Vegetable Grafting

Principles and Practices

This book is dedicated to the memory of our friend Prof. Dr Jung-Myung Lee who was a pioneer in the field of vegetable grafting and gave a substantial contribution to the first chapter of this book.

Vegetable Grafting

Principles and Practices

Giuseppe Colla

Department of Agricultural and Forestry Sciences
University of Tuscia
Italy

Francisco Pérez-Alfocea

Department of Plant Nutrition
Centro de Edafología y Biología Aplicada del Segura (CEBAS)
Consejo Superior de Investigaciones Científicas (CSIC)
Campus Universitario de Espinardo
Spain

Dietmar Schwarz

Leibniz Institute of Vegetable and Ornamental Crops
Germany

CABI

CABI is a trading name of CAB International

CABI
Nosworthy Way
Wallingford
Oxfordshire OX10 8DE
UK

Tel: +44 (0)1491 832111
Fax: +44 (0)1491 833508
E-mail: info@cabi.org
Website: www.cabi.org

CABI
745 Atlantic Avenue
8th Floor
Boston, MA 02111
USA

Tel: +1 (617)682-9015
E-mail: cabi-nao@cabi.org

A catalogue record for this book is available from the British Library, London, UK.

Library of Congress Cataloging-in-Publication Data

Names: Colla, Giuseppe, 1972- | Pérez-Alfocea, F. (Francisco) | Schwarz, Dietmar, 1956- Title: Vegetable grafting : principles and practices / [edited by] Giuseppe Colla, Department of Agricultural and Forestry Sciences, University of Tuscia, Viterbo, Italy, Francisco Pérez-Alfocea, Department of Plant Nutrition, CEBAS-CSIC, Murcia, Spain, Dietmar Schwarz, Leibniz Institute of Vegetable and Ornamental Crops, Grossbeeren, Germany.
Description: Wallingford, Oxfordshire, UK : CAB International, [2017] | Includes bibliographical references and index.
Identifiers: LCCN 2016057959 | ISBN 9781780648972 (hbk : alk. paper)
Subjects: LCSH: Vegetables. | Grafting. | Rootstocks.
Classification: LCC SB324.7 .V43 2017 | DDC 635--dc23
LC record available at https://lccn.loc.gov/2016057959

ISBN-13: 978 1 78064 897 2

Commissioning editor: Rachael Russell
Editorial assistant: Emma McCann
Production editor: Shankari Wilford

Typeset by SPi, Pondicherry, India
Printed and bound by CPI Group (UK) Ltd, Croydon, CR0 4YY

First printed in 2017. Transferred to Print on Demand in 2019.

Contents

Contributors

Alfonso Albacete, Department of Plant Nutrition, Centro de Edafología y Biología Aplicada del Segura (CEBAS), Consejo Superior de Investigaciones Científicas (CSIC), Campus Universitario de Espinardo, Murcia 30100, Spain. E-mail: alfmoreno@cebas.csic.es

Muhammad Azher Nawaz, College of Horticulture and Forestry Sciences, Huazhong Agricultural University/Key Laboratory of Horticultural Plant Biology, Ministry of Education, Wuhan, PR China; Department of Horticulture, University College of Agriculture, University of Sargodha, Sargodha, Pakistan. E-mail: azher490@hotmail.com

Penelope J. Bebeli, Laboratory of Plant Breeding and Biometry, Department of Crop Science, Agricultural University of Athens, Athens, Greece. E-mail: pbeb14@gmail.com

Meni Ben-Hur, Institute of Soil, Water and Environmental Sciences, Volcani Center, Agricultural Research Organization, Bet Dagan 50250, Israel. E-mail: meni@volcani.agri.gov.il

Zhilong Bie, College of Horticulture and Forestry Sciences, Huazhong Agricultural University/Key Laboratory of Horticultural Plant Biology, Ministry of Education, Wuhan, PR China. E-mail: biezhilong@hotmail.com

Angeles Calatayud, Departamento de Horticultura, Instituto Valenciano de Investigaciones Agrarias (IVIA), Valencia, Spain. E-mail: calatayud_ang@gva.es

Roni Cohen, Department of Vegetable Crops, Newe Ya'ar Research Center, Agricultural Research Organization, Ramat Yishai 30095, Israel. E-mail: ronico@volcani.agri.gov.il

Giuseppe Colla, Department of Agricultural and Forestry Sciences, University of Tuscia, 01100 Viterbo, Italy. E-mail: giucolla@unitus.it

Sarah Jane Cookson, UMR Ecophysiologie et Génomique Fonctionnelle de la Vigne, Institut des Sciences de la Vigne et du Vin (ISVV), INRA/Université de Bordeaux, Bordeaux, France. E-mail: scookson@bordeaux.inra.fr

J. Anja Dieleman, Wageningen UR Greenhouse Horticulture, PO Box 644, 6700 AP Wageningen, The Netherlands. E-mail: anja.dieleman@wur.nl

Ian C. Dodd, Lancaster Environment Centre, Lancaster University, Lancaster LA1 4YQ, UK. E-mail: i.dodd@lancaster.ac.uk

Aviv Dombrovsky, Institute of Plant Protection, Volcani Center, Agricultural Research Organization, Bet Dagan 50250, Israel. E-mail: aviv@volcani.agri.gov.il

Menahem Edelstein, Department of Vegetable Crops, Newe Ya'ar Research Center, Agricultural Research Organization, Ramat Yishai 30095, Israel. E-mail: medelst@volcani.agri.gov.il

Pilar Errea, Unidad de Hortofruticultura, Centro de Investigación y Tecnología Agroalimentaria de Aragón (CITA), Instituto Agroalimentario de Aragón – IA2 (CITA Universidad de Zaragossa), Avda Montañana 930, 50059 Zaragoza, Spain. E-mail: perrea@aragon.es

Carmina Gisbert, Instituto de Conservación y Mejora de la Agrodiversidad (COMAV), Universitat Politecnica de Valencia, Camino de Vera s/n, 46022 Valencia, Spain. E-mail: cgisbert@btc.upv.es

Francesco Giuffrida, Dipartimento di Agricoltura, Alimentazione e Ambiente (Di3A), Università degli Studi di Catania, Via Valdisavoia 5, 95123 Catania, Italy. E-mail: francesco.giuffrida@unict.it

Yuan Huang, College of Horticulture and Forestry Sciences, Huazhong Agricultural University/Key Laboratory of Horticultural Plant Biology, Ministry of Education, Wuhan, PR China. E-mail: huangyuan@mail.hzau.edu.cn

Jan Janse, Wageningen UR Greenhouse Horticulture, PO Box 644, 6700 AP Wageningen, The Netherlands. E-mail: jan.janse@wur.nl

Eyal Klein, Hishtil Nurseries Ltd, 22nd Yarden St., Moshav Nehalim 4995000, Israel. E-mail: kleine@hishtil.com

Amnon Koren, Hishtil Nurseries Ltd, 22nd Yarden St., Moshav Nehalim 4995000, Israel. E-mail: amkor52@gmail.com

Pradeep Kumar, Division of ILUM&FS (Division II), ICAR-Central Arid Zone Research Institute, Jodhpur (Rajasthan) 342003, India. E-mail: pradeephort@gmail.com

Marios C. Kyriacou, Department of Vegetable Crops, Postharvest Technology Laboratory, Agricultural Research Institute, PO Box 22016, 1516 Nicosia, Cyprus. E-mail: m.kyriacou@ari.gov.cy

Jung-Myung Lee, formerly of Department of Horticultural Biotechnology, Kyung Hee University, Seoul, Republic of Korea.

Cherubino Leonardi, Dipartimento di Agricoltura, Alimentazione e Ambiente (Di3A), Università degli Studi di Catania, Via Valdisavoia 5, 95123 Catania, Italy. E-mail: cherubino.leonardi@unict.it

Frank J. Louws, Center for Integrated Pest Management, Department of Plant Pathology, North Carolina State University, Raleigh, NC 27606, USA. E-mail: frank_louws@ncsu.edu

Isabel Mourão, CIMO, Escola Superior Agrária, Instituto Politécnico de Viana do Castelo, Refóios 4990-706 Ponte de Lima, Portugal. E-mail: isabelmourao@esa.ipvc.pt

Georgia Ntatsi, Department of Crop Science, Agricultural University of Athens, Iera Odos 75, 11855 Athens, Greece. E-mail: gntatsi@aua.gr

Gölgen B. Oztekin, Department of Horticulture, Faculty of Agriculture, Ege University, 35100 Bornova-Izmir, Turkey. E-mail: golgen.oztekin@ege.edu.tr

Ivan Paponov, NIBIO, Norwegian Institute of Bioeconomy Research, NO-4353 Klepp Station, Norway. E-mail: ivan.paponov@bioforsk.no

Francisco Pérez-Alfocea, Department of Plant Nutrition, Centro de Edafología y Biología Aplicada del Segura (CEBAS), Consejo Superior de Investigaciones Científicas (CSIC), Campus Universitario de Espinardo, Murcia 30100, Spain. E-mail: alfocea@cebas.csic.es

Maria Belen Pico, Instituto de Conservación y Mejora de la Agrodiversidad (COMAV), Universitat Politecnica de Valencia, Camino de Vera s/n, 46022 Valencia, Spain. E-mail: mpicosi@btc.upv.es

Ana Pina, Unidad de Hortofruticultura, Centro de Investigación y Tecnología Agroalimentaria de Aragón (CITA), Instituto Agroalimentario de Aragón – IA2 (CITA Universidad de Zaragossa), Avda Montañana 930, 50059 Zaragoza, Spain. E-mail: apina@aragon.es

Youssef Rouphael, Department of Agricultural Sciences, University of Naples Federico II, Via Università 100, 80055 Portici, Italy. E-mail: joerouphael@yahoo.com

Dimitrios Savvas, Department of Crop Science, Agricultural University of Athens, Iera Odos 75, 11855 Athens, Greece. E-mail: dsavvas@aua.gr

Dietmar Schwarz, Leibniz Institute for Vegetable and Ornamental Crops, Theodor Echtermeyer Weg 1, 14979 Großbeeren, Germany. E-mail: schwarz@igzev.de

Andrew J. Thompson, Cranfield Soil and AgriFood Institute, School of Water, Energy and Environment, Cranfield University, Cranfield, Bedfordshire MK43 0AL, UK. E-mail: a.j.thompson@cranfield.ac.uk

Allessandra Trinchera, Consiglio per la Ricerca in Agricoltura e L'analisi Dell'economia Agraria, Centro di Ricerca per lo Studio delle Relazioni tra Pianta e Suolo (CRA-RPS), Rome, Italy. E-mail: alessandra.trinchera@entecra.it

Jan Henk Venema, University of Groningen, Genomics Research in Ecology & Evolution in Nature (GREEN) – Plant Physiology, GELIFES, PO Box 11103, 9700 CC Groningen, The Netherlands. E-mail: j.h.venema@rug.nl

Halit Yetişir, Department of Horticultural Science, Agricultural Faculty, Erciyes University, Kayseri, Turkey. E-mail: yetişir1@yahoo.com

Preface

Although grafting has been practised on fruit trees for thousands of years, the commercial application of grafting on vegetables constitutes a relatively recent innovation in most countries. After more than 50 years of vegetable crop improvement, dedicated principally to selecting for above-ground traits, scientists now perceive root system engineering as an opportunity for integrating dynamic novel approaches in fostering sustainable vegetable production under changing environmental conditions, while minimizing the demand for new resources. Introduction of excellent rootstocks possessing multiple resistances and efficient grafting systems will greatly encourage the extended application of vegetable grafting all over the world. Although the benefits of using grafted transplants are now fully recognized worldwide, the need to enlighten the scientific basis of rootstock–scion interactions under variable environmental pressures remains vital for extracting grafting-mediated crop improvement. This has prompted the COST Action FA1204 entitled 'Vegetable grafting to improve yield and fruit quality under biotic and abiotic stress conditions' aimed at systematizing research findings (http://www.vegetablegrafting.unitus.it). The COST action allowed the development of a multidisciplinary network of partners targeting the root system and employing rootstock breeding to unravel the mechanisms behind rootstock-mediated crop improvement: the enhancement of productivity and fruit quality, and the sustainability of vegetable crops under multiple and combined stresses. The current book is the major output of the COST Action and contains nine chapters drawing on the 2012–2016 activities of four Working Groups (WGs) dealing with 'Genetic resources and rootstock breeding' (WG1), 'Rootstock–scion interactions and graft compatibility' (WG2), 'Rootstock-mediated resistance to biotic and abiotic stresses' (WG3) and 'Rootstock-mediated improvement of fruit quality' (WG4). While recent advances of scientific knowledge constitute the core of this COST book, valuable practical information is also provided on rootstock–scion combinations, on applicable grafting methods, on the establishment of grafted transplants and on recommendations for the use of grafted plants as an effective tool for sustainable vegetable production.

This book could not have been produced without the dedication and help of many, and we would like to thank the authors and co-authors who contributed to the compiled chapters. However, we would also like to express our appreciation to a large number of scientists and experts who served as reviewers and contributed to improving the quality of the book. Finally, we would like to thank the COST Association in Brussels (Belgium) for funding COST Action FA1204 and providing additional financial support for publishing the current book.

We planned and compiled this book as a collection of scientific information and as a practical tool aimed at both the people involved in the commercial production and cultivation of grafted plants, as well as researchers interested in an understanding of the science and technology behind a grafted plant. We hope all readers benefit from this book and we remain open to ideas and proposals on how to amend a future edition.

Giuseppe Colla, University of Tuscia, Viterbo, Italy
Francisco Pérez-Alfocea, CEBAS-CSIC, Murcia, Spain
Dietmar Schwarz, Leibniz Institute for Vegetable and Ornamental Crops, Großbeeren, Germany
July 2016

Acknowledgements

This article is based upon work from COST Action FA1204, supported by COST (European Cooperation in Science and Technology).

COST is a pan-European intergovernmental framework. Its mission is to enable break-through scientific and technological developments leading to new concepts and products and thereby contribute to strengthening Europe's research and innovation capacities. It allows researchers, engineers and scholars to jointly develop their own ideas and take new initiatives across all fields of science and technology, while promoting multi- and interdisciplinary approaches. COST aims at fostering a better integration of less research intensive countries to the knowledge hubs of the European Research Area. The COST Association, an International not-for-profit Association under Belgian Law, integrates all management, governing and administrative functions necessary for the operation of the framework. The COST Association has currently 36 Member Countries (www.cost.eu).

COST is supported by the EU Framework Programme Horizon 2020

The authors appreciate and thank the reviewers of the book chapters as follows:

Chapter 1

Marios C. Kyriacou, Department of Vegetable Crops, Agricultural Research Institute; Luigi Morra, Council for Agricultural Research and Economics, Research Institute for Cereals and Industrial Crops; Youssef Rouphael, Department of Agricultural Sciences, University of Naples Federico II; Min Wei, College of Horticulture, Shandong Agricultural University.

Chapter 2

Ahmet Balkaya, Department of Horticulture, University of Ondokuz Mayis; María José Díez, Instituto de Conservación y Mejora de la Agrodiversidad Valenciana (COMAV), Universitat Politècnica de València; Willem van Dooijeweert, Centre for Genetic Resources, The Netherlands, and Wageningen University and Research Centre; Harry S. Paris, Department of Vegetable Crops, Newe Ya'ar Research Center; Andrea Mazzucato, Department of Agricultural and Forestry Sciences, University of Tuscia.

Chapter 3

David Herzog, Crop Coordinator Tomato & Rootstock Solanaceae, Rijk Zwaan Ibérica S.A. Ctra Viator-PJ. Mami; Andrea Mazzucato, Department of Agricultural and Forestry Sciences, University of Tuscia.

Chapter 4

Hakan Aktas, Horticulture Department, Agriculture Faculty, Suleyman Demirel University; Vicent Arbona, Ecofisiologia i Biotecnologia, Dept Ciències Agràries i del Medi Natural, Universitat Jaume I; Eloise Foo, School of Biological Sciences, University of Tasmania.

Chapter 5

Marco Landi, Department of Agriculture, Food & Environment, University of Pisa; Dieter Treutter (deceased), Institute of Fruit Science, Center of Life Science, Technische Universität München.

Chapter 6

Maria Lodovica O. Gullino, Centro di Competenza per l'Innovazione in Campo Agro-ambientale (AGROINNOVA), Università degli Studi di Torino; Jaacov Katan, Department of Plant Pathology and Microbiology, Robert H. Smith Faculty of Agriculture, Food & Environment.

Chapter 7

Ravindra Mohan Bhatt, Division of Plant Physiology and Biochemistry, Indian Institute of Horticultural Research; Menahem Edelstein, Department of Vegetable Crops, Newe Ya'ar Research Center; Yuan Huang, College of Horticulture and Forestry Sciences, Huazhong Agricultural University/Key Laboratory of Horticultural Plant Biology; Damianos Neocleous, Soil Science Section, Agricultural Research Institute; Alberto Pardossi, Department of Agriculture,

Food and Environment (DAFE), University of Pisa; Dimitrios Savvas, Department of Crop Science, Agricultural University of Athens; Xin Zhao, Horticultural Sciences Department, University of Florida.

Chapter 8

Angelika Krumbein, Leibniz Institute of Vegetable and Ornamental Crops; Anastasios S. Siomos, Department of Horticulture, Aristotle University of Thessaloniki.

Chapter 9

Jorge Barmaimón, La Sala Nurseries; Zhilong Bie, College of Horticulture and Forestry Sciences, Huazhong Agricultural University/Key Laboratory of Horticultural Plant Biology; Giovanna Causarano, Centro Seia; Francesco Di Gioia, Department of Horticultural Sciences, University of Florida/Institute of Food and Agricultural Sciences; Yung-Myung Lee, Department of Horticultural Biotechnology, Kyung Hee University; Juan José Magán, Estación Experimental Cajamar-Las Palmerillas; Leo Sabatino, Department of Agricultural and Forest Sciences, University of Palermo.

1 Introduction to Vegetable Grafting

ZHILONG BIE,[1*] MUHAMMAD AZHER NAWAZ,[1,2] YUAN HUANG,[1] JUNG-MYUNG LEE[3] AND GIUSEPPE COLLA[4]

[1]*Huazhong Agricultural University/Key Laboratory of Horticultural Plant Biology, Wuhan, PR China;* [2]*Department of Horticulture, University College of Agriculture, University of Sargodha, Sargodha, Pakistan;* [3]*formerly of Kyung Hee University, Seoul, Republic of Korea;* [4]*University of Tuscia, Viterbo, Italy*

1.1 Importance and Use of Vegetable Grafting

1.1.1 Historical perspective

Grafting is the art of joining together two plant parts (a rootstock and a scion) by means of tissue regeneration, in which the resulting combination of plant parts achieves physical reunion and grows as a single plant (Janick, 1986). It is a centuries-old technique but a relatively new one in vegetable cultivation. Various references to fruit grafting appear in the Bible and in ancient Greek and Chinese literature, suggesting that grafting was used in Europe, the Middle East and Asia by the 5th century BC (Melnyk and Meyerowitz, 2015). Grafting occurs commonly in nature, and the observation of natural grafts may have inspired human use of this technique in horticulture thousands of years ago (Mudge *et al.*, 2009).

Grafting of fruit trees has been practised for thousands of years, but in vegetables this technique is a relatively new one. Self-grafting was used as a technique to produce large-sized gourd fruits, as reported in a Chinese book written in the 5th century and a Korean book written in the 17th century (Lee and Oda, 2003). However, commercial grafting of vegetables only originated in the early 20th century with the aim of managing soilborne pathogens (Louws *et al.*, 2010).

Scientific vegetable grafting was first launched in Japan and Korea in the late 1920s by grafting watermelon on to gourd rootstocks to avoid soilborne diseases (Ashita, 1927; Yamakawa, 1983). This new technique was disseminated to farmers in Japan and Korea by the agricultural extension workers. In the early 1930s, the commercial use of grafted transplants was started in Japan by grafting

* biezhilong@hotmail.com

watermelon on to bottle gourd (*Lagenaria siceraria* (Mol.) Standl) and summer squash (*Cucurbita moschata* Duch.) to induce resistance to Fusarium wilt (Oda, 2002; Sakata *et al.*, 2007, 2008). Grafting of cucumber to reduce soilborne diseases and to enhance scion vigour is believed to have started in the 1920s but was not applied on a commercial scale until the 1960s (Sakata *et al.*, 2008).

Among the Solanaceae crops, aubergine (*Solanum melongena* L.) was first grafted on to scarlet aubergine (*Solanum integrifolium* Lam.) in the 1950s (Oda, 1999). Similarly, grafting of tomato (*Solanum lycopersicum* L.) was started in the 1960s (Lee and Oda, 2003). In the 1950s, the rapid development of protected cultivation with the use of greenhouses or tunnels for offseason vegetable production and intensive cropping patterns changed the existing crop rotation system; consequently, farmers became dependent on grafting to control soilborne pathogens and other pests (Kubota *et al.*, 2008; Lee *et al.*, 2010).

Scientific studies investigating and developing rootstocks was initiated in the 1960s in Korea. By 1990, the percentage of grafted Solanaceae and Cucurbitaceae (e.g. cucumber, melon, aubergine, tomato) had increased to 59% in Japan and 81% in Korea (Lee, 1994). Currently, most greenhouse-cultivated cucurbits are grafted in China, Japan, Korea, Turkey and Israel, while grafted vegetables are cultivated on a commercial scale in more than 20 countries worldwide (Table 1.1).

1.1.2 Purpose and scope

Although vegetable grafting in ancient times was intended mainly to produce large-sized gourds for rice storage (Hong, 1710; PSNCK, 1982), it expanded rapidly in many countries to control soilborne pathogens (e.g. root-knot nematodes) and foliar pathogens, to enhance plant vigour, to extend the harvesting period, to increase yield and fruit quality, to prolong postharvest life, to increase nutrient uptake, to allow tolerance to low and high temperatures, to cope with salinity and heavy-metal stress, and to increase tolerance to drought and waterlogging (Table 1.2; *see* Chapters 6 and 7, this volume).

As well as myriad applications in advancing sustainable crop production, grafting can be used as a tool in both breeding and research. Recently, a group of researchers from Germany working on tobacco published a unique way of producing new allohexaploid tobacco species by using the graft site as propagation

Table 1.1. Main countries where grafted vegetables are produced and/or cultivated on a commercial scale.

Continent	Countries
East Asia	China, Japan, Korea, the Philippines
Europe	Spain, Italy, the Netherlands, France, Greece, Cyprus, Belgium, Portugal, Germany, Croatia, Bosnia and Herzegovina
Middle East and North Africa	Turkey, Israel, Morocco, Egypt, Iran, Algeria
Americas	Mexico, Canada, the USA, Argentina

Table 1.2. Benefits of vegetables grafting.

Benefit	Crop	Reference
Disease resistance to soilborne pathogens and foliar pathogens	Tomato, watermelon, aubergine, artichoke, cucumber, pepper, melon	Black *et al.* (2003); Bletsos *et al.* (2003); Bletsos (2005, 2006); Sakata *et al.* (2006, 2007, 2008); King *et al.* (2008); Lee *et al.* (2010); Louws *et al.* (2010); Kousik *et al.* (2012); Jang *et al.* (2012); Temperini *et al.* (2013); Gilardi *et al.* (2013a,b); Vitale *et al.* (2014); Arwiyanto *et al.* (2015); Miles *et al.* (2015); Shibuya *et al.* (2015); Suchoff *et al.* (2015)
Nematode resistance	Tomato	Dong *et al.* (2007); Lee *et al.* (2010); Louws *et al.* (2010)
Salt tolerance	Cucumber, pepper, watermelon, tomato	Huang *et al.* (2009); Colla *et al.* (2010, 2012, 2013); Huang *et al.* (2010, 2013a); Lee *et al.* (2010); Schwarz *et al.* (2010); Fan *et al.* (2011); Yang *et al.* (2012, 2013); Wahb-Allah (2014); Penella *et al.* (2015); Xing *et al.* (2015)
High- and low-temperature tolerance	Tomato, pepper, cucumber	Venema *et al.* (2008); Li *et al.* (2008); Lee *et al.* (2010); Schwarz *et al.* (2010); López-Marín *et al.* (2013)
Drought tolerance	Pepper, tomato	Lee *et al.* (2010); Schwarz *et al.* (2010); Penella *et al.* (2014); Wahb-Allah (2014)
Flooding tolerance	Tomato	Lee *et al.* (2010); Bhatt *et al.* (2015)
Nutrient uptake	Watermelon, tomato, melon	Kim and Lee (1989); Ruiz *et al.* (1997); Lee *et al.* (2010); Colla *et al.* (2010b, 2011); Huang *et al.* (2013b, 2016a,b); Schwarz *et al.* (2013); Huang *et al.* (2016a,b); Nawaz *et al.* (2016)
Yield increase	Watermelon, melon cucumber, tomato, aubergine, pepper, artichoke	Jeong (1986); Ruiz *et al.* (1997); Nisini *et al.* (2002); Colla *et al.* (2008); Huang *et al.* (2009); Lee *et al.* (2010); Gisbert *et al.* (2011); Moncada *et al.* (2013); Tsaballa *et al.* (2013); Temperini *et al.* (2013)
Fruit quality improvement	Tomato, cucumber, aubergine, pepper, melon, watermelon	Jeong (1986); Proietti *et al.* (2008); Huang *et al.* (2009); Lee *et al.* (2010); Rouphael *et al.* (2010); Gisbert *et al.* (2011); Zhao *et al.* (2011); Condurso *et al.* (2012); Krumbein and Schwarz (2013); Moncada *et al.* (2013); Tsaballa *et al.* (2013); Verzera *et al.* (2014); Kyriacou *et al.* (2016)
Scion vigour improvement	Cucumber	Jeong (1986); Lee *et al.* (2010)
Reproductive growth promotion	Cucumber	Jeong (1986); Lee *et al.* (2010)
Shelf-life/ postharvest life improvement	Melon	Zhao *et al.* (2011)

Continued

Table 1.2. Continued.

Benefit	Crop	Reference
Heavy metals/ organic pollutants tolerance	Cucumber, tomato	Rouphael *et al.* (2008); Lee *et al.* (2010); Schwarz *et al.* (2010); Zhang *et al.* (2010a,b, 2013); Kumar *et al.* (2015a,b)
Extension of harvesting period	Cucumber	Jeong (1986); Itagi (1992); Ito (1992); Lee *et al.* (2010)
Weed control/ management	–	Dor *et al.* (2010); Louws *et al.* (2010)
Production of new species (tetraploid)	Tobacco	Fuentes *et al.* (2014)

material *in vitro* (Fuentes *et al.*, 2014); in this case, grafting can be seen as a breeding tool to generate novel genetic combinations – in a process that is conceptually similar to protoplast fusion – by hybridization at the cellular level, bypassing sexual compatibility barriers (*see* Chapter 3, this volume). Independent breeding for rootstock and scion traits can also make 'trait stacking' in breeding programmes generally easier. Researchers have used reverse genetics and grafting to investigate root-to-shoot signalling: by grafting genetically defined and distinct rootstocks and scions, it is possible to assign the origins of physiological functions to one or the other, and to study the movement between roots and shoots of specific biomolecules (e.g. phytohormones, metabolites, small RNAs) and the consequent effects on root and shoot phenotypes (*see* Chapter 4, this volume). For example, the extent and impact of abscisic acid (Holbrook *et al.*, 2002) and cytokinin (Ghanem *et al.*, 2011) in root-to-shoot signalling became apparent through grafting with mutant and transgenic rootstocks. Also, by grafting hundreds of different rootstock genotypes from a genetically defined population to a common scion, it is possible to identify new genetic loci and processes that control rootstock traits by forward genetics (Asins *et al.*, 2015); this can advance scientific understanding and provide molecular markers for rootstock breeding. Finally, fundamental studies can also be carried out in the graftable dicotyledonous model plant *Arabidopsis thaliana*, such as how the graft junction is formed and the fundamental process of vascular regeneration following wounding (Melnyk *et al.*, 2015).

In short, grafting is being used in different ways for economic, societal and environmental benefits, and to extend the depth of knowledge about fundamental process in plant science.

Commercial grafting is currently practised in watermelon (*Citrullus lanatus* (Thunb.) Matsum. & Nakai), melon (*Cucumis melo* L.), cucumber (*Cucumis sativus* L.), pumpkin (*C. moschata*), bitter gourd (*Momordica charantia* L.), tomato (*S. lycopersicum*), aubergine (*S. melongena*) and pepper (*Capsicum annuum* L.). However, grafting may also be used in other vegetables, such as artichoke (*Cynara cardunculus* subsp. *scolymus* (L.) Hegi) grafted on to cardoon (Temperini *et al.*, 2013), or in different *Phaseolus vulgaris* L. graft combinations (Cichy *et al.*, 2007), where the yield increased significantly compared with non-grafted plants (*see* Chapter 9, this volume).

From the grower's point of view, yield is the most important factor in the economics of farming. Grafting directly increases yield by invigorating scions, increasing resource use efficiency (e.g. water, fertilizer), and extending the harvest period. Additionally, it helps to reduce the costs involved in plant protection measures compared with the use of self-rooted vegetables. An overview of reported percentage yield increases among various vegetables is summarized in Table 1.3.

1.1.3 Growing areas and plantlet production

Statistics about the cultivation and use of grafted vegetables worldwide are difficult to obtain and are often not updated as the use of grafted vegetables continues to increase. The trend for grafted vegetable production varies widely from country to country, and even within a country. The largest market for grafted vegetable crops is East Asia because of the high concentration of cucurbits in general and the high concentration of grafted plants in particular. For example, 99% of watermelons are grafted in Korea, 94% in Japan, and 40% in China. In contrast, solanaceous crops are less frequently grafted: about 60–65% of tomatoes and aubergines, and 10–14% of peppers. Under protected cultivation, the percentage is higher, and almost all cucumbers, watermelons, and tomato are grafted under these conditions. A similar high ratio of grafted plants compared with non-grafted can be found in Mediterranean countries, particularly those with high production areas, such as Spain, Italy, Turkey and Israel. In the Netherlands, nearly all the tomatoes produced in soil-less culture are grafted on to vigorous rootstocks to increase or at least secure the yield. In France, tomatoes and aubergines in particular are grafted to enhance resistance to soilborne pathogens and nematodes. Currently, grafting is expanding in many countries worldwide, particularly in eastern Europe, North and South America, India and the Philippines. The market for grafted vegetables in North America was first advanced in Canada by the Dutch, who introduced

Table 1.3. Yield increase of grafted plants in comparison with non-grafted or self-grafted plants in different vegetable crops.

Vegetable	Yield increase (%)	Reference
Melon	3.4–92	Ruiz *et al.* (1997); Lee *et al.* (2010); Condurso *et al.* (2012); Verzera *et al.* (2014); Salar *et al.* (2015); Han *et al.* (2015); Mohammadi *et al.* (2015); Esmaeili *et al.* (2015)
Watermelon	22.7–43.0	Mohamed *et al.* (2014); Soteriou *et al.* (2015)
Cucumber	8.8–57.0	Rouphael *et al.* (2008); Colla *et al.* (2012, 2013); Farhadi and Malek (2015); Gao *et al.* (2015)
Pepper	9.2	Jang *et al.* (2012)
Aubergine	27.7	Gisbert *et al.* (2011)
Tomato	5.4–80.3	Chung and Lee (2007); Schwarz *et al.* (2013); Wahb-Allah (2014); Bhatt *et al.* (2015); Boncato and Ellamar (2015); Suchoff *et al.* (2015)
Artichoke	21.7	Temperini *et al.* (2013)

grafting in tomato to increase yield under greenhouse conditions. Similarly, in Mexico grafting was introduced first in tomato and later in other vegetable crops.

Canada and Mexico currently have several large-scale grafting nurseries producing millions of grafted plants annually. As import of tomato plants from Mexico is prohibited in the USA, Canadian nurseries have been the only source of grafted plants until recently. More recently, nurseries in the USA have started grafting vegetables, and a large international nursery announced its plan to build the first North American operation. In most countries where grafted vegetables are cultivated, plantlets are produced by commercial nurseries based on growers' needs. In China alone, more than 1500 commercial nurseries are producing grafted transplants. In a small number of countries, particularly with small farm sizes and poor growers, grafted transplants are self-produced by the growers or imported from neighbouring countries. International trading of grafted vegetable transplants is rapidly increasing, but the majority of grafted transplants are still produced by the grower. In nurseries with a high production volume, fully automatic machines (grafting robots) may also be used, as occurs in the Netherlands and Korea.

1.2 The Process of Vegetable Grafting

The grafting process comprises four main steps (Fig. 1.1): (i) selection of the rootstock and scion cultivars; (ii) plantlet production and creation of the physical union by physical manipulation; (iii) healing of the graft union; and (iv) acclimatization of the grafted plants (Lee and Oda 2003; Lee *et al.*, 2010).

1.2.1 Selection of rootstock and scion cultivars

Selection of the correct rootstock and scion cultivars is a critical step for the success of grafted vegetable production. The seed of the scion cultivar is selected on the basis of purity, viability, yield, fruit quality and market demand. Rootstock cultivars are selected based on purity, viability, resistance to diseases, compatibility with the scion cultivar, and adaptability to local soil and environmental conditions.

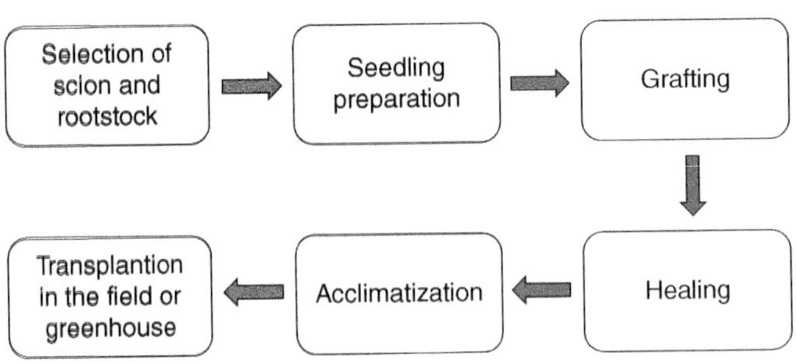

Fig. 1.1. Production process of grafted vegetable plantlets.

The public sector and private seed companies have introduced a number of high-yielding scion and rootstock cultivars with desired characteristics, such as resistance to diseases and nematodes, and tolerance to salinity, drought, flood, heat and chilling stress. Seed companies are aware that it is important that the rootstocks they breed do not impair fruit taste or quality. Thus, growers have a wide range of cultivars from which they can select rootstocks for Cucurbitaceae and Solanaceae crops in accordance with their own requirements. Cucurbit rootstock breeding work is concentrated mainly in China, Korea and Japan. The number of registered cucurbit rootstocks is increasing continuously because of the increased popularity of cultivation of grafted plants (Kato and Lou 1989; Ko, 1999; Lee *et al.*, 2008). According to one estimate in China, over 600 Cucurbitaceae rootstocks are in trials at various stages, although only a few are released each year (King *et al.*, 2010).

1.2.2 Overview of grafting methods

Different grafting methods are selected depending on the type of crop, the farmer's previous technical experience, personal choice, the number of grafts required, the purpose of grafting, plantlet production (own use or commercial), access to labour, and the availability of machinery and infrastructural facilities (Lee *et al.*, 2010). In general, grafting methods can be divided into two categories: (i) manual grafting, where most of the grafting process is performed manually; and (ii) mechanical grafting, where machines (robots) are used to carry out the main grafting processes.

Although many machines and grafting robots have been developed, manual grafting is still the most popular and widely used method (Lee *et al.*, 2010). In manual grafting, a number of methods are used for the process and described in detail in the following sections (*see* Plate 1).

Hole insertion

Hole insertion grafting is preferred for grafted watermelon transplant production in many areas because the size of watermelon seedlings is relatively small compared with the rootstock (bottle gourd or squash). In this method, rootstock seeds are sown 7–8 or 3–4 days earlier than the watermelon seeds for the bottle gourd and squash rootstock seeds, respectively. Grafting is performed 7–8 days after watermelon seed sowing. In the case of tomato and aubergine, rootstock seeds are sown 5–10 days before sowing the scion seeds, and grafting is performed 20–25 days after sowing the scion seeds (Lee *et al.*, 2010). At the time of grafting, both rootstock and scion seedlings should be uniform, healthy and vigorous.

For graft operation, the true leaves and growing point of the rootstock are carefully and completely removed just above (Plate 1a) or sometimes below (Plate 1b) the cotyledonary leaves, and a slanting hole is made with a wooden or plastic gimlet. The hypocotyl portion of the scion is prepared by making a slanting cut with a tapering end for easy insertion (*see* Plate 1a). Care must be taken during insertion of the scion into the rootstock to avoid insertion of the scion into the rootstock hypocotyl cavity, because this will strongly affect the reunion, and at a later point, adventitious roots from the scion will grow through the rootstock cavity and reach the soil, thus ultimately minimizing the purpose of using a rootstock.

This method is very popular in China, because it results in a strong union and vascular connection compared with the tongue grafting approach, and additional labour for clipping, transplanting, cutting and clip removal is not required (Oda, 1994). However, sometimes parallel growth of the rootstock along with that of the newly grafted scion starts just above the cotyledonary leaves, so these offshoots need to be removed. Plate 2 shows an example of grafted watermelon transplants produced by the hole insertion method.

Tongue grafting

A rootstock and scion of equal size are used for tongue approach to grafting. Therefore, in order to attain plants of a uniform size, the seeds of scion cultivars (e.g. watermelon, cucumber, melon) are sown 5–7 days earlier compared with the rootstock seeds. The growing point and the true leaves are carefully and completely removed from the rootstock and a downward-slanting cut is made in the hypocotyl while an upward-slanting cut is made on the hypocotyl of the scion. The angle of cut is made at 30–40° in relation to the perpendicular axis. When removing the growing point of the rootstock, one cotyledonary leaf is often also removed to ensure complete removal of the growing point and to avoid crowding in limited growing space. The cut region on the scion is inserted into the rootstock (*see* Plate 1c), after which specially designed grafting clips are placed at the graft point to hold it firmly in place. The grafted rootstock and scion are immediately planted in pots. The plants are placed under partial shade conditions for 1–2 days after which they can be placed under normal greenhouse conditions. After 10–12 days, the lower hypocotyl portion of the scion of several plants is cut to see the response of the plants; if the graft union has been successful, the scions will continue to grow, but if the union is incomplete or partially complete, the scions will wilt or show restricted growth. These results are used to judge the overall response of the plants. The clips are removed before transplanting the plants.

An experienced person can perform about 800 grafts per day (Lee *et al.*, 2010). Therefore, this ancient method of vegetable grafting is rarely used by professional seedling growers compared with other methods as more labour is required for cutting the scion again for testing and adding the clips (which also need to be removed before transplanting) and more space is needed to grow the plantlets. Moreover, frequent rooting from the scion occurs under field conditions, especially if the plantlets are planted deep in the ground (Lee, 1994).

Splice grafting

This method, also known as tube grafting or one-cotyledon splice grafting, is the most widely used and preferred method by growers and commercial grafted transplant producers. It can be performed in most vegetables by hand or by machines/robots. In this method, the growing point and one cotyledonary leaf of the rootstock are removed by performing a slanted cut (35–45°) and a prepared scion is matched to it (*see* Plate 1d, e and j). This method is popular in cucurbits and solanaceous crops. Grafted watermelon transplants processed by this method are shown in Plate 3) and pepper transplants in Plate 4). To hold the graft site, a grafting clip, pin or tube (an elastic tube with a slit on one side) is used. For Solanaceae crops, grafting is performed at lower epicotyls of the rootstock.

The main features of the hole insertion, tongue and splice grafting methods are compared in Table 1.4.

Cleft grafting

Cleft grafting is also called apical or wedge grafting. In this method, the rootstock seedling is decapitated and a 0.5–1.5 cm vertical cut is made in the centre of the stem along the stem axis. The scion is pruned to one to three true leaves, and the lower stem end is given a slanted cut from both sides to form a wedge. This wedge-shaped scion is inserted into the slit made on the rootstock and a clip is placed to hold the rootstock and scion together (*see* Plate 1f and g). Various types of grafting clips in different sizes are available for this purpose.

In this method, the scion is tightly held by the rootstock compared with other methods, so clips may not be necessary and grafting tape, wax tape or Parafilm

Table 1.4. Comparison of the different grafting methods.

Hole insertion grafting	Tongue grafting approach	Splice grafting, tube grafting or one-cotyledon splice grafting
Strong vascular connection	Weak vascular connection	Very strong vascular connection
Trained workers required to accomplish the task	Less trained workers can successfully graft by this method	Trained workers required to accomplish the task
Grafting machine not available	Grafting machine available	Grafting machine available
Grafting clips not required	Grafting clips required	Grafting clips, pins and tubes required to hold the graft union
Labour for clipping, transplanting, cutting and clip removal not required	Labour for clipping, transplanting, cutting and clip removal not required	Labour for clipping, transplanting, cutting and clip removal may be required
Offshoot removal required	Offshoots removal required	Offshoots removal often not required
Scion does not need to be planted with the rootstock during the healing process	Scion needs to be planted with the rootstock during the healing process	Scion does not need to be planted with the rootstock during the healing process
Less space required during the healing process	More space required during the healing process	More space required during the healing process
Less labour-intensive	Most labour-intensive	More labour-intensive
Environmental control (high humidity) required during the healing process	No strict environmental control (high humidity) required during the healing process	Careful environmental control (high humidity) required during the healing process
Can be used mainly in cucurbits	Can be used in both solanaceous crops and cucurbits	Can be used in solanaceous crops, cucurbits and other minor crops

may be used to hold the scion. However, the grafting process takes more time than splice grafting, and sometimes rootstock stems split completely during the grafting process. Cleft grafting is relatively difficult to perform in vegetables compared with woody tree species, so this method is confined to several Solanaceae crops such as aubergine and chilies (*see* Plate 5) (Lee *et al.*, 2010; Johnson *et al.*, 2011), while it is rarely used in cucurbits.

Pin grafting

Pin grafting is similar to splice grafting, the only difference being that specially designed pins are used instead of clips to fix the grafted position of the scion and rootstock (*see* Plate 1h and i). These pins are made of a natural ceramic material so that they can remain within the plant without causing problems. The Takii Seed Company in Japan has designed ceramic pins of 15 mm length and 0.5 mm width with a hexagonal cross-section. This method saves time and labour, as clips need removal while pins do not. However, the ceramic pins are costly and are used only once as they are not removed, while clips can be reused for the grafting process. Recently, it has been found that rectangular-shaped bamboo or wooden sticks can be used as a replacement for ceramic pins. Careful environmental control is necessary for the success of a graft union. Rootstock suckers/offshoots may emerge during the healing stage of this method, or even under field conditions, and need removing.

Mechanical grafting

Grafting machines or robots are increasingly being used for the grafting process. The first robotic 'one-cotyledon grafting system' was developed in the 1980s by Iam Brain in Japan for cucurbit vegetables. The prototype was developed in 1987 and then adapted in 1989 (Ito, 1992; Kubota *et al.*, 2008). It takes 4.5 s to make a graft, and the success rate is 95%. The technologies used in this robot were shared with agriculture machinery companies and a prototype semi-automatic grafting system was developed in Korea. Several grafting robots were developed by the Rural Development Administration of Korea and were provided to plug seedling nursery growers at a relatively low price. By 2001, three grafting robots had been developed in Korea. The simple and economical grafting machine developed by the Yopoong Company was provided for local growers and has been exported to Asian countries for more than a decade. Another semi-automatic grafting machine was developed by a private company in Korea and provided to growers. This semi-automatic multifunctional machine was adopted by many countries because of its reasonable price, adjustability and convenient handling (*see* Plate 14)

In the Netherlands, a fully automatic grafting robot with a capacity of 1000 grafts h^{-1} has been developed and used for tomato; similarly, another fully automatic grafting robot has been developed in Japan with a capacity of 750 grafts h^{-1} and a success rate of 90%. Currently, a total of six models of semi- or fully automatic grafting robots are available in the market; three of these models have been developed in Japan, and one each in Korea, the Netherlands and Spain, as described in Table 1.5. According to a published report (http://jhawkins54.typepad.com/files/vegetable-grafting-1.pdf, accessed 24 November 2016), at least two companies are producing mechanically grafted transplants in the USA. The use

Table 1.5. Features of grafting robots available in different countries. (From http://cals.arizona.edu/grafting/grafting-robots, accessed 24 November 2016.)

Make	Country of origin	Distribution	Suitability	Properties/characteristics/ specifications
Helper Robotech (semi-automated machine)	Korea	Distributed to Asia, Europe and North America	Cucurbits and tomato	The first model that can graft both cucurbits and tomato. Widely marketed in Asia and North America. Produces 650–900 grafts h^{-1} at ≥95% success rate. Needs two to three workers to operate the machine.
Iseki (semi-automated machine)	Japan	Distributed to Asia and Europe	Cucurbits	Introduced to the Asian and European market. One machine has been introduced in the USA for trial use. Produces 900 grafts h^{-1} at ≥95% success rate. Needs two to three workers to operate the machine.
Iseki (semi-automated machine)	Japan	Distributed to Asia	Tomato and aubergine	Produces 800 grafts h^{-1} at ≥95% success rate. Seedling size required for grafting was too large for Japanese standard, limiting the market. However, the seedling size is acceptable for USA standard. Needs two to three workers to operate the machine.
Iseki (fully automated machine)	Japan	–	Cucurbits	Introduced in Japanese market in 2009. Produces 800 grafts h^{-1} at ≥95% success rate. A tomato model is also under development at IAM BRAIN[a]. Only one person needed operate the machine.
ISO Group (fully automated machine)	The Netherlands	–	Tomato and aubergine	Introduced in 2009. Produces 1000 grafts h^{-1}. A semi-automated model is also available that requires manual feeding of plants into the system.
Conic System (semi-automated machine)	Spain	–	Tomato	A semi-automated robot to cut tomato scions and rootstocks at a selected angle. Produces 400–600 grafts h^{-1}. Only one person needed to operate the machine.

[a]Institute of Agricultural Machinery Bio-oriented Technology Research Advancement Institution, Japan.

of grafting machines and robots is increasing; Korea exported 32 grafting robots to different countries around the world between 2011 and 2013.

1.2.3 Preference of grafting method for different species

Curcubits

In Japan, grafted watermelon transplants are produced mainly by hole insertion grafting, while for cucumber, individual farmers produce grafted transplants for their own use by the tongue grafting approach. Commercial growers prefer splice grafting for curcubits (Lee *et al.*, 2010). In a recent study conducted in Egypt, Mohamed *et al.* (2014) concluded that, for watermelon, grafted transplants produced using the tongue grafting approach were better compared with those produced by hole insertion and splice grafting under the prevailing cultural and environmental conditions. A table-top grafting machine is available for the tongue grafting approach and is small, handy and easy to operate. However, the tongue grafting approach is considered a simple and basic grafting method, so these machines are now used only by a limited number of growers on a small scale in Korea.

Solanaceous crops

Individual farmers produce grafted aubergine transplants by a number of grafting methods of their own choice, while commercial growers generally adopt splice grafting. According to a published report (http://jhawkins54.typepad.com/files/vegetable-grafting-1.pdf, accessed 24 November 2016), cleft grafting is used routinely for the production of grafted tomato, pepper and aubergine transplants, although Johnson *et al.* (2011) reported that splice grafting is the most commonly used method (95%) for aubergine and tomato in the USA. Generally, less experienced and small farmers prefer the tongue grafting approach, while professional and commercial plantlet producers use splice grafting for most of their grafted vegetables. The quality of grafted transplants produced by splice grafting is considered to be better than those produced by tongue grafting (Lee *et al.*, 2010).

Other crops

Grafting is also reported in other vegetables, although only with local importance (*see* Chapters 2 and 9, this volume). Temperini *et al.* (2013) grafted artichoke (*C. cardunculus* subsp. *scolymus*) on to cardoon by cleft grafting and observed that grafting increased the yield. Similarly, Chinese cabbage (*Brassica rapa* ssp. *pekinensis* (Lour.) Kitam, inbred line) was grafted on to three *Brassica* rootstocks (mustard, turnip and broccoli) by cleft grafting to perform gene expression studies (Mun *et al.*, 2015). Although there is limited data on grafting in other vegetables, cleft grafting seems to be the most appropriate method.

1.2.4 Post-graft healing environment

Figure 1.2 presents a typical time line for graft production (http://www.ces.ncsu.edu/fletcher/programs/ncorganic/research/grafting_techniques.pdf,

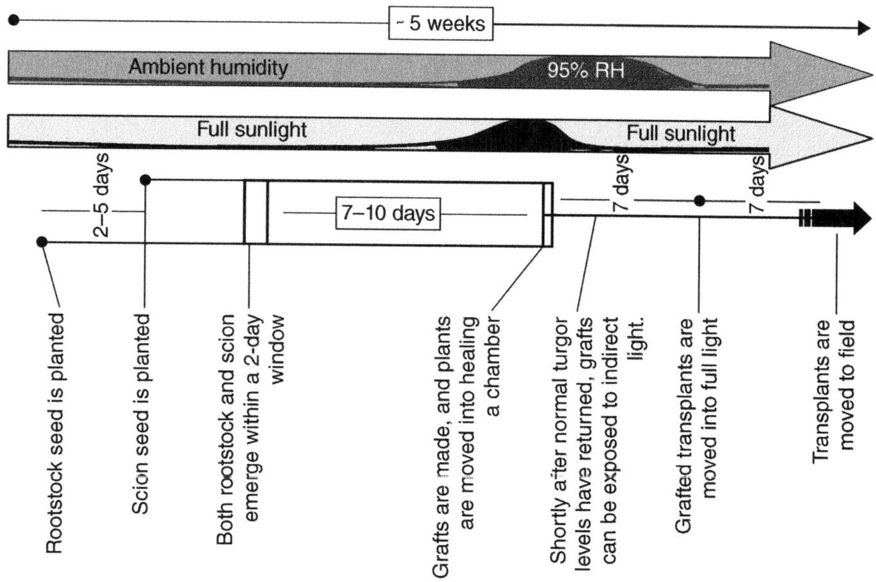

Fig. 1.2. A typical time line for graft production. RH, relative humidity. (Courtesy of C. Rivard and F. Louws, North Carolina State University, Raleigh, USA).

accessed 24 November 2016). Proper care of newly grafted transplants is necessary to secure a higher success rate for the grafting process. In the case of hole insertion grafting, splice grafting, cleft grafting and pin grafting, a very high humidity (95%) is required during the first 48 h when the temperature should be maintained at 27–28°C (82°F). Later, the plantlets can be shifted to a normal greenhouse environment (Guan and Zhao, 2014). Loss of water from the scion during the first 2 days may lead to wilting of the scion and ultimately failure of the grafting process; therefore, a high humidity is required to prevent water loss. Normally, grafted transplants are covered for 5–7 days after grafting with black plastic sheeting or 0.01 mm black polyethylene film to increase humidity, reduce light intensity and promote the healing process (Denna, 1962). However, prolonged covering of grafted transplants leads to unfavourable stem elongation and the plants become spindly, so proper care should be taken to avoid this.

Experienced growers can use plastic tunnels or chambers as healing chambers. However, on a smaller scale, farmers can use specially designed healing containers. In commercial nurseries, the grafted transplants are placed on greenhouse benches and the trays are sealed with 0.01 mm polythene film for 5–7 days to raise the humidity. Partial shading during the day time helps to improve the results. Several types of healing growth chambers with sophisticated controls (temperature and humidity) have been designed by different companies and are used by commercial nurseries in Japan, Korea (Lee *et al.*, 1998, 2008; Kawai *et al.*, 1996), China, Spain and the USA. Dong *et al.* (2015) reported that use of a healing room is becoming increasingly common in China and a 95% grafting success can be obtained on commercial scale using this method.

In a recent report, Li *et al.* (2015) found that the health and vigour of watermelon transplants grafted on to pumpkin and bottle gourd were affected by different light sources (combinations of light controlled by light emitting diodes or LEDs). They obtained healthy and vigorous plantlets under a light source with a red/blue ratio of 7/3 compared with white fluorescent light only. Although they did not comment on the effect of the light source on the healing process, it was apparent that different light sources affected the healing of the graft union. This needs further investigation.

Plants grafted using the tongue approach can attain a high graft success rate without strict humidity control. However, exposure of plantlets to direct sunlight during the healing process should be avoided.

1.3 Problems Associated with Vegetable Grafting

Various problems are associated with the production and management of grafted transplants. The technique is labour-intensive and specialized trained workers are required. It also requires time management for sowing of the rootstock and scion seeds, a controlled environment for graft healing, and efficient grafting machines and robots. Overgrowth of transplants under field conditions may occur, and the yield and quality of scion fruit may also be significantly affected (Huang *et al.*, 2015). Sometimes rootstock–scion incompatibility is observed during the initial stages or after transplantation under field conditions. Careful selection of rootstock and scion combinations is required depending on the prevailing soil and environmental conditions of the area. Both rootstock and scion seeds are required, and hybrid and special types of seed can be costly. The rootstock suckers/offshoots that develop during the healing process or under field conditions (after transplanting) need removal. Moreover, grafting can increase the risk of pathogen spread, especially for seedborne pathogens (e.g. bacterial canker caused by *Clavibacter michiganensis* subsp. *michiganensis* in tomato, bacterial fruit blotch caused by *Acidovorax citrulli* in watermelon and melon, charcoal rot caused by *Macrophomina phaseolina* in melon and bottle gourd, and tomato mosaic virus and pepino mosaic virus infections in tomato) in the nursery. This is due to the use of two seeds for producing a grafted plant and to the use of cutting instruments in the grafting process. For the above reasons, it is important to adopt procedures for preventing the spread of pathogens in the nursery by using seeds that have been certified free of pathogens, and by the periodical disinfection of cutting instruments, the use of clean clothing and disinfected hands by the grafting workers, the periodical disinfection of grafting areas and plant growing environments, and the continuous monitoring of the phytosanitary status of seedlings. Despite vegetable grafting providing many job opportunities for the workforce, researchers have identified some problems directly related to the health of nursery workers. Manual grafting is the leading grafting method (Lee *et al.*, 2010), and workers performing grafting within a greenhouse and growth chamber face the problems of heat stress and discomfort, especially during April–June, September and October (Marucci *et al.*, 2012). Although the working conditions can be adjusted to a certain degree by cooling pads, fans and covering sheets, better facilities (air-conditioned environments) are still required for the welfare of workers. However, the intensity of these problems can be reduced considerably by careful management practices.

1.4 Conclusions

Commercial vegetable grafting has been practised for decades and the area used for grafted vegetables is increasing continuously. The main objective of grafting remains an increase in yield, especially under the high pressure from soilborne pathogens and nematodes and unfavourable environmental conditions (e.g. sub- and supra-optimal temperatures, salinity, drought). Future research should contribute towards improving grafting technologies and nursery management practices to ensure high-quality grafted transplants for growers. Nursery production and management is labour-intensive. To solve this problem, scientists must focus on developing and popularizing facilities, equipment and grafting robots to increase the efficacy of grafting and reduce labour costs. The trend for plug plantlet nurseries is increasing in developing countries. Thus, the use of precise seeders, carrier vehicles, germination rooms, plant growth chambers and acclimatization facilities for grafted transplants should lead to great improvement.

To improve seed germination, uniformity and seedling vigour, seed-priming techniques warrant attention. Similarly, storage technology for grafted transplants demands the consideration of researchers. The development of databases, software, mobile applications and crop models related to grafted vegetables will assist nursery managers and farming communities in the selection of suitable scion and rootstock cultivars, and will also provide guidelines for optimal management practices. Although some problems associated with grafting of vegetables remain, these are outweighed by the benefits attained through grafting, so this technique will continue to proliferate and be adopted worldwide.

References

Arwiyanto, T., Lwin, K., Maryudani, Y. and Purwantoro, A. (2015) Evaluation of local *Solanum torvum* as a rootstock to control *Ralstonia solanacearum* in Indonesia. *Acta Horticulturae* 1086, 101–106.

Ashita, E. (ed.) (1927) *Grafting of Watermelons* Korea (Chosun). *Agricultural Newsletter* 1, 9 (in Japanese).

Asins, M.J., Raga, V., Roca, D., Belver, A. and Carbonell, E.A. (2015) Genetic dissection of tomato rootstock effects on scion traits under moderate salinity. *Theoretical and Applied Genetics* 128, 667–679.

Bhatt, R.M., Upreti, K.K., Divya, M.H., Bhat, S., Pavithra, C.B. and Sadashiva, A.T. (2015) Interspecific grafting to enhance physiological resilience to flooding stress in tomato (*Solanum lycopersicum* L.). *Scientia Horticulturae* 182, 8–17.

Black, L., Wu, D., Wang, J., Kalb, T., Abbass, D. and Chen, J. (2003) Grafting tomatoes for production in the hot-wet season. AVRDC International Cooperators' Guide. AVRDC publication number 03-551. Available at: http://www.ces.ncsu.edu/fletcher/programs/ncorganic/research/grafting.pdf (accessed 30 April 2016).

Bletsos, F.A. (2005) Use of grafting and calcium cyanamide as alternatives to methyl bromide soil fumigation and their effects on growth, yield, quality and Fusarium wilt control in melon. *Journal of Phytopathology* 153, 155–161.

Bletsos, F.A. (2006) Grafting and calcium cyanamide as alternatives to methyl bromide for greenhouse eggplant production. *Scientia Horticulturae* 107, 325–331.

Bletsos, F., Thanassoulopoulos, C. and Roupakias, D. (2003) Effect of grafting on growth, yield, and Verticillium wilt of eggplant. *HortScience* 38, 183–186.

Boncato, T. and Ellamar, J. (2015) Off-season tomato production: a new technology in Tarlac Province of Philippines. *Acta Horticulturae* 1086, 261–267.

Chung, H.D. and Lee, J.M. (2007) Rootstocks for grafting. In: *Horticulture in Korea*. Korean Society for Horticultural Science Technical Bulletin, Korea, pp. 162–167.

Cichy, K.A., Snapp, S.S. and Kirk, W.W. (2007) Fusarium root rot incidence and root system architecture in grafted common bean lines. *Plant Soil* 300, 233-244.

Colla, G., Rouphael, Y., Cardarelli, M., Temperini, O., Rea, E., Salerno, A. and Pierandrei F. (2008). Influence of grafting on yield and fruit quality of pepper (*Capsicum annuum* L.) grown under greenhouse conditions. *Acta Horticulturae* 782, 359–363.

Colla, G., Rouphael, Y., Leonardi, C. and Bie, Z. (2010b) Role of grafting in vegetable crops grown under saline conditions. *Scientia Horticulturae* 127, 147–155.

Colla, G., Cardona Suarez, C.M., Cardarelli, M. and Rouphael, Y. (2010c) Improving nitrogen use efficiency in melon by grafting. *Hortscience* 45, 559–565.

Colla, G., Rouphael, Y., Mirabelli, C. and Cardarelli, M. (2011) Nitrogen-use efficiency traits of mini-watermelon in response to grafting and nitrogen-fertilization doses. *Journal of Plant Nutrition and Soil Science* 174, 933–941.

Colla, G., Rouphael, Y., Leonardi, C. and Bie, Z. (2010a) Role of grafting in vegetable crops grown under saline conditions. *Scientia Horticulturae* 127, 147–155.

Colla, G., Rouphael, Y., Rea, E. and Cardarelli, M. (2012) Grafting cucumber plants enhance tolerance to sodium chloride and sulfate salinization. *Scientia Horticulturae* 135, 177–185.

Colla, G., Rouphael, Y., Jawad, R., Kumar, P., Rea, E. and Cardarelli, M. (2013) The effectiveness of grafting to improve NaCl and $CaCl_2$ tolerance in cucumber. *Scientia Horticulturae* 164, 380–391.

Condurso, C., Verzera, A., Dima, G., Tripodi, G., Crinò, P., Paratore, A. and Romano, D. (2012) Effects of different rootstocks on aroma volatile compounds and carotenoid content of melon fruits. *Scientia Horticulturae* 148, 9–16.

Denna, D.W. (1962) A simple grafting technique for cucurbits. *Proceedings of the American Society of Horticultural Sciences USA* 81, 369–370.

Dong, D., Cao, Z., Wang, X., Hu, J. and Gullino, M.L. (2007) Effect of nematode resistant rootstocks on growth characteristics and yields of tomato. *Acta Horticulturae Sinica* 34, 1305–1308.

Dong, W., Zhou, Z.C., Bu, Y.L., Zhuo, J.Q., Chen, L.Z. and Li, Y.Z. (2015) Research and application of grafted seedlings healing room. *Acta Horticulturae* 1086, 51–57.

Dor, E., Alperin, B., Wininger, S., Ben-Dor, B., Somvanshi, V.S., Koltai, H., Kapulnik, Y. and Hershenhorn, J. (2010) Characterization of a novel tomato mutant resistant to the weedy parasites *Orobanche* and *Phelipanche* spp. *Euphytica* 171, 371–380.

Esmaeili, M., Salehi, R., Taheri, M.R., Babalar, M. and Mohammadi, H. (2015) Effect of different nitrogen rates on fruit yield and quality of grafted and non-grafted muskmelon. *Acta Horticulturae* 1086, 255–260.

Fan, M., Bie, Z., Krumbein, A. and Schwarz, D. (2011) Salinity stress in tomatoes can be alleviated by grafting and potassium depending on the rootstock and K-concentration employed. *Scientia Horticulturae* 130, 615–623.

Farhadi, A. and Malek, S. (2015) Evaluation of graft compatibility and organoleptic traits of greenhouse cucumber seedlings grafted on different rootstocks. *Acta Horticulturae* 1086, 219–224.

Fuentes, I., Stegemann, S., Golczyk, H., Karcher, D. and Bock, R. (2014) Horizontal genome transfer as an asexual path to the formation of new species. *Nature* 511, 232–235.

Gao, Y., Tian, Y., Liang, X. and Gao, L. (2015) Effects of single-root-grafting, double-root-grafting and compost application on microbial properties of rhizosphere soils in Chinese protected cucumber (*Cucumis sativus* L.) production systems. *Scientia Horticulturae* 186, 190–200.

Ghanem, M.E., Albacete, A., Smigocki, A.C., Frébort, I., Pospíilová, H., Martínez-Andújar, C., Acosta, M., Sánchez-Bravo, J., Lutts, S., Dodd, I.C. and Pérez-Alfocea, F. (2011) Root-synthesized cytokinins improve shoot growth and fruit yield in salinized tomato (*Solanum lycopersicum* L.) plants. *Journal of Experimental Botany* 62, 125–140.

Gilardi, G., Gullino, M.L. and Garibaldi, A. (2013a) Critical aspects of grafting as a possible strategy to manage soil-borne pathogens. *Scientia Horticulturae* 149, 19–21.

Gilardi, G., Baudino, M., Moizio, M., Pugliese, M., Garibaldi, A. and Gullino, M.L. (2013b) Integrated management of *Phytophthora capsici* on bell pepper by combining grafting and compost treatment. *Crop Protection* 53, 13–19.

Gisbert, C., Prohens, J., Raigón, M.D., Stommel, J.R. and Nuez, F. (2011) Eggplant relatives as sources of variation for developing new rootstocks: effects of grafting on eggplant yield and fruit apparent quality and composition. *Scientia Horticulturae* 128, 14–22.

Guan, W. and Zhao, X. (2014) *Techniques for Melon Grafting*. Document no. HS1257, Horticultural Sciences Department, University of Florida, Florida.

Han, X.Y., Huang, Q.W., Kong, X.Y., Luo, F., Wang, S., Wan, S.L. and Lin, D.P. (2015) Effects of different stocks on the development and fruit quality of Hami melon. *Acta Horticulturae* 1086, 237–246.

Holbrook, N.M., Shashidhar, V.R., James, R.A. and Munns, R. (2002) Stomatal control in tomato with ABA-deficient roots: response of grafted plants to soil drying. *Journal of Experimental Botany* 53, 1503–1514.

Hong, M.S. (1710) Forest economics. 1, 38–39 (see PSNCK (1982) for translated version).

Huang, Y., Tang, R., Cao, Q. and Bie, Z. (2009) Improving the fruit yield and quality of cucumber by grafting onto the salt tolerant rootstock under NaCl stress. *Scientia Horticulturae* 122, 26–31.

Huang, Y., Bie, Z., He, S., Hua, B., Zhen, A. and Liu, Z. (2010) Improving cucumber tolerance to major nutrients induced salinity by grafting onto *Cucurbita ficifolia*. *Environmental and Experimental Botany* 69, 32–38.

Huang, Y., Bie, Z., Liu, P., Niu, M., Zhen, A., Liu, Z., Lei, B., Gu, D., Lu, C. and Wang, B. (2013a) Reciprocal grafting between cucumber and pumpkin demonstrates the roles of the rootstock in the determination of cucumber salt tolerance and sodium accumulation. *Scientia Horticulturae* 149, 47–54.

Huang, Y., Li, J., Hua, B., Liu, Z., Fan, M. and Bie, Z. (2013b) Grafting onto different rootstocks as a means to improve watermelon tolerance to low potassium stress. *Scientia Horticulturae* 149, 80–85.

Huang, Y., Kong, Q.S., Chen, F. and Bie, Z. (2015) The history, current status and future prospects of the vegetable grafting in China. *Acta Horticulturae* 1086, 31–39.

Huang, Y., Zhao, L.Q., Kong, Q.S., Cheng, F., Niu, M.L., Xie, J.J., Nawaz, M.A. and Bie, Z.L. (2016a) Comprehensive mineral nutrition analysis of watermelon grafted onto two different rootstocks. *Horticultural Plant Journal* 2, 105–113.

Huang, Y., Jiao, Y., Nawaz, M.A., Chen, C., Liu, L., Lu, Z., Kong, Q., Cheng, F. and Bie, Z. (2016b) Improving magnesium uptake, photosynthesis and antioxidant enzyme activities of watermelon by grafting onto pumpkin rootstock under low magnesium. *Plant and Soil* doi: 10.1007/s11104-016-2965-3.

Itagi, T. (1992) Status of transplant production systems in Japan and new grafting techniques. In: *Symposium on Protected Horticulture*. Horticultural Experimental Station, Rural Development Administration, Suwon, Republic of Korea, pp. 32–67 (in Korean).

Ito, T. (1992) Present state of transplant production practices in Japanese horticultural industry. In: Kurata, K. and Kozai, T. (eds) *Transplant Production System*. Kluwer Academic Publishers, Yokohama, Japan, pp. 65–82.

Jang, Y., Yang, E., Cho, M., Um, Y., Ko, K. and Chun, C. (2012) Effect of grafting on growth and incidence of phytophthora blight and bacterial wilt of pepper (*Capsicum annuum* L.). *Horticulture, Environment, and Biotechnology* 53, 9–19.

Janick, J. (1986) *Horticultural Science*, 4th edn. W.H. Freeman & Co., New York.

Jeong, S.J. (1986) Effects of several rootstocks on the growth, fruit yield and quality of two cucumber cultivars. Institute of Food Development, Kyung Hee University, Suwon, Korea. *Research Collection* 7, 67–76 (in Korean with English summary).

Johnson, S., Kreider, P. and Miles, C. (2011) Vegetable grafting eggplants and tomatoes. Washington State University Extension Fact Sheet, FS052E, Washington State University, Washington, DC.

Kato, T. and Lou, H. (1989) Effect of rootstock on the yield, mineral nutrition and hormone level in xylem sap in eggplant. *Journal of the Japanese Society for Horticultural Science* 58, 345–352.

Kawai, J., Onouti, N., Niwa, K. and Ito, T. (1996) The development of grafts for seedlings raised in cellular pots and simple acclimating equipment in tomato plants. Aichi Agricultural Research Center, Japan. *Research Bulletin* 28, 149–155 (in Japanese with English summary).

Kim, S.E. and Lee, J.M. (1989) *Effect of Rootstocks and Fertilizers on the Growth and Mineral Contents in Cucumber* (*Cucumis sativus* L.). Institute of Food Development, Kyung Hee University, Suwon, Korea. *Research Collection* 10, 75–82 (in Korean with English summary).

King, S.R., Davis, A.R., Liu, W.G. and Levi, A. (2008) Grafting for disease resistance. *HortScience* 43, 1673–1676.

King, S.R., Davis, A.R., Zhang, X. and Crosby, K. (2010) Genetics, breeding and selection of rootstocks for Solanaceae and Cucurbitaceae. *Scientia Horticulturae* 127, 106–111.

Kyriacou, M.C., Soteriou, G.A., Rouphael, Y., Siomos, A.S. and Gerasopoulos, D. (2016). Configuration of watermelon fruit quality in response to rootstock-mediated harvest maturity and postharvest storage. *Journal of the Science of Food and Agriculture* 96, 2400–2409.

Ko, K.D. (1999) Response of cucurbitaceous rootstock species to biological and environmental stresses. PhD thesis, Seoul National University, Seoul, Korea.

Kousik, C.S., Donahoo, R.S. and Hassell, R. (2012) Resistance in watermelon rootstocks to crown rot caused by *Phytophthora capsici*. *Crop Protection* 39, 18–25.

Krumbein, A. and Schwarz, D. (2013) Grafting: a possibility to enhance health-promoting and flavour compounds in tomato fruits of shaded plants? *Scientia Horticulturae* 149, 97–107.

Kubota, C., McClure, M.A., Kokalis-Burelle, N., Bausher, M.G. and Rosskopf, E.N. (2008) Vegetable grafting: history, use, and current technology status in North America. *HortScience* 43, 1164–1669.

Kumar, P., Lucini, L., Rouphael, Y., Cardarelli, M., Kalunke, R.M., and Colla, G. (2015a) Insight into the role of grafting and arbuscular mycorrhiza on cadmium stress tolerance in tomato. *Frontiers in Plant Science* 6, 477.

Kumar, P., Rouphael, Y., Cardarelli, M., and Colla, G. (2015b) Effect of nickel and grafting combination on yield, fruit quality, antioxidative enzyme activities, lipid peroxidation, and mineral composition of tomato. *Journal of Plant Nutrition and Soil Science* 178, 848–860.

Lee, J.M. (1994) Cultivation of grafted vegetables. I. Current status, grafting methods, and benefits. *HortScience* 29, 235–239.

Lee, J.M. and Oda, M. (2003) Grafting of herbaceous vegetable and ornamental crops. *Horticultural Reviews* 28, 61–124.

Lee, J.M., Bang, H.J. and Ham, H.S. (1998) Grafting of vegetables. *Journal of the Japanese Society for Horticultural Science* 67, 1098–1114.

Lee, J.M., Kubota, C., Tsao, S.J., Vinh, N.Q., Huang, Y. and Oda, M. (2008) Recent progress in vegetable grafting. In: *International Workshop on Development and Adaptation of Green Technology for Sustainable Agriculture and Enhancement of Rural Entrepreneurship*. IRRI, Los Banos, Laguna, Philippines, p. 21.

Lee, J.M., Kubota, C., Tsao, S.J., Bie, Z., Echevarria, P.H., Morra, L. and Oda, M. (2010) Current status of vegetable grafting: diffusion, grafting techniques, automation. *Scientia Horticulturae* 127, 93–105.

Li, J.-Y., Tian, H.-X., Li, X.-G., Meng, J.-J., and He, Q.-W. (2008) Higher chilling-tolerance of grafted-cucumber seedling leaves upon exposure to chilling stress. *Agricultural Sciences in China* 7, 570–576.

Li, X.E., Liu, F.Y., Huang, Y., Kong, Q.S., Wan, Z.J., Li, X. and Bie, Z. (2015) Growth and physiology of grafted watermelon seedlings as affected by different light sources. *Acta Horticulturae* 1086, 59–64.

López-Marín, J., González, A., Pérez-Alfocea, F., Egea-Gilabert, C. and Fernández, J.A. (2013) Grafting is an efficient alternative to shading screens to alleviate thermal stress in greenhouse-grown sweet pepper. *Scientia Horticulturae* 149, 39–46.

Louws, F.J., Rivard, C.L. and Kubota, C. (2010) Grafting fruiting vegetables to manage soil borne pathogens, foliar pathogens, arthropods and weeds. *Scientia Horticulturae* 127, 127–146.

Marucci, A., Pagniello, B., Monarca, D., Colantoni, A., Biondi, P. and Cecchini, M. (2012) The heat stress for workers during vegetable grafting in greenhouses. In: *International Conference RAGUSA SHWA*. Ragusa, Italy, pp. 321–328.

Melnyk, C.W. and Meyerowitz, E.M. (2015) Plant grafting. *Current Biology* 25, R183–R188.

Melnyk, C.W., Schuster, C., Leyser, O. and Meyerowitz, E.M. (2015) A developmental framework for graft formation and vascular reconnection in *Arabidopsis thaliana*. *Current Biology* 25, 1306–1318.

Miles, C., Wimer, J. and Inglis, D. (2015) Grafting eggplant and tomato for Verticillium wilt resistance. *Acta Horticulturae* 1086, 113–118.

Mohamed, F.H., Abd El-Hamed, K.E., Elwan, M.W.M. and Hussien, M.N.E. (2014) Evaluation of different grafting methods and rootstocks in watermelon grown in Egypt. *Scientia Horticulturae* 168, 145–150.

Mohammadi, H., Salehi, R. and Esmaeili, M. (2015) Yield and fruit quality of grafted and non-grafted muskmelon (*Cucumis melo* L.) affected by planting density. *Acta Horticulturae* 1086, 247–254.

Moncada, A., Miceli, A., Vetrano, F., Mineo, V., Planeta, D. and D'Anna, F. (2013) Effect of grafting on yield and quality of eggplant (*Solanum melongena* L.). *Scientia Horticulturae* 149, 108–114.

Mudge, K., Janick, J., Scofield, S. and Goldschmidt, E.E. (2009) A history of grafting. *Horticultural Reviews* 35, 437–493.

Mun, M., Yi, H., Nou, I. and Hur, Y. (2015) Grafting-induced gene expression change in *Brassica rapa* leaves is different from fruit trees. *Plant Breeding and Biotechnology* 3, 67–76.

Nawaz, M.A., Imtiaz, M., Kong, Q., Fei, C., Ahmed, W., Huang, Y. and Bie, Z. (2016). Grafting: a technique to modify ion accumulation in horticultural crops. *Frontiers in Plant Science* 7, 1457.

Nisini, P.T., Colla, G., Granati, E., Temperini, O., Crino, P., and Saccardo, F. (2002). Rootstock resistance to fusarium wilt and effect on fruit yield and quality of two muskmelon cultivars. *Scientia Horticulturae* 93, 281–288.

Oda, M. (1994) Effects of uniconazole and gibberellic acid application on elongation of hypocotyls and internodes in figleaf gourd for rootstock. *Japan Agricultural Research Quarterly* 28, 195–199.

Oda, M. (1999) Grafting of vegetables to improve greenhouse production. *Food and Fertilizer Technology Center Extension Bulletin* 480, 1–11.

Oda, M. (2002) Grafting of vegetable crops. *Scientific Report of the Graduate School of Agriculture and Biological Sciences, Osaka Prefecture University* 54, 49–72.

Penella, C., Nebauer, S.G., Bautista, A.S., López-Galarza, S. and Calatayud, A. (2014) Rootstock alleviates PEG-induced water stress in grafted pepper seedlings: physiological responses. *Journal of Plant Physiology* 171, 842–851.

Penella, C., Nebauer, S.G., Quinones, A., Bautista, A.S., López-Galarza, S. and Calatayud, A. (2015) Some rootstocks improve pepper tolerance to mild salinity through ionic regulation. *Plant Science* 230, 12–22.

Proietti, S., Rouphael, Y., Colla, G., Cardarelli, M., De Agazio, M., Zacchini, M., Rea, E., Moscatello, S. and Battistelli, A. (2008) Fruit quality of mini-watermelon as affected by grafting and irrigation regimes. *Journal of the Science of Food and Agriculture* 88, 1107–1114.

PSNCK (1982) *Forest Economics*. Rev. Korean edition of original writings of M.S. Hong. Proceedings of the Society of National Culture of Korea, Korea.

Rouphael, Y., Cardarelli, M., Rea, E. and Colla, G. (2008) Grafting of cucumber as a means to minimize copper toxicity. *Environmental and Experimental Botany* 63, 49–58.

Rouphael, Y., Schwarz, D., Krumbein, A. and Colla, G. (2010) Impact of grafting on product quality of fruit vegetables. *Scientia Horticulturae* 127, 172–179.

Ruiz, J.M., Belakbir, A., Lopez-Cantarero, I. and Romero, L. (1997) Leaf-macronutrient content and yield in grafted melon plants. A model to evaluate the influence of rootstock genotype. *Scientia Horticulturae* 71, 227–234.

Sakata, Y., Sugiyama, M., Ohara, T. and Morishita, M. (2006) Influence of rootstocks on the resistance of grafted cucumber (*Cucumis sativus* L.) scions to powdery mildew (*Podosphaera xanthii* U. Braun & N. Shishkoff). *Journal of Japanese Society for Horticultural Sciences* 75, 135–140.

Sakata, Y., Ohara, T. and Sugiyama, M. (2007) The history and present state of the grafting of cucurbitaceous vegetables in Japan.. *Acta Horticulturae* 731, 159–170.

Sakata, Y., Ohara, T. and Sugiyama, M. (2008) The history of melon and cucumber grafting in Japan. *Acta Horticulturae* 767, 217–228.

Salar, N., Salehi, R. and Delshad, M. (2015) Effect of grafting and nitrogen application on yield and fruit quality of grafted and non-grafted melon. *Acta Horticulturae* 1086, 225–230.

Schwarz, D., Rouphael, Y., Colla, G. and Venema, J.H. (2010) Grafting as a tool to improve tolerance of vegetables to abiotic stresses: thermal stress, water stress and organic pollutants. *Scientia Horticulturae* 127, 162–171.

Schwarz, D., Öztekin, G.B., Tüzel, Y., Brückner, B. and Krumbein, A. (2013) Rootstocks can enhance tomato growth and quality characteristics at low potassium supply. *Scientia Horticulturae* 149, 70–79.

Shibuya, T., Itagaki, K., Wang, Y. and Endo, R. (2015) Grafting transiently suppresses development of powdery mildew colonies, probably through a quantitative change in water relations of the host cucumber scions during graft healing. *Scientia Horticulturae* 192, 197–199.

Soteriou, G.A., Kyriacou, M.C., Siomos, A.S. and Gerasopoulos, D. (2015) Rootstock-mediated effects on watermelon ripening behavior and fruit physicochemical and phytochemical composition. *Acta Horticulturae* 1079, 707–714.

Suchoff, D., Gunter, C., Schulthesis, J. and Louws, F.J. (2015) On farm grafted tomato trial to manage bacterial wilt. *Acta Horticulturae* 1086, 119–127.

Temperini, O., Calabrese, N., Temperini, A., Rouphael, Y., Tesi, R., Lenzi, A., Carito, A. and Colla, G. (2013) Grafting artichoke onto cardoon rootstocks: graft compatibility, yield and verticillium wilt incidence. *Scientia Horticulturae* 149, 22–27.

Tsaballa, A., Athanasiadis, C., Pasentsis, K., Ganopoulos, I., Nianiou-Obeidat, I. and Tsaftaris, A. (2013) Molecular studies of inheritable grafting induced changes in pepper (*Capsicum annuum*) fruit shape. *Scientia Horticulturae* 149, 2–8.

Venema, J.H., Dijk, B.E., Bax, J.M., Hasselt, P.R.V. and Elzenga, J.T.M. (2008) Grafting tomato (*Solanum lycopersicum*) onto the rootstock of a high-altitude accession of *Solanum habrochaites* improves suboptimal-temperature tolerance. *Environmental and Experimental Botany* 63, 359–367.

Verzera, A., Dima, G., Tripodi, G., Condurso, C., Crinò, P., Romano, D., Mazzaglia, A., Lanza, C.M., Restuccia, C. and Paratore, A. (2014) Aroma and sensory quality of honeydew melon fruits (*Cucumis melo* L. subsp. *melo* var. inodorus H. Jacq.) in relation to different rootstocks. *Scientia Horticulturae* 169, 118–124.

Vitale, A., Rocco, M., Arena, S., Giuffrida, F., Cassaniti, C., Scaloni, A., Lomaglio, T., Guarnaccia, V., Polizzi, G., Marra, M. and Leonardi, C. (2014) Tomato susceptibility to Fusarium crown and root rot: effect of grafting combination and proteomic analysis of tolerance expression in the rootstock. *Plant Physiology and Biochemistry* 83, 207–216.

Wahb-Allah, M.A. (2014) Effectiveness of grafting for the improvement of salinity and drought tolerance in tomato (*Solanum lycopersicon* L.). *Asian Journal of Crop Science* 6, 112–122.

Xing, W.W., Li, L., Gao, P., Li, H., Shao, Q.S., Shu, S., Sun, J. and Guo, S.R. (2015) Effects of grafting with pumpkin rootstock on carbohydrate metabolism in cucumber seedlings under $Ca(NO_3)_2$ stress. *Plant Physiology and Biochemistry* 87, 124–132.

Yamakawa, B. (1983) Grafting. In: Nishi (ed.) *Vegetable Handbook*. Yokendo Book Company, Tokyo, Japan, pp. 141–153 (in Japanese).

Yang, Y., Wang, L., Tian, J., Li, J., Sun, J., He, L., Guo, S. and Tezuka, T. (2012) Proteomic study participating the enhancement of growth and salt tolerance of bottle gourd rootstock-grafted watermelon seedlings. *Plant Physiology and Biochemistry* 58, 54–65.

Yang, Y., Lu, X., Yan, B., Li, B., Sun, J., Guo, S. and Tezuka, T. (2013) Bottle gourd rootstock-grafting affects nitrogen metabolism in NaCl-stressed watermelon leaves and enhances short-term salt tolerance. *Journal of Plant Physiology* 170, 653–661.

Zhang, Z.-K., Liu, S.-Q. Hao, A.-Q. and Liu, S.-H. (2010a) Grafting increases the copper tolerance of cucumber seedlings by improvement of polyamine contents and enhancement of antioxidant enzymes activity. *Agricultural Sciences in China* 9, 985–994.

Zhang, Z.-K., Zhag, H.L., Huang, Z.-J., Chen, K. and Liu, S.-Q. (2010b) Grafting enhances copper tolerance of cucumber through regulating nutrient uptake and antioxidative system. *Agricultural Sciences in China* 9, 1758–1770.

Zhang, Z.-K., Li, H., He, H.-J. and Liu, S.-Q. (2013) Grafting raises the Cu tolerance of cucumber through protecting roots against oxidative stress induced by Cu stress. *Journal of Integrative Agriculture* 12, 815–824.

Zhao, X., Guo, Y., Huber, D.J. and Lee, J. (2011) Grafting effects on postharvest ripening and quality of 1-methylcyclopropene-treated muskmelon fruit. *Scientia Horticulturae* 130, 581–587.

2 Genetic Resources for Rootstock Breeding

Maria Belen Pico,[1*] Andrew J. Thompson,[2] Carmina Gisbert,[1] Halit Yetişir[3] and Penelope J. Bebeli[5]

[1]Universitat Politècnica de Valencia, Valencia, Spain; [2]Cranfield University, Bedfordshire, UK; [3]Erciyes University, Kayseri, Turkey; [5]Agricultural University of Athens, Athens, Greece

2.1 Genetic Diversity

2.1.1 Diversity in the Cucurbitaceae family

The Cucurbitaceae comprises about 960 species (Jeffrey, 2005; Schaefer *et al.*, 2009). Three genera, *Cucumis*, *Citrullus* (Tribe Benincaseae) and *Cucurbita* (Cucurbiteae), include the most widely cultivated cucurbits worldwide. In addition to these, there are other notable cucurbits belonging to regionally cultivated genera, such as *Benincasa*, *Lagenaria* (Benincaseae), *Sicana* (Cucurbiteae), *Momordica* (Momordiceae), *Luffa*, *Sechium*, *Sicyos*, and *Trichosanthes* (Sicyeae) (Bates *et al.*, 1995; Behera *et al.*, 2008; Janick and Paull, 2008; Schaefer and Renner, 2011). Most are not only cultivated as crops but also used as rootstocks for the same or a different species. Knowledge of their origin and diversification, their distribution and ecology, and the genetic relationships among the different species is essential for selecting the potentially best rootstocks for specific scions. The family is known to have an Asian origin, but currently approximately 40% of the Cucurbitaceae species are endemic in the South American continent, while the rest occur in Africa, Asia, Australia and Europe (Schaefer *et al.*, 2009; Renner and Pandey, 2013).

The genus *Cucumis* is considered to be of Asian origin (Schaefer *et al.*, 2009). *Cucumis melo* ($2n = 24$), which is split into two subspecies (*melo* and *agrestis*), is the most variable species in the genus. It could have originated in Asia and then reached Africa (Sebastian *et al.*, 2010). The South Asian species *Cucumis trigonus* and *Cucumis callosus* probably represent the wild progenitors and are fully crossable with melon. Wild melons are also found in East and West Africa, and

* mpicosi@btc.upv.es

Central Asia. The primary and secondary centres of diversity for this species are located from eastern Asia to the Mediterranean Sea. *Cucumis sativus* (2n = 14) was probably domesticated on the Indian subcontinent (Sebastian *et al.*, 2010). China is considered a secondary centre of diversity for this species. *C. sativus* is less variable than melons, but houses several botanical varieties, including the cultivated cucumber and the wild var. *hardwickii*, cross-compatible with *C. sativus*. Additionally, the genus *Cucumis* includes a number of other species, such as *Cucumis africanus*, *Cucumis anguria*, *Cucumis dipsaceus*, *Cucumis ficifolius*, *Cucumis hystrix*, *Cucumis metuliferus*, *Cucumis myriocarpus*, *Cucumis picrocarpus*, *Cucumis pustulatus*, *Cucumis zeyheri*. Severe cross-breeding barriers prevent interspecific crosses in this genus (Kirkbride, 1993; Matsumoto *et al.*, 2012).

It is currently generally accepted that watermelon, *Citrullus* spp. (2n = 22), originated in Africa, while China is considered a secondary centre of diversity. Archaeological remains and historical records, some over 4000 years old, and the current presence of wild and primitive watermelons in the region suggest that the dessert watermelon originated in north-eastern Africa (Paris, 2015). *Citrullus lanatus* has been considered to contain three subspecies: *lanatus* (citron watermelon), *mucosospermus* (egusi watermelon) and *vulgaris* (dessert watermelon). Although cross-compatibility among these three taxa is quite high, difficulties have been encountered in crossing them, and recently it was proposed that they should be considered as separate species (Chomicki and Renner, 2015). Other *Citrullus* species are *Citrullus rehmii* and *Citrullus ecirrhosus*, endemic to desert regions of Namibia. There is yet another non-cultivated species from southern Africa, *Citrullus naudinianus*. Successful cross-breeding of *C. ecirrhosus* and *C. lanatus* has been reported (Levi *et al.*, 2005; Dane and Liu, 2007).

Benincasa hispida (2n = 24) and *Lagenaria siceraria* (2n = 22) are two other regionally important species of the tribe Benincaseae. *B. hispida* has its centre of diversity in the Indo-China region (Sureja *et al.*, 2006; Pandey *et al.*, 2008). *L. siceraria* is native to Africa, but was used by diverse human cultures across Eurasia, the Pacific Islands, and the New World during pre-Columbian times (Decker-Walters *et al.*, 2004; Erickson *et al.*, 2005). Recent phylogenetic studies indicated that *Lagenaria* was brought into North America from Asia (Kistler *et al.*, 2014). Wild *Lagenaria* spp., such as *Lagenaria sphaerica*, *Lagenaria abyssinica* and *Lagenaria breviflora*, are distributed in northern Africa. Successful crosses of *L. siceraria* with some of these wild types have been reported (Decker-Walters *et al.* 2004; Levi *et al.*, 2009; Bhawna *et al.*, 2014).

Pumpkin and squash species, *Cucurbita* spp. (2n = 40) originated in Mexico, other parts of Central America and South America. The most economically important species of the genus are *Cucurbita pepo*, *Cucurbita moschata* and *Cucurbita maxima* (Ferriol and Picó, 2008). *Cucurbita argyrosperma* and *Cucurbita ficifolia* are not as widely cultivated. Each cultivated species was domesticated independently in distinct regions of the American continent. A high level of variation has been reported in *C. pepo*, *C. maxima* and *C. moschata*, whereas lower levels have been found in *C. ficifolia* and *C. argyrosperma* (Ferriol and Picó, 2008). *Cucurbita* has also non-cultivated species, including the xerophytic *Cucurbita digitata*, *Cucurbita palmata*, *Cucurbita cordata*, *Cucurbita foetidissima*, *Cucurbita radicans* and *Cucurbita pedatifolia* and the mesophytic species *Cucurbita okeechobeensis* and

Cucurbita lundelliana (Sanjur *et al.*, 2002; Zheng *et al.*, 2013). These wild and cultivated species have diverse crossability relationships. Among the cultivated species, *C. maxima* is partially crossable with *C. moschata*, *C. argyrosperma*, *C. ficifolia* and *Cucurbita ecuadorensis*; *C. moschata* crosses with *C. argyrosperma* and *C. lundelliana*; and *C. ficifolia* crosses with *C. lundelliana*, *C. foetidissima* and *C. pedatifolia* (Lira *et al.*, 2009; Zheng *et al.*, 2013). Another less known species in the same tribe as Cucurbiteae is *Sicana odorifera* (2n = 40). This is a large, herbaceous perennial vine, native to tropical South America, grown as an ornamental plant and for its non-bitter, edible fruits, which resemble squash (Janick and Paull, 2008; Schaefer and Renner, 2011).

The genus *Momordica* comprises 59 species (47 in Africa and 12 in Asia and Australia) growing in rainforest, deciduous forest, bushland, savannah or grassland. Most species are perennial climbers, but two species are small shrubs and two species are annuals (Schaefer and Renner, 2011). They are probably of Indian origin, and are widely cultivated in India, China, Malaysia, tropical and subtropical Africa, and South America. *Momordica charantia* (2n = 22) is the most important species (Schaefer and Renner, 2011). Two main varieties have been reported: var. *charantia* and var. *muricata*. Several studies describe the existence of high diversity levels for this species in Indian landraces, and a secondary centre of diversity is located in China (Dalamu *et al.*, 2012).

The genera *Luffa* (2n = 26) and *Sechium* (2n = 28) comprise very popular vegetables in the tropics. *Luffa* originated in the Old World or Australia, and then reached the New World from Africa (Sirohi *et al.*, 2005; Sebastian *et al.*, 2010). This genus comprises several species from the Old World (*Luffa aegyptiaca*, *Luffa acutangula*, *Luffa echinata*, *Luffa graveolens*, *Luffa hermaphrodita*, *Luffa tuberosa* and *Luffa umbellata*) and from the New World (*Luffa quinquefida*, *Luffa operculata* and *Luffa astorii*) (Pandravada *et al.*, 2014). *L. aegyptiaca* and *L. acutangula* are the two main cultivated species, especially in Asia, India and Africa. Interspecific crosses allowing gene flow have been reported between these two species (Marr *et al.*, 2005). In the genus *Sechium*, the most important species is *Sechium edule*. This is native to Central America, and Mexico is considered the main centre of diversity (Newstrom, 1991). The wild ancestor of the cultivated forms is *S. edule* ssp. *sylvestre*, which is endemic to Mexico (Cross *et al.*, 2006). Within the same tribe, *Sicyos angulatus* (2n = 24) is native to the Americas but has spread as an invasive plant in Asia and Europe, as it is adapted to wet soils and tolerant to low root temperatures (Kurokawa *et al.*, 2009). Another less known species in the same tribe is *Trichosanthes cucumerina* (2n = 22). This originated in India where it is quite popular (Schaefer and Renner, 2011) as a quick-growing annual plant adapted to humid tropics. It is also cultivated in South-east Asia, northern Australia, tropical Africa and America.

2.1.2 Diversity in the Solanaceae family

The Solanaceae family consists of approximately 90 genera and 3000–4000 species (Knapp *et al.*, 2004). It contains economically important food crops cultivated worldwide, such as potato (*Solanum tuberosum*), tomato (*Solanum lycopersicum*), aubergine (*Solanum melongena*) and pepper (*Capsicum annuum*). This family includes other important crops such as tobacco (*Nicotiana tabacum*) and ornamental and poisonous species. Tomato and pepper originated in the New World, while aubergine has its origins in the Old World. The commercially grafted crops tomato, aubergine and pepper all have the same basic chromosome number (2n = 24), and cultivated genetic material, as well as wild relatives of these three crops, has been used for grafting. Therefore, knowledge of the existing resources and their genetic relationships will assist in rootstock breeding programmes.

Half of the Solanaceae species, approximately 1400, belong to the genus *Solanum*, which exhibits wide genetic and phenotypic variation (Knapp *et al.*, 2004, 2013). The highest number of *Solanum* spp. can be found in South America, and secondary centres of diversity are in North and Central America, Asia and Africa. The tomato was classified in 1754 as *Lycopersicon esculentum* Mill. But, following the Linnaean nomenclature, was reclassified in 2006 as *S. lycopersicum*. It is an herbaceous perennial that is cultivated as an annual (Peralta *et al.*, 2006). Its centre of origin is in the Andean region where its wild relatives are found (Díez and Nuez, 2008). It belongs to the section *Lycopersicon*, which is distributed from the high Andes to Equador, Peru and Chile, and the Galapagos Islands (Peralta *et al.*, 2005). *S. lycopersicum* can be divided into two botanical varieties: var. *cerasiforme* and var. *lycopersicum*. *S. lycopersicum* var. *cerasiforme* is phylogenetically positioned between *S. lycopersicum* var. *lycopersicum* and *Solanum pimpinellifolium* (Blanca *et al.*, 2012, 2015). *S. lycopersicum* var. *cerasiforme* is sometimes considered to be synonymous with 'cherry tomato', but this is not supported by molecular phylogenetic data, as many 'cherry tomato' accessions are not considered to be true *S. lycopersicum* var. *cerasiforme*, but may be admixtures from interbreeding of *S. pimpinellifolium* and *S. lycopersicum* var. *cerasiforme* (Blanca *et al.* 2015). There are 12 closest wild relatives of tomato that belong to the same section *Lycopersicum* (Peralta *et al.*, 2008). Their geographical distribution, habitat, mating systems and genetic polymorphisms have been reviewed by Bauchet and Causse (2012). For example, *S. pimpinellifolium*, referred to as 'currant tomato', is native to south Equador and north Peru, and is the only relative with small red edible fruits (Peralta *et al.*, 2008); it has recently been characterized as an endangered species (Bauchet and Causse, 2012). The most variable wild species are *Solanum chilense*, *Solanum peruvianum*, *Solanum pennellii* and *Solanum habrochaites*; of these, *S. chilense* is native to southern Peru and northern Chile in a wide range of habitats and altitudes (0–3000 m above sea level), whereas *S. peruvianum* is native to central Peru and northern Chile, and *S. pennellii* is native to Peru. Among the wild relatives, the most commonly utilized in tomato rootstock breeding is *S. habrochaites*, which grows at high altitudes (up to 3600 m above sea level) in the Peruvian Andes in xeric habitats and at low temperatures (Venema *et al.*, 1999). Within this section *Lycopersicon*, most species can be intercrossed but with varying degrees of difficulty, as described by Stevens and Rick (1986). In addition there are two further sections considered to be close wild relatives of cultivated tomato: the section *Lycopersicoides* containing *Solanum lycopersicoides* and *Solanum sitiens*, which have been successfully been crossed with *S. lycopersicum* on occasions (Ji *et al.*, 2004; Chetelat, 2016), and the section *Juglandifolia* containing the two species *Solanum juglandifolium* and *Solanum ochranthum*, which, although morphologically more distinct and not sexually compatible with *S. lycopersicum*, are graft-compatible with *S. lycopersicum* hybrids (Chetelat and Petersen, 2003).

Common aubergine (*S. melongena*) is an important vegetable consumed worldwide. It belongs to the largest subgenus of the genus *Solanum*, that is, *Leptostemonum*, the 'spiny solanums', which includes the cultivated species *S. melongena* and *Solanum macrocarpon* (section *Melongena*), and *Solanum aethiopicum* (section *Oliganthes*), as well as many wild relatives. *S. melongena* is of Asian origin, with at least two domestication events in India and in southern China/South-east Asia (Meyer *et al.*, 2012), while many of its wild relatives are grown in Africa and the Middle East. The taxonomy of aubergine relatives remains a challenge and a controversial matter (Daunay, 2008; Meyer *et al.*, 2012, 2013; Samuels, 2013). The progenitor of *S. melongena* is probably a complex of wild species known as *Solanum incanum* (Frary *et al.*, 2007) that can be found in northern and eastern Africa and the Middle East (Daunay, 2008; Meyer *et al.*, 2012). *S. incanum* and *S. melongena* produce fertile hybrids (Gisbert *et al.*, 2011a).

S. macrocarpon is an important vegetable in sub-Saharan Africa (Yang and Ojiewo, 2013), grown for its fruits and leaves, and is known to exhibit tolerance to drought (Rotino *et al.*, 2014). The fruity types can be found in Suriname, South America and the Caribbean. It has four cultivar groups: Mukuno, Nabingo, Uganda and one semi-wild group (Ebert, 2013).

S. aethiopicum (scarlet aubergine) is a hypervariable tropical cultivated species native to sub-Saharan Africa, grown for its fruits and leaves, and is now found in many parts of the world; it is divided into four cultivar groups, Shum, Kumpa, Gilo and Aculeatum (Lester, 1986; Yang and Ojiewo, 2013). Its taxonomy is complex, and there are many alternative species names, but it is useful in this context to point out that the Aculeatum group is also known as *Solanum integrifolium*, and there are several rootstock studies that use the latter name. Many *S. aethiopicum* accessions carry resistance to Fusarium and bacterial wilt, as well as to fruit and shoot borers and spider mites (Kashyap *et al.*, 2003; Toppino *et al.*, 2008), and some display immunity against tobamoviruses (tobacco mosaic virus, tomato mosaic virus and pepper mild mottle virus) (Tzortzakakis *et al.*, 2006). Gilo, Kumba and Shum are important cultivar groups for food in tropical Africa and can easily be found in markets (Domingos *et al.*, 2016). There are also intermediate forms between them, as well as between *S. aethiopicum* and its wild ancestor *Solanum anguivi*, which is widely distributed in Africa and in the Arabian Peninsula. Crosses are easy between *S. anguivi* and all cultivar groups of *S. aethiopicum*, and are exploited for developing cultivars resistant to *Ralstonia solanacearum* (bacterial wilt) and other diseases (Ebert, 2013). Interspecific hybridization between *S. macrocarpon*, *S. aethiopicum* and *S. melongena* has been achieved to transfer disease and insect pest resistances from African to Asian aubergine (Prohens *et al.*, 2012; Ebert, 2013).

One of the most promising aubergine relatives as a rootstock is *Solanum torvum*, a relatively unknown wild species found especially in Mexico and South America, which is considered an invasive weed. It has been identified as a potential source of resistance to Verticillium wilt, root-knot nematodes, mycoplasma, *R. solanacearum*, *Fusarium* and other soilborne diseases. It is also tolerant of salt and metals (Kashyap *et al.*, 2003; Clain *et al.*, 2004; Bebeli and Mazzucato, 2008; Bagnaresi *et al.*, 2013; Rotino *et al.*, 2014).

The genus *Capsicum* originated in Bolivia and consists of around 30 species, with five having been domesticated, namely *C. annuum* var. *annuum*, *Capsicum baccatum* var. *pendulum*, *Capsicum chinense*, *Capsicum frutescens* and *Capsicum pubescens* (Paran and Falik, 2011; Ebert, 2013). Some species are considered to be semi-domesticated, namely *C. annuum* var. *glabriusculum*, *C. baccatum* var. *baccatum* and var. *praetermissum*, *Capsicum cardenasii*, *Capsicum eximium* and *Capsicum tovarii* (Heiser, 1995; Ebert, 2013). The phylogenetic tree of the cultivated species consists of three branches. One branch contains *C. annuum* (var. *annuum* and var. *glabriusculum*), *C. chinense* and *C. frutescens*. These three species form the *annuum* complex. The other branch contains the *C. baccatum* var. *pendulum* varieties, which, along with the *praetermissum* varieties and *C. tovarii*, form the *baccatum* complex. The third branch comprises *C. pubescens*. This species along with *C. cardenasii* and *C. eximium* form the *pubescens* complex (Pickersgill, 1997; Albrecht *et al.*, 2012; Ibiza *et al.*, 2012). Crosses between species within each complex result in fertile hybrids. However, crosses between species from the *annuum* complex and *C. baccatum* are difficult and the hybrids are sterile. The degree of compatibility depends on which parent is the pollen donor, and also on the species involved. Some interspecific hybrids developed within the genus are *C. frutescens* × *C. annuum*, *C. frutescens* × *C. chinense*, and *C. frutescens* × *C. baccatum* (Sikora and Nowaczyk, 2014).

C. annuum originated in Mexico and Central America and is the most economically important *Capsicum* spp., producing sweet and pungent fruits. It is both a vegetable and a spice crop, and also has ornamental uses (Heiser, 1995; Bebeli and Mazzucato, 2008). It includes many diverse morphological types, ranging from small pungent types to large-fruited bell peppers. Most sweet pepper varieties are susceptible to damage by soil-borne diseases and flooding. Resistances are available in many of the other species; for example, *C. chinense*, found in the Amazonian basin, in central and north-eastern regions of Brazil and in the Caribbean, shows good resistance against anthracnose and various viruses, and is preferred by African farmers during the rainy season (Ebert, 2013); *C. baccatum*, found in South America, shows tolerance to powdery mildew and various viruses (Mo *et al.*, 2015); *C. pubescens*, which originated in the highlands of the Andes, is well adapted to low temperatures; and *C. frutescens*, from the lowlands of south-eastern Brazil to Central America (Heiser, 1995; Ebert, 2013), and *Capsicum chacoense*, native to South America, both have accessions resistant to the nematode *Meloidogyne incognita* (Oka *et al.*, 2004; Gisbert *et al.*, 2013).

2.2 Gene Bank Collections

2.2.1 Cucurbitaceae

Cucurbits have diversified around the world. A large number of genetic resources adapted to many different environmental and growing conditions can be found in different areas. Knowing the extant genetic diversity among cucurbits is important to optimize collection and conservation programmes. The availability of these genetic resources facilitates the ongoing efforts by plant breeders to develop new varieties and rootstocks for melon, cucumber, watermelon and squash. Among the initiatives to rationalize the genetic resources of cucurbits and optimize their use are the activities of the cucurbits working group of the European Cooperative Programme for Plant Genetic Resources (ECPGR) (Díez *et al.*, 2008). One of their activities has been to create the European Central Cucurbits Database (ECCUDB; http://www.comav.upv.es, accessed 24 November 2016). At the time of writing, the database contained passport information on 27,485 accessions of cucurbit species following the International Plant Genetic Resources Institute/Food and Agriculture Organization of the United Nations (IPGRI/FAO) Multicrop Passport Descriptors List (http://www.bioversityinternational.org/e-library/publications/detail/faoipgri-multi-crop-passport-descriptors-mcpd/, accessed 24 November 2016) and will be extended to include passport, characterization and evaluation data. The ECCUDB contains information about the cucurbit collections held in gene banks from southern Europe, including Mediterranean countries (Portugal, Spain, Italy, France), and also collections from Mediterranean Near East countries (Turkey and Israel), northern Europe (Austria, Germany, the Netherlands, the UK, Poland, Switzerland and Sweden), eastern Europe (Armenia, Bulgaria, Hungary, Romania, Czech Republic and Slovakia), and Russia and the previous Soviet Union republics (Azerbaijan, Georgia, Lithuania, Latvia and Ukraine). An additional resource, although not specific for cucurbits, is the European Genetic Resources Search Catalogue (EURISCO), providing information about *ex situ* plant collections maintained in Europe. The EURISCO database was first maintained by Bioversity International (Rome, Italy; http://www.bioversityinternational.org/, accessed 24 November 2016), but in April 2014 it was moved to the Leibniz Institute of Plant Genetics and Crop Plant Research (IPK; Gatersleben, Germany) and will incorporate phenotypic information in the future. The cucurbits collections in the main gene banks of the ECCUDB and other Asian and American gene banks and databases are shown in Table 2.1.

Other global gene banks that host large cucurbit collections are the National Plant Germplasm System (NPGS) of the Agricultural Research Service/US Department of Agriculture (ARS-USDA) (available in the Germplasm Resources Information Network GRIN; Table 2.1), including not only US accessions but also ones from Europe and Asia, representing the areas of origin and primary centres of diversity of most species, and the Centro Agronómico Tropical de Investigación y Enseñanza (CATIE) gene bank, which hosts one of the most interesting collections of *Cucurbita* and other American genera not represented in other collections (Table 2.1) (Ebert *et al.*, 2007). Asian and African germplasm from many of the origin and primary diversification centres of the different cucurbit species are available at the Asian Vegetable Research and Development Center (AVRDC; http://avrdc.org/, accessed 24 November 2016), recently renamed as the World Vegetable Center, at Tainan in Taiwan (see AVGRIS in Table 2.1).

Some of these databases, but not all, include characterization and evaluation data, along with taxonomic and ecological information of the accessions that could help to select new rootstocks. The database Genesys PGR (Table 2.1) provides some information specifically on rootstocks: *C. ficifolia* is used as the rootstock for cucumber, and *C. maxima*, *L. siceraria* and *L. aegyptiaca* are also listed as rootstocks.

Table 2.1. Status of the cucurbits collections in the main gene banks of the ECCUDB and other Asian and American gene banks and databases, 2016. Numbers indicate the number of accessions available for each genus. For many of these, and additional European collections, detailed information about cucurbit germplasm is included in EURISCO and in ECCUDB.

Institution	Genus				
	Citrullus	*Cucumis*	*Cucurbita*	*Lagenaria*	Other
Institute for Plant Genetic Resources 'K. Malkov', Bulgaria (data available in EURISCO)	190	1023	388	11	5 (*Luffa*)
Research Institute of Crop Production, Olomouc, Czech Republic (data available in GRIN Czech)	8	988	740	12	25 (*Luffa*, *Momordica*, *Benincasa*, *Echinocystis*)
Institute of Plant Genetics and Crop Plant Research, Gatersleben, Germany (data available in GBIS/I)	274	1148	1105	74	83 (*Luffa*, *Momordica*, *Benincasa*, *Sechium*, *Trichosanthes*, *Ecballium*; *Melothria*, *Echinocystis*, *Cyclanthera*)
Polytechnical University of Valencia, Spain (data available in COMAV Genebank)	251	761	772	40	4 (*Luffa*, *Momordica*)
Research Centre for Agrobotany, Tápiószele, Hungary (data available in EURISCO)	215	431	899	36	15 (*Luffa*, *Momordica*, *Benincasa*, *Trichosanthes*, *Bryonia*)
Centre for Genetic Resources (CGN), the Netherlands	–	1002	–		–
Plant Breeding and Acclimatization Institute National Research Institute, Poland (data available in EURISCO)	91	537	381		4 (*Luffa*, *Benincasa*)
Banco Portugues de Germoplasma Vegetal (BPGV), Portugal	32	184	305	11	
N.I. Vavilov Research Institute of Plant Industry (VIR), Russian Federation	2412	2998	5771	254	192 (*Luffa*, *Momordica*, *Benincasa*)
Aegean Agricultural Research Institute (AARI), Turkey	–	638	633		–
Selcuk University, Turkey	–	–	400		

Continued

Table 2.1. Continued.

Institution	Genus				
	Citrullus	*Cucumis*	*Cucurbita*	*Lagenaria*	Other
Ondokuz Mayıs University, Turkey	–	–	311		
Erciyes University, Turkey	–	–	–	322	
Çukurova University, Turkey	385	400	–		
Germplasm Resources Information Network (GRIN), USA	1467	3549	1773	706	649 (*Luffa*, *Momordica*, *Benincasa*)
Centro Agronómico Tropical de Investigación y Enseñanza (CATIE), Costa Rica	9	5	2231	133	32 (*Luffa*, *Momordica*, *Benincasa*, *Sicana*, *Cyclanthera*, *Melothria*, *Cionosicyos*)
AVRDC Vegetable Genetic Resources Information System (AVGRIS), Taiwan	57	644	326	346	1437 (*Luffa*, *Momordica*, *Benincasa*, *Trichosanthes*)
National Institute of Agrobiological Sciences (NIAS), Japan	229	1236	123	148	66 (*Luffa*, *Momordica*)
National Plant Genetic Resources Information System (NPGRIS), Taiwan	1911	1235	421	300	942 (*Luffa*, *Momordica*, *Benincasa*)

AARI: http://map.seedmap.org/solutions/conservation/seed-banks/aegean-agricultural-research-institute-aari/; AVGRIS: http://203.64.245.173/; BPGV: http://www.iniav.pt/noticias/banco-portugues-de-germoplasma-vegetal-(bpgv); CATIE: http://www.catie.ac.cr; CGN: http://cgngenis.wur.nl/; COMVAC, https://www.comav.upv.es/index.php/databasesgermplasm/databases/genebank-database; Çukurova University, http://www.cukurova.edu.tr; Erciyes University, http://www.erciyes.edu.tr; EURISCO, http://www.eurisco.ecpgr.org; GRIN, https://npgsweb.ars-grin.gov/gringlobal/taxon/taxonomysearch.aspx; GRIN Czech, https://grinczech.vurv.cz/gringlobal/search.aspx; GBIS/I, http://www.ipk-gatersleben.de/en/gbisipk-gaterslebendegbis-i/informationordering/; NIAS, http://www.gene.affrc.go.jp; NPGRIS, http://www.npgrc.tari.gov.tw/web/apec130e.jsp; Selcuk University, http://www.selcuk.edu.tr/; VIR, http://www.vir.nw.ru); Ondokuz Mayıs University, http://www.omu.edu.tr/en. Websites accessed 24 November 2016.

2.2.2 Solanaceae

In 2007, Robertson and Labate reported an estimated number of *Solanum* section *Lycopersicon* germplasm accessions of over 75,000 held in more than 120 countries, while according to Ebert in 2013, the number of seed bank entries for tomato worldwide was 84,279. Tomato, aubergine and pepper accessions can be searched in many databases and at Genesys PGR (Tables 2.2 and 2.3), the largest gateway through which users can discover material in gene banks around the world; at the time of writing it had 80,601 *Solanum* entries, including 20,171 *S. tuberosum* entries. The largest individual collection is at AVDRC with 8107 accessions (Ebert, 2013). The Tomato Genetic Resource Center (TGRC) at the Department of Vegetable Crops of the University of California, Davis, holds about 3700 accessions, of which 30% are wild and 30% are monogenic stocks (Bebeli and Mazzucato, 2008). The TGRC conserves 2489 *S. lycopersicum* accessions (Table 2.2), as well as accessions of tomato wild relatives classified in the *Solanum* sections *Lycopersicon*, *Lycopersicoides* and *Juglandifolia*, as described above.

In western Europe, the largest collection of tomato accessions (4046 accessions) is held at the gene bank of the IPK at Gatersleben in Germany (Table 2.2). Other important gene banks are in the Russian Federation at the N.I. Vavilov All-Russian Scientific Research Institute of Plant Industry, in the Philippines at the National Plant Genetic Resources Laboratory, and in Japan at the National Institute of Agrobiological Sciences (NIAS) gene bank (Ebert, 2013). A more complete list with links on tomato genetic resources is available at the Tomato Genetic Resource Center (Table 2.2).

Similarly to the cucurbits, the Solanaceae working group of the ECPGR provides useful links about Solanaceae genetic resources (http://www.ecpgr.cgiar.org/working-groups/solanaceae/, accessed 24 November 2016). Within this working group, an online searchable central tomato database has been developed, hosted at the Centre for Genetic Resources in the Netherlands (Table 2.2), with the tomato germplasm collected and preserved by ECPGR members and associated countries. The database has 21,332 accessions, with the majority represented by 20,213 accessions of *S. lycopersicum*; *S. pimpinellifolium* ranks second, followed by *S. habrochaites*, *S. peruvianum* and *S. pennellii*. The database also includes hybrids. The rest are, using the new classification (Peralta *et al.*, 2008), *S. cheesmaniae*, *S. chilense*, *Solanum chmielewskii*, *Solanum corneliomulleri*, *Solanum neorickii* and *S. lycopersicoides*. The database has been developed according to the IPGRI/FAO Multicrop Passport Descriptors List and Bioversity International.

Genetic resources of aubergine are comprised of both *S. melongena* cultivated forms and related wild species. The genetic resources of *S. melongena* comprise a wealth of landraces, open-pollinated cultivars and hybrids that present intraspecific diversity, related cultivated species, and wild and weedy species. Ebert (2013) estimated the aubergine collections worldwide at a total of 21,616 accessions, with AVRDC having the largest number of accessions at 3524, followed by the National Bureau of Plant Genetic Resources (NBPGR) in India and NIAS in Japan (Ebert, 2013). Other important gene banks for aubergine accessions are in the

Table 2.2. Status of the *Solanum* collections in the main gene banks and databases, 2016. Numbers indicate the number of accessions for each species or class.

Institution	*S. lycopersicum*[a]	Other tomato-related species[b]	*S. melongena*	Other aubergine-related species[c]
Research Institute of Crop Production, Olomouc, Czech Republic (data available in GRIN Czech)	1,850	24	30	–
Institute of Plant Genetics and Crop Plant Research (IPK), Gatersleben, Germany (data available in GBIS/I)	4,046	290	113	17
Polytechnical University of Valencia, Spain (data available in COMAV Genebank)	2,361	144	181	51
Centre for Genetic Resources, the Nederlands (CGN)	1,222	66	373	107
N.I. Valilov Research Institute of Plant Industry (data available in Genesys PGR)	4,216	24	564	–
Aegean Agricultural Research Institute, Turkey (AARI)	1,066	–	350	–
Alata Horticultural Research Institute, Turkey (AHRI)	1,480	–	–	–
West Mediterranean Agricultural Research Institute (BATEM), Turkey	250	–	250	–
Germplasm Resources Information Network, USA (GRIN)	10,566	990	1,205	137
AVRDC Vegetable Genetic Resources Information System, Taiwan (AVGRIS)	6,575	610	2,214	680
Tomato Genetic Resource Center (TGRC)	2,489	594	–	–

Continued

Table 2.2. Continued.

Institution	*S. lycopersicum*[a]	Other tomato-related species[b]	*S. melongena*	Other aubergine-related species[c]
National Institute of Agrobiological Sciences, Japan (NIAS)	710	29	424	24
National Plant Genetic Resources Information System, Taiwan (NPGRIS)	7,123	4	3,419	39

AHRI, http://arastirma.tarim.gov.tr/alata; BATEM, http://arastirma.tarim.gov.tr/batem; PGR GENESYS, https://www.genesys-pgr.org/; TGRC, http://tgrc.ucdavis.edu/. Websites accessed 24 November 2016. See Table 2.1 for other website details.

[a]A search was carried out for *S. lycopersicon* and synonyms.

[b]The other related species include *S. pimpinellifolium*, *S. habrochaites* (syn. *S. hirsutum*), *S. pennellii*, *S. chilense* and *S. cheesmaniae* and synonyms.

[c]The other related species include *S. aethiopicum*, *S. torvum*, *S. macrocarpon*, *S. sisymbriifolium* and *S. incanum*.

USA, Bangladesh, the Philippines and the Netherlands (Ebert, 2013). Regarding databases, Genesys PGR lists 5474 accessions of *S. melongena*, and also dozens of *S. torvum*, *S. aethiopicum*/ *S. integrifolium*, *S. macrocarpon*, *S. anguivi* and other accessions of aubergine wild relative species. The ECPGR Eggplant Database at the Botanical and Experimental Garden, Radboud University, Nijmegen, the Netherlands (http://www.ru.nl/bgard/databases/ecpgr-solanaceae/, accessed 24 November 2016), has been developed within the framework of the EGGplant genetic resources NETwork (EGGNET), a project funded by the European Union up to 2005 (http://www.bgard.science.ru.nl/eggnet/eggnet01.html, accessed 24 November 2016).

According to Ebert (2013), there are 73,572 *Capsicum* accessions worldwide, and around 10% of the *Capsicum* germplasm, comprising accessions from five species (*C. annuum*, *C. baccatum*, *C. chinense*, *C. frutescens* and *C. pubescens*), is conserved in AVRDC (Table 2.3). Other countries with large collections of accessions are the USA (4698 accessions), Mexico (4661 accessions), India (3835 accessions), Brazil (2321 accessions) and Japan (2271 accessions) (Ebert, 2013). In western Europe, IPK in Germany holds the highest number of *Capsicum* accessions (Table 2.3). Regarding databases, Genesys PGR has 24,823 entries for *Capsicum* of which 18,525 are *C. annuum*, 1002 are *C. baccatum*, 1310 are *C. chinense*, 1436 are *C. frutescens* and 180 are *C. pubescens*. The European Database for Pepper is being developed by the Aegean Agricultural Research Institute (AARI), Izmir, Turkey (Table 2.3).

In general, gene banks maintaining solanaceous resources and associated databases have very limited descriptions of the use of these genetic materials as rootstocks, as is also the case with cucurbits. However, Genesys PGR (Table 2.2) provides some information: four accessions of *S. habrochaites* × *S. lycopersicum*,

Table 2.3. Status of the *Capsicum* collections in the main gene banks and databases, 2016. Numbers indicate the number of accessions for each species.

Institution	*C. annuum*	*C. chinense*	*C. baccatum*	*C. frutescens*	*C. pubescens*
Research Institute of Crop Production, Olomouc, Czech Republic (data available in GRIN Czech)	593	–	–	5	–
Institute of Plant Genetics and Crop Plant Research (IPK), Gatersleben, Germany (data available in GBIS/I)	1207	55	51	179	25
Polytechnical University of Valencia, Spain (data available in COMAV Genebank)	1177	7	19	12	4
Research Centre for Agrobotany, Tápiószele, Hungary (data available in EURISCO)	1104	2	11	18	2
Centre for Genetic Resources(CGN), the Netherlands	804	117	38	43	8
Plant Breeding and Acclimatization Institute National Research Institute, Poland (data available in EURISCO)	138	–	–	–	–
N.I. Vavilov Research Institute of Plant Industry, Russian Federation (data available in Genesys PGR)	1322	–	–	2	–
Aegean Agricultural Research Institute, Turkey	483	14	13	37	–

Continued

Table 2.3. Continued.

Institution	*C. annuum*	*C. chinense*	*C. baccatum*	*C. frutescens*	*C. pubescens*
West Mediterranean Agricultural Research Institute (BATEM), Turkey	450	–	–	–	–
Alata Horticultural Research Institute, Turkey	2069	3	–	2	1
Germplasm Resources Information Network (GRIN), USA	4054	517	442	672	133
Centro Agronómico Tropical de Investigación y Enseñanza (CATIE), Costa Rica	340	51	30	262	6
National Institute of Agrobiological Sciences (NIAS), Japan	248	16	32	4	3
National Plant Genetic Resources Information System (NPGRIS), Taiwan	4161	299	271	362	20
AVRDC Vegetable Genetic Resources Information System (AVGRIS), Taiwan	5488	505	388	741	30
Capsicum Active Germplasm Bank of Embrapa Temperate Agriculture (Southern Brazil)[a]	265 landraces representing these five species				

See Tables 2.1 and 2.2 for website details.
[a]Neitzke *et al.* (2011).

four accessions of *Solanum* section *Lycopersicon* and four accessions of *S. melongena*, most of them developed in the Netherlands, are described as rootstocks, and their resistance to different soilborne pathogens is noted. The TGRC provides vigorous hybrid rootstocks for use in multiplying difficult germplasm (Chetelat and Petersen, 2003).

2.3 Current Usage of Genetic Material in Rootstocks

2.3.1 Rootstocks for cucurbit production

The primary purpose of grafting cucurbits worldwide has been to provide resistance to soilborne diseases and nematodes, especially after the loss of methyl bromide as a soil fumigant and the appearance of pathogen resistance to commonly used pesticides (Davis *et al.*, 2008; King *et al.*, 2008, 2010; Lee *et al.*, 2010). The use of grafting has expanded to address a wide variety of fungal diseases (e.g. Fusarium wilt, vine decline, Verticillium wilt, gummy stem blight) and to enhance whole-plant biotic and abiotic stress responses. Grafting is also an environmentally friendly strategy, as grafting commercial varieties on to disease-resistant rootstocks is a potential alternative to chemical means of soilborne disease management. It can also facilitate the use of wastewater as an alternative water source for irrigation, increasing the tolerance of the scions to salt, heavy metals and high boron concentrations commonly found in this water.

The production of grafted cucurbits first began in Japan and Korea in the late 1920s with watermelon grafted on to squash rootstocks (Davis *et al.*, 2008). The use of grafted transplants has increased hugely since then. Currently, it is more common in watermelons and cucumbers and has not yet gained widespread acceptance in melons or squashes. The choice of rootstock has a vital role in the success of grafting. Good rootstocks should be compatible with the scions and provide the desired traits without a reduction in fruit yield and quality. Several rootstocks have been used for cucurbit grafting with varying degrees of success (e.g. *Cucurbita, Lagenaria, Benincasa, Luffa, Trichosanthes, Sicyos, Cucumis, Citrullus*); most enhance scion growth and yield under stressful conditions, but some lack tolerance to specific stresses, and others can have a negative impact on fruit quality (Rouphael *et al.*, 2010, 2012).

The various species and their hybrids currently used as rootstocks for cucurbits, their properties and uses are described in the following sections and summarized in Table 2.4.

Cucurbita *as rootstocks*

The most popular rootstocks for cucurbit crops, principally watermelons, melons and cucumbers, belong to the genus *Cucurbita*. Interspecific crosses are an effective way to create new genotypes that combine favourable traits from different parents. *C. maxima* shows variable degrees of compatibility with several other *Cucurbita* spp., being used as a bridge for interspecific crosses. *C. moschata* is a vigorous species reported to be resistant to biotic and abiotic stresses (Paris and Kabelka, 2009). The hybrid combination *C. maxima* × *C. moschata* has been exploited as a favourable source of rootstocks (Colla *et al.*, 2010; see also Chapter 3, this volume), with most of the currently used commercial rootstocks coming from breeding work performed in far-eastern countries (China, Japan and Korea). New compatible combinations of *C. maxima* × *C. moschata* are being identified by screening alternative germplasm collections of these species (Karaagac and Balkaya, 2013). *C. maxima* × *C. moschata* hybrids provide non-specific but efficient protection to a wide range of soilborne diseases and tolerance to some

Table 2.4. Examples of Cucurbitaceae rootstock–scion combinations for diverse purposes.

Rootstock	Scion Me	Scion Wm	Scion Cb	Purpose	Reference
Cucumis melo: var. *flexuosus*, var. *conomon*, var. *agrestis*, Asian landraces	X			Improved tolerance to *Monosporascus* vine decline	Fita *et al.* (2007); Jang *et al.* (2014)
				Tolerance to nematodes	Ito *et al.* (2014)
				Fruit quality	Condurso *et al.* (2012); Verzera *et al.* (2014)
				Salt tolerance	Dasgan *et al.* (2015)
				Tolerance to *Fusarium oxysporum*	Lee and Oda (2003)
Cucumis metuliferus	X			Management of *Meloidogyne incognita*	Sigüenza *et al.* (2005); Kokalis-Burelle and Rosskopf (2011); Guan *et al.* (2014); Selvi *et al.* (2013); Punithaveni *et al.* (2015)
Cucumis africanus, *Cucumis myriocarpus*, *Cucumis ficifolius*, *Cucumis anguria*		X		Management of *M. incognita*	Pofu *et al.* (2011); Pofu *et al.* (2013)
				Tolerance to Fusarium wilt	Trionfetti Nisini *et al.* (2002)
Cucumis maxima × *Cucumis moschata*	X	X	X	Improved tolerance to vine decline diseases	Jifon and Crosby (2008)
				Improved tolerance to salinity	Orsini *et al.* (2013); Rouphael *et al.* (2012)
				Improved tolerance to boron	Edelstein *et al.* (2011)
				Improved tolerance to Fusarium wilt and fruit quality	Zhou *et al.* (2014))
				Improved response in nematode-infested soil	Goreta Ban *et al.* (2014)
				Management of *Verticillium*	Buller *et al.* (2013)
				Higher yield, larger fruit	Zhang *et al.* (2014)
				Water use efficiency	Kivi *et al.* (2014)
				Nitrogen use efficiency	Salar *et al.* (2015)
				Didymella bryoniae tolerance	Silva *et al.* (2012)

Continued

Table 2.4. Continued.

Rootstock	Scion			Purpose	Reference
	Me	Wm	Cb		
				Macrophomina wilt tolerance	Cohen *et al.* (2012)
C. ficifolia		X	X	Improved cold tolerance	Bulder *et al.* (1991); Zhou *et al.* (2009); Li *et al.* (2014a)
				Improved salt tolerance	Huang *et al.* (2010)
				Increase yield	Cheshmehmanesh *et al.* (2004); Hoyos Echebarría *et al.* (2001); Hernández-González *et al.* (2014)
				M. incognita tolerance	Punithaveni *et al.* (2015)
C. moschata	X		X	Improved salt tolerance	Liu *et al.* (2012)
				M. incognita tolerance	Selvi *et al.* (2013)
				Tolerance to *F. oxysporum*	Traka-Mavrona *et al.* (2000)
				Tolerance to high pH	Roosta and Karimi (2012)
Curcubita pepo		X		Improved vegetative growth (but reduced yield)	Bekhradi *et al.* (2011)
Curcubita argyrosperma			X	Increased yield	Hernández-González *et al.* (2014)
Cucurbita martinezii		X		Grafting incompatibility	Huh *et al.* (2003)
Lagenaria siceraria		X	X	Tolerance to Fusarium wilt	Yetişir *et al.* (2007); Karaca *et al.* (2012)
				Tolerance to flooding	Yetişir *et al.* (2006)
				Improved tolerance to *Phytophthora capsici*	Kousik *et al.* (2012)
				Tolerance to salinity	Yang *et al.* (2013)
				Management of Verticillum wilt	Buller *et al.* (2013)
				Tolerance to nematodes	Ozarslandan *et al.* (2011)
				Tolerance to powdery mildew	Kousik *et al.* (2008)
				Fruit volatiles, quality and yield	Petropoulos *et al.* (2012); Guler *et al.* (2014)
				Plant development and fruit quality	Petropoulos *et al.* (2014); Bekhradi *et al.* (2011)
Luffa cylindrica	X	X	X	Improved heat tolerance	Li *et al.* (2014a,b); Yetişir and Sari (2004); Galatti *et al.* (2013)

Continued

Table 2.4. Continued.

Rootstock	Scion: Me	Wm	Cb	Purpose	Reference
Sicyos angulatus			X	Improved cold tolerance	Zhang *et al.* (2008); Bulder *et al.* (1991)
Citrullus spp.		X	X	Reduced negative impact on fruit quality; *M. incognita* tolerance	Huh *et al.* (2003); Selvi *et al.* (2013); Punithaveni *et al.* (2015)
Citrullus lanatus var. *citroides*	X	X		Tolerance to Fusarium wilt	Keinath and Hassell (2014)
				Tolerance to nematodes	Thies *et al.* (2010); Thies *et al.* (2015a)
				Melon necrotic spot virus tolerance	Huitrón *et al.* (2007)
Benincasa hispida	X		X	Tolerance to Fusarium wilt	Trionfetti Nisini *et al.* (2002)
				Tolerance to nematodes	Ito *et al.* (2014); Amin and Mona (2014); Galatti *et al.* (2013)
				Tolerance to Verticillium wilt	Wimer *et al.* (2014)
Trichosanthes cucumerins	X			Tolerance to nematodes	Ito *et al.* (2014)
Sicana odorifera	X			Tolerance to nematodes	Ito *et al.* (2014)
Momordica charantia			X	Tolerance to nematodes	El-Eslamboly and Deabes (2014)

Me, melon; Wm, watermelon; Cb, cucumber.

abiotic stresses. They have been used to improve tolerance to Fusarium and Verticillium wilts, and to vine decline diseases (caused by *Monosporascus* spp. and other fungal pathogens) of watermelons, melons and cucumbers. These diseases constitute major limitations for cucurbit production in many regions worldwide. For example, melons and watermelons grafted on to hybrid *Cucurbita* rootstocks exhibited dramatically reduced incidences of Fusarium wilt and increased yields (Keinath and Hassell, 2014; Zhou *et al.*, 2014). Late-season vine decline can be circumvented in watermelon crops by grafting to vigorous hybrid rootstocks that improve the capacity for water uptake (Davis *et al.*, 2008; Jifon and Crosby, 2008). An improved response of grafted watermelons to Verticillium wilt has been also reported (Buller *et al.*, 2013). A better response to salinity, due to improvements in the exclusion of Na^+ and the uptake of K^+ and to a more efficient control of stomatal functions, has been found in melons grafted on to *C. maxima* × *C. moschata* (Orsini *et al.*, 2013). In addition, the use of hybrid *Cucurbita* rootstocks improves melon and cucumber photosynthetic capacity under salt stress and consequently crop performance (Rouphael *et al.*, 2012). Grafting melons that are boron sensitive on to boron-tolerant *Cucurbita* hybrids can be used to prevent boron damage (Edelstein *et al.*, 2011).

Apart from their tolerance to biotic and abiotic stress, hybrid *Cucurbita* rootstocks are preferred because they show better emergence performance, even at low temperatures, than other rootstocks, and they develop long and thick hypocotyls that facilitate grafting (Yetişir and Sari, 2004). However, this excess of vigour can result in unequal stem diameters at the graft union, reducing the survival of grafted plantlets, and can later cause a delay in flowering if cultivation is not managed properly; it can also prolong the ripening process (Soteriou *et al.*, 2014, 2016).

Less frequent is the use of single non-hybrid *Cucurbita* species as rootstocks. Selected accessions of *C. ficifolia*, *C. moschata*, *C. maxima*, *C. argyrosperma* and *C. pepo* have been assayed (Goreta Ban *et al.*, 2014). For example, *C. moschata* accessions can reduce the deleterious effects of salt stress on cucumber plants (Liu *et al.*, 2012; Roosta and Karimi, 2012), and yield can be increased in cucumber grafted on to *C. argyrosperma* (Hernández-González *et al.*, 2014). Landraces of these species can be selected on the basis of favourable traits that facilitate the grafting process, such as tolerance to the stem canker caused by *Didymella* (Keinath, 2014a,b).

The single *Cucurbita* sp. used most for grafting, mainly for cucumber, is *C. ficifolia* (figleaf gourd) as it provides superior cold tolerance. The mechanisms by which cold-tolerant *C. ficifolia* rootstocks induced scion tolerance have been studied (Li *et al.*, 2014a). *C. ficifolia* rootstocks also improve cucumber tolerance to salinity (Huang *et al.*, 2010).

Single species have been reported to have less graft compatibility and lower post-grafting survival ratios, and result in lower yields than hybrid rootstocks (Traka-Mavrona *et al.*, 2000; Yetişir and Sari, 2004; Bekhradi *et al.*, 2011). However, there is not enough information to know whether this is a general effect. Therefore, further screenings of germplasm collections of these species and of wild *Cucurbita* that have not been assayed to date are necessary to identify promising rootstocks or parental line candidates.

Cucurbita rootstocks face several limitations, and these are common to most of the accessions in use. Although they have been reported to reduce nematode damage to the scion due to increased vigour (Goreta Ban *et al.*, 2014), they are still susceptible to nematodes and are not suitable under high-infestation conditions, as their use may lead to a build-up of nematode populations in soils. Another challenge in using *Cucurbita* rootstocks is their susceptibility to powdery mildew; when combined with powdery mildew-susceptible scions, minor disease outbreaks can quickly become a serious problem, impacting on yield and fruit quality. Resistance to powdery mildew has been reported in some accessions of *C. moschata* and *C. pepo*, and in wild species such as *C. okeechobeensis*, *C. ecuadorensis* and *C. lundelliana* (Paris and Kabelka, 2009). The use of some of these resistant accessions, as well as intra- or interspecific hybrids between them, can be useful to overcome these limitations. There are only a few reports describing the use of wild *Cucurbita* spp. as rootstocks: Huh *et al.* (2003) report a high level of grafting incompatibility of *Cucurbita martinezii* (syn. *C. okeechobeensis* subsp. *martinezii*) with watermelon; this compatibility problem can be minimized by generating interspecific wild × cultivated crosses; for example, the hybrid *C. ecuadorensis* × *C. maxima* was found to have high levels of grafting compatibility with melon cultivars (Nunes *et al.*, 2013).

Negative effects on quality have been reported in plants grafted on to *Cucurbita* rootstocks (*see* Chapter 7, this volume), including modifications of fruit size or fruit shape, a decrease in sugars and acidity, an increase in vitrescence, and a change in fruit volatiles and colour (Davis *et al.*, 2008; Rouphael *et al.*, 2010). These effects are mainly dependent on the scion × rootstock interaction and on modified ripening behaviour but are also influenced by crop management practices (Bekhradi *et al.*, 2011; Petropoulos *et al.*, 2014; Soteriou *et al.*, 2014), which need to be adapted to suit the graft combination to achieve the best outcome.

Closely related to the genus *Cucurbita* is the cassabanana, *S. odorifera*, which has been assayed as a rootstock for melon to protect against nematodes (Ito *et al.*, 2014). However, its potential for grafting melons and other cucurbits has not yet been exploited.

Lagenaria *and* Benincasa *as rootstocks*

Bottle gourd (*L. siceraria*) has been used as a rootstock since the 1920s. Because of the quality problems associated with *Cucurbita* spp. and interspecific hybrids, bottle gourd became a preferred rootstock, mainly for watermelon. It was reported that *L. siceraria* is useful against soilborne diseases and low soil temperatures (Davis *et al.*, 2008; Bekhradi *et al.*, 2011). However, in comparison with *Cucurbita* rootstocks, *L. siceraria* accessions have been observed to have poor seedling emergence and poor grafted plantlet establishment at low temperatures (H. Yetişir, unpublished data), which may make them less appropriate for early growing cycles of watermelon, although the fact that they are less vigorous can help them finish their fruiting-ripening cycle earlier.

This species develops thinner and shorter hypocotyls than *Cucurbita*, which makes grafting more difficult but can result in higher survival rates due to more similar stem diameters at the graft union (Yetişir and Sari, 2004). This reduced vigour is also associated with fewer reported negative effects on fruit quality and flowering, although some scion–rootstock combinations have been reported to affect watermelon and cucumber fruit quality (Karaca *et al.*, 2012; Petropoulos *et al.*, 2012).

Evaluation of the rootstock potential of *L. siceraria* germplasm has been conducted regarding the graft compatibility, plant growth and resistance to Fusarium wilt, and some promising accessions have been selected (Yetişir *et al.*, 2007; Keinath and Hassell, 2014). Similarly to *Cucurbita*, bottle gourd rootstocks are useful to manage Verticillium wilt in watermelon (Buller *et al.*, 2013), and they enhance watermelon salt tolerance (Yang *et al.*, 2013). However, bottle gourd rootstocks are reported to provide less tolerance to nematodes than *Cucurbita* hybrids (Goreta Ban *et al.*, 2014), whose moderate degree of tolerance is probably due to their high vigour and rapid root growth (Ozarslandan *et al.*, 2011). Additionally, resistance to Phytophthora crown rot caused by *Phytophthora capsici* has been found in *Lagenaria* rootstocks, whereas high levels of susceptibility were exhibited by *Cucurbita* hybrids and watermelon rootstocks (Kousik *et al.*, 2012). Crown rot-resistant bottle gourd rootstocks may be more appropriate than *Cucurbita* in areas where *P. capsici* is a recurring problem.

Screenings of bottle gourd germplasm collections have reported that some accessions have certain levels of tolerance to flooding, *Didymella bryoniae* (gummy stem blight), powdery mildew, whitefly and zucchini yellow mosaic virus (Yetişir *et al.*, 2006; Kousik *et al.*, 2008; Levi *et al.*, 2009; Rouphael *et al.*, 2010; Keinath, 2014a, b). This information is essential for the selection of superior *L. siceraria* rootstock lines with enhanced resistance to diseases and insect pests of cucurbit crops. Variability also exists between rootstocks concerning their effects on fruit quality (Guler *et al.*, 2014). The potential of *Lagenaria* spp. other than *L. siceraria* as rootstocks remains largely underexploited.

B. hispida, a close relative of bottle gourd with similar rootstock characteristics, has been tested as a rootstock for melons or watermelons, conferring resistance to *Fusarium* spp., *Verticillium* spp. and nematodes (Ito *et al.*, 2014; Wimer *et al.*, 2014). However, grafting on to *B. hispida* negatively influenced both yield and fruit quality (Trionfetti Nisini *et al.*, 2002). Yetişir and Uygur (2009) reported that *B. hispida* is not a suitable rootstock for watermelon under saline conditions.

Luffa, Trichosanthes, Sicyos *and* Momordica *as rootstocks*

Luffa cylindrica (loofah) is a heat-tolerant cucurbit species that has been used mainly to graft cucumbers. It shares characteristics with *L. siceraria*, displaying low-vigour emergence of seedlings prior to grafting – and also low vigour in the grafted transplants – at low temperatures, and thin and short hypocotyls; as a rootstock, *L. cylindrica* is only suitable when high temperatures occur. Grafting cucumbers on to luffa rootstocks significantly alleviated heat-induced growth reduction; these rootstocks induced significant changes in the transcripts of stress-responsive and defence-related genes through root–shoot communication, especially at stressful growth temperatures (Li *et al.*, 2014b). This species has been also reported as nematode tolerant and useful for grafting melons (Galatti *et al.*, 2013). Other less well known cucurbits that have been experimentally assayed as new rootstocks are *T. cucumerinus*, used to provide nematode resistance to melons (Ito *et al.*, 2014), and *S. angulatus*, used to provide watermelons and cucumbers with a greater tolerance to low root temperature than that provided by *C. ficifolia* (Zhang *et al.*, 2008). Both *Trichosanthes* and *Sicyos* spp. have shown important shortcomings that prevent their general use as rootstocks, such as germination difficulties, grafting incompatibility and yield reductions (Bulder *et al.*, 1991), although such problems can potentially be overcome through suitable breeding programmes.

M. charantia has also been used as a rootstock for cucumbers, with which it shows high graft compatibility. Grafted cucumbers had a significant increment in yield and vegetative growth, even in soil infested with nematodes, although this species gave a lower survival rate compared with other rootstocks under study (*Cucurbita* sp. hybrids, *C. moschata*, *C. ficifolia*, *C. metuliferus*, *Lagenaria* spp.) (El-Eslamboly and Deabes, 2014).

Cucumis *as rootstocks*

One of the main limitations of some of the previously described rootstocks is their excessively high vigour, which appears to be associated with reduced survival of grafted transplants, delayed flowering and ripening, and negative impacts on fruit quality (mainly in watermelon and melon). The use of rootstocks belonging to the same species as the scions could minimize the compatibility and quality problems, but this strategy is problematic in melons because of a lack of disease-resistant germplasm (Pitrat, 2008). However, screenings of large germplasm collections have identified some promising accessions of *C. melo* that could be used to provide disease resistance without impacting on quality; for example, two accessions from Brazil (CNPH 01-962 and CNPH 01-963) of *C. melo* subsp. *melo*, cultivar group *conomon*, were found to be resistant to *M. incognita* (Ito *et al.*, 2014). Condurso *et al.* (2012) and Verzera *et al.* (2014) compared two *Fusarium oxysporum*-resistant *C. melo* rootstocks with *Cucurbita* hybrids as rootstocks of honeydew melon and found less effect on the aroma and carotenoid contents of the fruit with the *C. melo* rootstocks. In addition to Fusarium wilt and nematode tolerance, *C. melo* rootstocks can be an alternative to *Cucurbita* genotypes for providing tolerance against Monosporascus root rot and vine decline and to increase the marketable yield without a reduction in fruit quality (Fita *et al.*, 2007; Jang *et al.*, 2014). One disadvantage of using intraspecific grafting is that it may be more labour-intensive to identify unwelcome rootstock grow-outs due to the similarity of leaf morphology between scion and rootstock.

As well as intraspecific variation, intragenus variation can also be exploited in melon grafting. There are numerous *Cucumis* spp., and some have been reported to be highly resistant to nematodes, Fusarium wilt, powdery mildew and/or downy mildew. One of the most promising *Cucumis* rootstocks is *C. metuliferus* (Kokalis-Burelle and Rosskopf, 2011; Guan *et al.*, 2014). Melons grafted on to *C. metuliferus* accessions have been reported to be resistant to nematodes, with reduced levels of root galling, improved shoot mass and significantly lower nematode levels at harvest; the same melons grafted on to *C. moschata* rootstocks gave similar results except that the final nematode levels were higher in comparison with *C. metuliferus* rootstocks and were similar to the non-grafted plants (Sigüenza *et al.*, 2005). Evaluation and characterization of the diversity in *C. metuliferus* have been performed, allowing the selection of the most favourable types (Weng, 2010). Although *C. metuliferus* is genetically closer to melon, it is also a useful rootstock for cucumber (El-Eslamboly and Deabes, 2014), but is graft-incompatible with watermelon, having negative impacts on productivity and fruit quality in some cases (Huh *et al.*, 2003). For cucumber, *Cucumis sativus* var. *hardwickii* has also been assayed as a rootstock (Nienhuis and Lower, 1980). Recenly, *Cucumis pustulatus* has been reported as a suitable rootstock for cucumber, melon and watermelon (Liu *et al.*, 2015). Some accessions of this species have resistance to root-knot nematode and Fusarium wilt.

Other *Cucumis* spp. useful as sources of resistance to nematodes and/or *Fusarium* are *C. ficifolius*, *C. zeyheri*, *C. africanus*, *C. anguria* and and *C. myriocarpus* (Trionfetti Nisini *et al.*, 2002). These rootstocks have lower compatibility levels with watermelon, but it has proved possible to improve compatibility by optimizing the stem diameters at the grafting union using different seedling tray sizes, seed priming and sequential planting (Pofu *et al.*, 2011, 2013). One way of increasing vigour is by using interspecific hybrids, but hybridization between different *Cucumis* spp. is extremely difficult, but some successful crosses have been reported. For example, Cáceres *et al.*, (2016) obtained two interspecific hybrids between *C. ficifolius* × *C. anguria* and *C. ficifolius* × *C. myriocarpus* that can be assayed as putative rootstocks for melon. The generation of intraspecific hybrids between genetically distant accessions can be a useful alternative. More studies are needed to demonstrate the advantages in terms of fruit yield and quality before *Cucumis* spp. rootstocks can be used widely in commercial melon production.

Citrullus *as rootstocks*

Similar to the situation found with melons, the main shortcomings of the previously described rootstocks for watermelon (*C. lanatus*) are fruit quality problems. Scions grafted on to accessions of the same species, their wild relatives or species from the same genus are expected to have fewer horticultural problems related to scion–rootstock compatibility and fruit quality. Development of *Citrullus* rootstocks for watermelon production may, therefore, have some advantages.

Some studies have evaluated disease-resistant *Citrullus* germplasm as putative watermelon rootstocks, finding high levels of grafting compatibility, shoot growth and fruit development comparable to rootstocks of *Cucurbita* and *Lagenaria*, and less effect on fruit quality, thus confirming the possibility of producing high-quality watermelons by using disease-resistant watermelon rootstocks (Huh *et al.*, 2003).

One of the most promising sources of rootstocks for watermelon is the citron (*C. lanatus* var. *citroides*) (Fredes *et al.*, 2016) because it has significantly less galling than *Cucurbita* hybrids and bottle gourd rootstocks (Thies and Levi, 2007; Thies *et al.*, 2008), and indeed it has proved useful for managing root-knot nematodes in watermelon (Thies *et al.*, 2010, 2015a, b). The use of citron as a rootstock for watermelon also enhanced the tolerance of the scion to Fusarium wilt, although not as effectively as either interspecific hybrid squash or bottle gourd rootstocks (Keinath and Hassell, 2014). Citron rootstocks were also more susceptible to *D. bryoniae* than *Cucurbita* or *Lagenaria* rootstocks (Keinath, 2014b).

Recent screenings of exotic *Citrullus* collections against soilborne diseases and their horticultural characterization have provided new and promising rootstock sources for watermelon (Cohen *et al.*, 2014; Edelstein *et al.*, 2014). Other *Citrullus* spp. such as *Citrullus colocynthis* and *C. ecirrhosus* have not been exploited and may be a source of interesting traits, such as drought tolerance (Jarret, 2014).

2.3.2 Rootstocks for production of solanaceous crops

The main solanaceous crops grafted are tomato, aubergine and pepper. The first and main reason for grafting these crops was the same as for cucurbits, namely resistance to soilborne pathogens and nematodes (King *et al.*, 2008), particularly after the ban of methyl bromide as a soil fumigant (Miguel, 2004). Grafting in solanaceous crops has been used to provide scion resistance or tolerance not only to fungal diseases (e.g. *F. oxysporum*, *Verticillium dahliae* and *Verticillium alboatrum*, *Sclerotium rolfsii*, *Pyrenochaeta lycopersici*), bacterial diseases (*R. solanacearum*) and viruses such as pepino mosaic virus, but also to abiotic stresses such as high levels of soil boron and salinity, low and high temperatures, and flooding (Davis *et al.*, 2008; King *et al.*, 2010; Lee *et al.*, 2010; Schwarz *et al.*, 2010a; Miguel *et al.*, 2011). Grafting has also been exploited for its impact on managing foliar or other soilborne pests (e.g. insects, mites, parasitic plants such as *Orobanche* sp.), where it often showed positive results (Louws *et al.*, 2010).

The production of grafted solanaceous crops started with aubergine in the 1950s, followed by tomato and pepper in the 1960s. Currently, of the solanaceous crops, tomato has the greatest area of production on rootstocks because it has the largest overall production, and a large proportion of the high-value protected crops for the fresh market are grafted, although grafting is not yet commercial practice for field-grown processing tomato, a lower value crop. In addition to resistance to abiotic and biotic stresses, vigour is a very important trait in tomato, as it boosts yield, especially towards the end of long-season glasshouse crops, and this is a key reason that nearly 100% of glasshouse tomatoes in northern Europe are grafted. Commercial rootstocks are often named to indicate vigour and strength, with names such as Big Power, Beauforte, He-Man, Armstrong and Stallone, suggestive of their ability to alter the balance of vegetative and reproductive growth to optimize the cropping system.

The properties and uses of various species and their hybrids currently used as rootstocks are described below.

Solanum *section* Lycopersicon *as rootstocks*

Rootstocks derived from this section of the genus *Solanum* are the most popular for tomato production and are also used widely for aubergine production (e.g. Ioannou, 2001). By far the most common commercial rootstocks from this section are *S. lycopersicum* × *S. habrochaites* F_1 hybrids, but open-pollinated *S. lycopersicum* and its intraspecific hybrids have also been used (Table 2.5). These (e.g. the inbred 'KVFN' and the F_1 hybrid 'Maxifort') have multiple resistances including to corky root, *Verticillium* spp., *Fusarium* spp. and nematodes.

S. habrochaites is native to cloud forests of the western Andean slopes up to elevations of 3600 m above sea level, and is well known for its cold tolerance, maintaining its growth rate at lower temperatures. It can confer cold tolerance to the grafted scion (Venema *et al.*, 1999, 2005, 2008); this trait is of great importance for the success of early-season crops and for limiting the energy used for heating in northern European glasshouse tomato production. The species is also particularly vigorous, with a large sprawling habit, growing up to 6 m wide and up to 1 m high, and with a corolla of up to 5 cm (Peralta *et al.*, 2008); this generally high vigour may be an important factor in its success in rootstock hybrids. Although it is common knowledge that the majority of commercial tomato rootstocks are *S. lycopersicum* × *S. habrochaites* hybrids, the development of the most successful rootstock cultivars is proprietary information and is not in the public domain. The authors are not aware of any published systematic studies to evaluate the success of different hybrids rootstocks conducted in the public domain, and therefore the scientific basis of the success of the *S. lycopersicum* × *S. habrochaites* hybrids is not well understood, and the extent to which other wild species are utilized in commercial tomato rootstock breeding is unknown. For example, a recent study showed that a commercial hybrid rootstock, 'Jjak Kkung', improved water conservation in tomato, but the origins of the hybrid have not been published (Nilsen *et al.*, 2014). Despite this lack of genetic information, many experimenters have demonstrated

Table 2.5. Examples of Solanaceae rootstock–scion combinations for diverse purposes.

Rootstock species	Purpose	Reference(s)
Grafted to tomato scion		
Solanum lycopersicum	Improved tolerance to alkalinity	Mohsenian *et al.* (2012); Mohsenian and Roosta (2015)
	Improved tolerance to thermal shock	Rivero *et al.* (2003)
	Improved vigour and virus tolerance	Lee *et al.* (2010)
S. lycopersicum × *S. lycopersicum*	Improved fruit quality	Barrett *et al.* (2012); Krumbein and Schwartz (2013)
	Improved *Ralstonia solanacearum* and *Meloidogyne incognita* resistance	Kunwar *et al.* (2015)
	Improved macronutrient uptake	Leonardi and Giuffrida (2006)
Solanum habrochaites	Low-temperature tolerance	Venema *et al.* (2008)
	Corky root disease resistance	Lee *et al.* (2010))
	Tomato yellow leaf curl virus resistance	Mahmoud (2014)
S. lycopersicum × *S. habrochaites*	Salinity tolerance improvement	di Gioia *et al.* (2013); Giuffrida *et al.* (2014)
	Yield and fruit quality improvement	Turhan *et al.* (2011); Barrett *et al.* (2012); Gajc-Wolska *et al.* (2015)
	Colletotrichum coccodes management	Garibaldi *et al.* (2008)
	Improved macronutrient uptake	Leonardi and Giuffrida (2006); Savvas *et al.* (2009)
	Sclerotium rolfsii and southern root-knot nematode management	Rivard *et al.* (2010)
	Abiotic stresses tolerance	Schwarz *et al.* (2010b)
	Soilborne disease management	Rivard and Louws (2008)
	Increased vigour and yield	Djidonou *et al.* (2013); Higashide *et al.* (2014)
S. lycopersicum × *Solanum cheesmaniae*	Fruit quality improvement, better performance and productivity under saline conditions	Albacete *et al.* (2009); Flores *et al.* (2010)
S. lycopersicum × *Solanum pimpinellifolium*	Improving agronomic water-use efficiency	Cantero-Navarro *et al.* (2016)
	Tolerance to potassium (K^+) or phosphorus (P) deficits	Albacete *et al.* (2015b)

Continued

Table 2.5. Continued.

Rootstock species	Purpose	Reference(s)
	Improved tolerance to drought, salinity and soil impedance	Albacete *et al.* (2015a)
S. lycopersicum var. *cerasiforme*	Soilborne pathogens, foliar pathogens, arthropods and weed management	Alvarez-Hernandez *et al.* (2009); Cortez-Madrigal (2012)
Solanum melongena	Flooding stress resilience enhancement	Bhatt *et al.* (2015)
	Improved tolerance to alkalinity	Mohsenian and Roosta (2015)
	R. solanacearum management	Singh *et al.* (2014)
	Low- and high-temperature tolerance	Lee *et al.* (2010)
Solanum laciniatum	Water-logging resistance	Lee *et al.* (2010)
Solanum chilense	Tomato yellow leaf curl virus management	Mahmoud (2014)
Solanum pennellii	Tomato yellow leaf curl virus management	Mahmoud (2014)
S. pimpinellifolium	Tomato yellow leaf curl virus management	Mahmoud (2014)
Solanum integrifolium (syn. *S. aethiopicum* gr. Aculeatum)	Fruit quality improvement	Lee *et al.* (2010)
Solanum luteum	Improved tolerance to alkalinity	Mohsenian and Roosta (2015)
Nicotiana tabacum	Improved tolerance to alkalinity	Mohsenian *et al.* 2012)
Datura patula	Improved tolerance to alkalinity	Mohsenian *et al.* (2012); Mohsenian and Roosta (2015)
Solanum sisymbrifolium	Improved disease resistance and fruit quality	Lee *et al.* (2010)
Solanum toxicarium	Improved disease resistance	Lee *et al.* (2010)
Solanum torvum	*R. solanacearum* management	Singh *et al.* (2014)
Solanum nigrum	Improved fruit quality	Lee *et al.* (2010)
Grafted to aubergine scion		
Solanum melongena × *S. melongena*	Compatibility and grafting success, nematode resistance, yield, fruit number and earliness	Gisbert *et al.* (2011a)
S. torvum	Fruit yield, apparent quality and proximate and mineral composition	Gisbert *et al.* (2011b)

Continued

Table 2.5. Continued.

Rootstock species	Purpose	Reference(s)
	Reduction in aubergine fruit cadmium (Cd) concentrations	Arao *et al.* (2008)
	pH, mechanical firmness and vitamin C content, and sensory parameters	Arvanitoyannis *et al.* (2005)
	Yield and quality	Moncada *et al.* (2013)
	Growth and yield and Verticillium wilt resistance of aubergine	Bletsos *et al.* (2003)
	Growth performance under saline stress conditions	Colla *et al.* (2010)
	Increased yields, marketable fruit, health qualities	Sabatino *et al.* (2013)
	Resistant to nematodes	Lee *et al.* (2010)
S. incanum	Graft success, nematode resistance, fruit yield and quality improvement	Gisbert *et al.* (2011a)
S. incanum × *S. melongena*	Graft success, nematode resistance, fruit yield and quality improvement	Gisbert *et al.* (2011a)
S. melongena × *S. aethiopicum*	Graft success, nematode resistance, fruit yield and quality improvement	Gisbert *et al.* (2011a)
S. macrocarpon	Fruit yield and quality improvement	Gisbert *et al.* (2011b)
S. integrifolium (syn. *S. aethiopicum* gr. Aculeatum)	Improved Fusarium wilt resistance	Iwamoto and Ezura (2006)
	Reduction of fruit Cd concentration	Arao *et al.* (2008)
	Improved resistance to *Verticillium dahliae*	Johnson *et al.* (2014)
	Vigour, fruit yield and quality improvement	Gisbert *et al.* (2011b)
S. sisymbriifolium	Improved growth, yield and resistance to Verticillium wilt	Bletsos *et al.* (2003)
	Fruit quality improvement	Arvanitoyannis *et al.* (2005)
S. torvum × *S. sanitwongsei*	Resistance to bacterial wilt	Lee *et al.* (2010)
S. integrifolium syn. S.aethiopicum gr. Aculeatum × *S. melongena*	High temperature tolerance	Lee *et al.* (2010)
S. lycopersicum	Improved resistance to Verticillium wilt	Liu *et al.* (2009)

Continued

Table 2.5. Continued.

Rootstock species	Purpose	Reference(s)
	Fruit quality improvement (pH, degrees Brix, acidity)	Khah (2011)
	Improvement of fruit yield, size and quality	Passam *et al.* (2005)
S. lycopersicum × *S. lycopersicum*	Fruit quality improvement	Kacjan Maršić *et al.* (2014)
	Grafting success, improved nematode resistance and fruit yield	Gisbert *et al.* (2011a)
S. habrochaites	Graft success, nematode resistance, fruit yield and quality improvement	Gisbert *et al.* (2011a)
S. lycopersicum × *S. habrochaites*	Na_2SO_4 salinity resistance	Giuffrida *et al.* (2014)
	Macronutrient uptake	Leonardi and Giuffrida (2006)
	V. dahliae Kleb.	Johnson *et al.* (2014)
	Graft success, nematode resistance, yield, fruit number and earliness	Gisbert *et al.* (2011a)
S. melongena	Reduction aubergine fruit Cd concentrations	Arao *et al.* (2008)
	Root-knot nematode and corky root resistance	Serges and Colombo (2005) (cited by Louws *et al.*, 2010)
	Grafting success, improved nematode resistance and fruit yield	Gisbert *et al.* (2011a)
Grafted to pepper scion		
Capsicum annuum	Tolerance to mild salinity	Lee *et al.* (2006); Penella *et al.* (2015)
	Tolerance to *Phytophthora nicotianae*	Saadoun and Allagui (2013)
	Changes in fruit shape	Tsaballa *et al.* (2013)
	Phytophthora blight, bacterial wilt and root-knot nematode tolerance	Jang *et al.* (2012); Ros-Ibáñez *et al.* (2014)
	Tolerance to soilborne pathogens	Morra and Bilotto (2006)
	P. nicotianae tolerance	Saadoun and Allagui (2013)
	M. incognita tolerance	Kokalis-Burelle *et al.* (2009)
	Yield and/or fruit quality	Attia *et al.* (2003); Yang *et al.* (2013); Doñas-Uclés *et al.* (2014)
C. annuum F_1	*Fusarium oxysporum* and *M. incognita* tolerance and Verticillium wilt tolerance	Geboloğlu *et al.* (2011)

Continued

Table 2.5. Continued.

Rootstock species	Purpose	Reference(s)
	High radiation, temperature tolerance	López-Marín *et al.* (2013)
	Control of *Phytophthora* blight	Gilardi *et al.* (2014)
	Nutritional quality improvement	Chávez-Mendoza *et al.* (2013)
	Yield and commercial quality	Doñas-Uclés *et al.* (2014)
Capsicum chinense	Tolerance to mild salinity	Penella *et al.* (2015)
Capsicum frutescens	Increased production under hot wet or hot dry conditions	Palada and Wu (2008)
Capsicum baccatum	Tolerance to mild salinity	Penella *et al.* (2015)
C. annuum × *C. chinense*	Superior growth and yield	Lee *et al.* (2010)
	Agronomic performance and fruit quality	Gisbert *et al.* (2010)
Capsicum chacoence	Root-knot nematode resistance	Oka *et al.* (2004)

the benefits of commercial *S. lycopersicum* × *S. habrochaites* hybrid rootstocks on tomato and aubergine production (Table 2.5) (e.g. Giuffrida *et al.*, 2014).

The F_1 hybrid *S. lycopersicum* VF36 × *S. pennellii* LA0716 has been reported to have very high vigour and resistance to common root diseases, including multiple races of Fusarium wilt, and it is recommended by the TGRC for seed production from tomato-like species such as *S. sitiens*, *S. juglandifolium* and *S. ochranthum*, which are difficult to grow to maturity in a glasshouse. This hybrid rootstock (TGRC accession LA4135) makes the scions more compact, allows them to flower earlier and prevents early death from root rots; it is also reported to be graft-compatible with aubergine and pepper (Chetelat and Petersen, 2003).

Finally, *C. annuum* has been grafted experimentally to a tomato (*S. lycopersicum*) rootstock (Rodriquez and Bosland, 2010), although the graft survival rate was very low. Attempts to graft pepper scions on aubergine rootstocks have been unsuccessful (Bletsos and Olympios, 2008).

Spiny Solanum *as rootstocks*

Aubergine and its wild relatives (together known as the 'spiny solanums') provide rootstocks for production of both tomato and aubergine itself (Table 2.5).

S. melongena is an alternative rootstock for cultivating susceptible tomatoes in some conditions; it was shown to provide a better protection than tomato rootstocks against bacterial wilt (*R. solanacearum*) and against flooding in the hot wet season in South-east Asia (Black *et al.*, 2003) and shows good graft compatibility with tomato.

S. torvum (turkey berry) is a wild species closely related to cultivated aubergine (*S. melongena*) and is native to India. It is used extensively as a rootstock for aubergine and tomato (Table 2.5) (Bletsos *et al.*, 2003; Lee *et al.*, 2010; Moncada *et al.*, 2013; Miceli *et al.*, 2014); graft compatibility is excellent with aubergine, and double-grafted plants are often used to give high yields (Miguel *et al.*, 2011), whereas compatibility with tomato is inferior in some cases but still practical (Petran and Hoover, 2014). *S. torvum* rootstocks provide a wide range of resistances to soilborne diseases including root-knot nematode (Boiteux and Charchar, 1996; Bagnaresi *et al.*, 2013), *V. dahliae* (Alconero *et al.*, 1988; Bletsos *et al.*, 2003) and notably complete resistance to bacterial wilt (Petran and Hoover, 2014; Singh *et al.*, 2014). In addition, *S. torvum* rootstocks generally give increased vigour as measured by plant height, main stem diameter and root system mass compared with non-grafted plants.

Information on the yield and quality of aubergines grafted on to *S. torvum* rootstock is conflicting, but generally it is considered to have a negative effect (Moncada *et al.*, 2013); for example, fruits from grafted plants were found to be darker with a less vivid colour and a lower total phenolic content compared with non-grafted plants. However, Lee *et al.* (2010) reported no effect of *S. torvum* rootstock on fruit sugar content, and, when used as a rootstock for Sicilian aubergine ecotypes, the fruit quality and levels of health-related metabolites increased (Gisbert *et al.*, 2011b; Sabatino *et al.*, 2013). In terms of abiotic stress tolerance, grafting *S. melongena* on to an *S. torvum* rootstock reduced the cadmium (Cd) concentration in aubergine fruits (Arao *et al.*, 2008; Yamaguchi *et al.*, 2011), and under soil moisture deficit increased the shoot and root dry mass, plant height, stem diameter, leaf relative water content, leaf chlorophyll content, and leaf soluble sugar and proline contents, relative to non-grafted plants (Zhou *et al.*, 2012).

Despite its many benefits, *S. torvum* exhibits slow and erratic germination and slow seedling growth, and the development of uniform stands of seedlings for grafting in large commercial nurseries is challenging. Propagation by cuttings from stock plants for subsequent grafting has been developed as an alternative (Miceli *et al.*, 2014), and research to improve germination protocols has been reported (Ranil *et al.*, 2015). Further improvements in *S. torvum* seedling establishment could be obtained through breeding and seed technologies. Finally, *S. torvum* is a federally regulated noxious weed whose entry and distribution within the USA is prohibited (USDA APHIS, 2011), but its use in Europe and Asia is not restricted.

Since the 1950s, *S. aethiopicum* has been used as a rootstock for tomato and common aubergine (Oda, 2006), and various scientific evaluations have been performed. Compared with tomato rootstocks or non-grafted tomato plants, two *S. aethiopicum* rootstocks increased fruit set, number and mass, and improved control of disease in tomato scions and the shelf-life of their fruit, but had no effect on degrees Brix or acidity (Nkansah *et al.*, 2013). In contrast, in comparison with tomato Hawaii 7998 rootstocks, *S. integrifolium* syn. *S. aethiopicum* gr. Aculeatum rootstocks reduced the growth and yield of tomato, while increasing blossom end rot and increasing fruit sugar and total soluble solids (Oda *et al.*, 1996). *S. aethiopicum* is reportedly the most commonly used rootstock for aubergine production in Japan (Iwamoto and Ezura 2006), but some combinations are reportedly poor rootstocks for aubergine (Gisbert *et al.*, 2011a). Specifically, some *S. aethiopicum* accessions have been noted for their resistance to three tobamoviruses (tobacco mosaic virus, tomato mosaic virus and pepper mild mottle virus; Tzortzakakis *et al.*, 2006) and to Fusarium wilt (Iwamoto and Ezura, 2006), but all the accessions tested were susceptible to root-knot nematodes (*Meloidogyne javanica* and *M. incognita*) (Tzortzakakis *et al.*, 2006). Germination of this species is considered to be more uniform than that of *S. torvum* (Ginoux and Laterrot, 1991).

The wild species *Solanum sisymbriifolium* (sticky nightshade) has small edible fruits and has been used as a rootstock for both tomato (Matsuzoe *et al.*, 1996, cited by Bletsos and Olympios, 2008) and aubergine (Bletsos *et al.*, 2003; Arvanitoyannis *et al.*, 2005) to provide vigour and resistance to *V. dahliae* (Alconero *et al.*, 1988), although *S. sisymbriifolium* was less resistant than *S. torvum* (Bletsos *et al.*, 2003). There were some negative effects on aubergine fruit quality reported, with fruits being less firm and less sweet when grafted on to *S. sisymbriifolium* rootstocks (Arvanitoyannis *et al.*, 2005)

F_1 aubergine hybrids and interspecific hybrids between aubergine and its relatives have been used as rootstocks (Lee *et al.*, 2010), and provide promising material for developing new rootstocks for aubergine production (Gisbert *et al.*, 2011a), despite some difficulties in hybridization (Calvo-Asensio *et al.*, 2014). The *S. melongena* × *S. incanum* and *S. melongena* × *S. aethiopicum* hybrids are vigorous rootstocks that enhance yield, fruit number and earliness, and show resistance to several diseases without negative effects on apparent fruit quality or composition; both hybrids exibited high germination (> 90%), unlike the erratic germination of *S. torvum*, and also gave excellent compatibility (100% graft success) with aubergine 'Black Beauty' scions (Gisbert *et al.*, 2011a).

Solanum sanitwongsei 'Karehen' was found to have excellent resistance to bacterial wilt, but as a rootstock it was not agronomically acceptable because of slow germination and slender seedling hypocotyls, which make grafting difficult, and because the rootstock conferred low vigour to the scion (Asao *et al.*, 2001). Attempts have therefore been made to utilize the bacterial wilt resistance in a hybrid with *S. torvum* (Lee *et al.*, 2010), *S. melongena* (Asao *et al.*, 2001) and *S. aethiopicum* (Iwamoto *et al.*, 2007).

Capsicum *as rootstocks*

As might be expected, *Capsicum* rootstocks are used mainly for pepper (*C. annuum*) production, and pepper grafting first began in Japan and Korea. Pepper is currently the least grafted among the solanaceous crops (Lee *et al.*, 2010), presumably because the available rootstocks provide relatively modest benefits; consequently, there is great interest in developing new *Capsicum* rootstocks to address the market opportunities. The primary purpose of grafting peppers worldwide has been to manage losses to soilborne pathogens of major economic importance such as *P. capsici*, pests including nematodes, and viruses such as tobacco mosaic virus or potato Y virus. In common with other crops, successful rootstocks for pepper production should also ideally have excellent scion compatibility, provide resistance to abiotic stresses, and improve fruit yield and quality (Gisbert *et al.*, 2010; Chávez-Mendoza *et al.*, 2013; Jang *et al.*, 2013).

The most popular rootstocks are cultivars or intraspecific hybrids of *C. annuum*, but accessions from the cultivated species *C. baccatum*, *C. chinense* and *C. frutescens*, and their interspecific hybrids with *C. annuum*, have also been tested as rootstocks for pepper scions (Lee *et al.*, 2010). It is hoped that the exploitation of *Capsicum* wild relatives such as *C. chacoense* may lead to the development of more-robust rootstocks (Oka *et al.*, 2004).

Breeding and selecting new rootstocks is necessary because pathogens may develop virulence capable of overcoming the resistance provided by a rootstock (Ros-Ibáñez *et al.*, 2014). As there is intraspecific variability for various agronomical and morphological traits in *C. annuum*, *C. chinense* and *C. frutescens*, screening many accessions of each species could assist in the selection of the right rootstock (Gisbert *et al.*, 2013). For example, evaluation of such accessions has identified *C. annuum* and *C. frutescens* genotypes that present moderate to high resistance to *M. incognita* and *M. javanica*, but susceptibility to *Meloidogyne enterolobii* (Oka *et al.*, 2004; Pinheiro *et al.*, 2015).

C. annuum varieties and intraspecific hybrids have been used as rootstocks for pepper scions for the management of *F. oxysporum* and *V. dahliae* and the nematode *M. incognita* (Geboloğlu *et al.*, 2011). The yield increase in the grafted plants with or without the presence of biotic stress showed that the right combination of scion–rootstock can be beneficial. Using *C. annuum* rootstocks improved the agronomic performance and marketable yield (Colla *et al.*, 2008), depending on the rootstock genotype used (Doñas-Uclés *et al.*, 2014). Comparative experiments have shown that *C. annuum* rootstocks were more compatible for pepper production in comparison with rootstocks of other *Capsicum* cultivated species, as determined from fruit yield and quality measurements (de Oliveira *et al.*, 2009). *C. annuum* rootstocks increased the nutritional fruit quality in terms of β-carotene, vitamin C and total antioxidant capacity (Chávez-Mendoza *et al.*, 2013). The wild serrano-type pepper SCM334 proved to be a successful rootstock against root rot and wilt (*Phytophthora nicotianae*) (Saadoun and Allagui, 2013).

The cultivated relative of sweet pepper, *C. baccatum*, has been evaluated as a rootstock in terms of graft compatibility and resistance to *Meloidogyne* nematodes; it showed resistance to *M. javanica* and susceptibility to *M. incognita* and *M. enterolobii* (Oka *et al.*, 2004; Pinheiro *et al.*, 2015). *C. baccatum* var. pendulum rootstock increased pepper fruit yield under salinity (Penella *et al.*, 2014, 2015), and *C. baccatum* rootstock lines provide the greatest tolerance to flooding damage, giving a survival rate greater than 45% (Palada and Wu, 2008). *C. annuum* × *C. baccatum* hybrid rootstocks have been commercially developed and patented (Hennart, 2014) to provide multiple disease resistances and a good balance between generative and vegetative vigour for high yields.

C. frutescens genotypes have been evaluated as rootstocks for sweet pepper and gave good resistance to *M. incognita* (Oka *et al.*, 2004; de Oliveira *et al.*, 2009; Pinheiro *et al.*, 2015) and *M. javanica*, but they were susceptible to *M. enterolobii* (Pinheiro *et al.*, 2015). *C. frutescens* rootstock provided tolerance to flooding damage (Palada and Wu, 2008), but provided lower fruit production and quality compared with *C. annuum* rootstocks (de Oliveira *et al.*, 2009). Intraspecific variability was shown in *C. frutescens*, with three out of four accessions being resistant to *M. incognita* (Gisbert *et al.*, 2013).

C. chinense genotypes have been evaluated as rootstocks to protect against *M. incognita* (Oka *et al.*, 2004; de Oliveira *et al.*, 2009; Gisbert *et al.*, 2013), and some studies showed intraspecific variability in resistance. However, one evaluation showed that all elite genotypes of *C. chinense* were susceptible to *M. ingognita*,

M. javanica and *M. enterolobii* (Pinheiro *et al.*, 2015). Nevertheless, *C. chinense* is considered to have potential to contribute to root-knot nematode management as a component in sustainable crop production systems (Kokalis-Burelle *et al.*, 2009). *C. chinense* has also been evaluated for its graft compatibility (Oka *et al.*, 2004), and one *Capsicum chinense* genotype was among the most tolerant tested under salinity conditions according to photosynthetic rate (Penella *et al.*, 2013). *C. annuum* × *C. chinense* hybrids have been reported to be in extensive use for green pepper production to increase growth and yield (Lee *et al.*, 2010).

Other Solanaceae species as rootstocks

Other miscellaneous rootstocks that have been used for grafting tomato are datura (*Datura patula*), orange nightshade (*Solanum luteum*) and tobacco (*N. tabacum*) (Lee *et al.*, 2010; Mohsenian *et al.*, 2012). Mohsenian *et al.* (2012) recommended *Datura* rootstocks to provide resistance to alkaline soils in tomato production.

2.4 Germplasm Collections and Grafting in Other Plant Families

2.4.1 *Cynara* gafting

Cynara cardunculus is a species of Mediterranean origin that includes three different taxa: the wild perennial cardoon (var. *sylvestris*), the cultivated or leafy cardoon (var. *altilis*) and the globe artichoke (var. *scolymus*) (Sonnante *et al.*, 2007). Globe artichoke is cultivated extensively around the world, with Italy being by far the top producer at 474,550 t in 2011 (FAOSTAT: http://faostat.fao.org/, accessed 24 November 2016); however, commercial varieties lack effective resistance to *V. dahliae*. The high grafting affinity between globe artichoke and wild and cultivated cardoon has stimulated interest in the potential of cardoon diversity for use as rootstocks; initial screening experiments identified *V. dahliae* resistance in some accessions of wild cardoon (Ciccarese *et al.*, 2012; Temperini *et al.* 2013; Trinchera *et al.*, 2013a, b). Cultivated cardoons used as rootstocks increased yield and reduced Verticillium wilt incidence on the scion (Ciccarese *et al.*, 2012). Development of rootstocks for globe artichoke is still in its infancy, but there are significant genetic resources available for further selection and breeding; a search of global genetic resources of the genus *Cynara* found 249 accessions, of which 199 were under the botanical species *C. cardunculus* var. *cardunculus* or *C. cardunculus*. For *Cynara scolymus*, the alternative name for globe artichoke, there were 23 accessions (http://www.genesys-pgr.org/, accessed 24 November 2016).

2.4.2 *Phaseolous* grafting

Since 2010, green beans have been extensively produced as a grafted crop in Portugal using rootstocks from *Phaseolus coccineus* and *Phaseolus vulgaris*; further discussion of this topic and the available genetic resources are given in Chapters 1 and 9 (this volume).

2.5 Conclusions

It is clear that the genetic resources available for the key grafted vegetable crops are very extensive, and they are securely held and accessible from germplasm centres around the world. The collection of molecular phylogenetic and DNA sequence information for these accessions will doubtless increase dramatically as genotyping and sequencing platforms become ever more powerful and economical. The remaining key challenges are to fully exploit the genetic information by discovery and incorporation of beneficial alleles into breeding lines, and to understand the myriad of interactions between loci, between genes and environment, and between rootstock and scion genotypes. The range of rootstock cultivars currently in widespread commercial use is extremely narrow and is much more limited than the diversity of scion cultivars. Chapter 3 (this volume) explores how genetic diversity can be exploited more deeply to create new rootstock cultivars, and describes the current and future breeding strategies applicable to rootstocks.

Acknowledgements

The authors thank the Spanish Ministerio de Economía y Competitividad (MINECO), project AGL2014-53398-C2-2-R, co-funded with FEDER funds.

References

Albacete, A., Martinez-Andujar, C., Ghanem, M.E., Acosta, M., Sanchez-Bravo, J., Asins, M.J., Cuartero, J., Lutts, S., Dodd, I.C. and Perez-Alfocea, F. (2009) Rootstock-mediated changes in xylem ionic and hormonal status are correlated with delayed leaf senescence, and increased leaf area and crop productivity in salinized tomato. *Plant, Cell and Environment* 32, 928–938.

Albacete, A., Andújar, C., Dodd, I., Giuffrida, F., Hichri, I., Lutts, S., Thompson, A. and Asins, M. (2015a) Rootstock-mediated variation in tomato vegetative growth under drought, salinity and soil impedance stresses. *Acta Horticulturae* 1086, 141–146.

Albacete, A., Andújar, C., Pérez-Alfocea, F., Lozano, J. and Asins, M. (2015b) Rootstock-mediated variation in tomato vegetative growth under low potassium or phosphorous supplies. *Acta Horticulturae* 1086, 147–152.

Albrecht, E., Zhang, D., Saftner, R.A. and Stommel, J.R. (2012) Genetic diversity and population structure of *Capsicum baccatum* genetic resources. *Genetic Resources and Crop Evolution* 59, 517–538.

Alconero, R., Robinson, R.W., Dicklow, B. and Shail, J. (1988) Verticillium wilt resistance in eggplant, related *Solanum* species, and interspecific hybrids. *HortScience* 23, 388–390.

Alvarez-Hernandez, J.C., Cortez-Madrigal, H., Garcia-Ruiz, I., Ceja-Torres, L.F. and Petrez-Dominguez, J.F. (2009) Incidence of pests in grafts of tomato (*Solanum lycopersicum*) on wild relatives. *Revista Colombiana De Entomologia* 35, 150–155.

Amin, A.W. and Mona, A.W. (2014) Protecting cucumber from *Meloidogyne incognita* using graft onto resistant cucurbit rootstocks and antagonistic marigold as an alternative to nematicide. *Pakistan Journal of Nematology* 32, 51–58.

Arao, T., Takeda, H. and Nishihara, E. (2008) Reduction of cadmium translocation from roots to shoots in eggplant (*Solanum melongena*) by grafting onto *Solanum torvum* rootstock. *Soil Science and Plant Nutrition* 54, 555–559.

Arvanitoyannis, I.S., Khah, E.M., Christakou, E.C. and Bletsos, F.A. (2005) Effect of grafting and modified atmosphere packaging on eggplant quality parameters during storage. *International Journal of Food Science and Technology* 40, 311–322.

Asao, H., Aral, S. and Hirai, M. (2001) Somatic hybridization between *Solanum melongena* L. (eggplant) and *Solanum sanitwongsei* Craib. In: Nagata, T. and Bajaj, Y.P.S. (eds) *Somatic Hybridization in Crop Improvement II. Biotechnology in Agriculture and Forestry 49*. Springer, Berlin, pp. 233–244.

Attia, M.F., Arafa, A.M., Moustafa, M.A. and Mohamed, M.A. (2003) Pepper grafting, a method of controlling soilborne diseases and enhancement of fruit yield: 1. Improvement of pepper resistance to Fusarium wilt. *Egyptian Journal of Phytopathology* 31, 151–165.

Bagnaresi, P., Sala, T., Irdani, T., Scotto, C., Lamontanara, A., Beretta, M., Rotino, G.L., Sestili, S., Cattivelli, L. and Sabatini, E. (2013) *Solanum torvum* responses to the root-knot nematode *Meloidogyne incognita*. *BMC Genomics* 14, 540.

Barrett, C.E., Zhao, X., Sims, C.A., Brecht, J.K., Dreyer, E.Q. and Gao, Z. (2012) Fruit composition and sensory attributes of organic heirloom tomatoes as affected by grafting. *HortTechnology* 22, 804–809.

Bates, D.M., Merrick, L.C. and Robinson, R.W. (1995) Minor cucurbits – *Benincasa*, *Lagenaria*, *Luffa*, *Sechium*, and other genera (Cucurbitaceae). In: Smartt, J. and Simmonds, N.W. (eds) *Evolution of Crop Plants*, 2nd edn. Longman Scientific, Harlow, UK, pp. 104–111.

Bauchet, G. and Causse, M. (2012) Genetic diversity in tomato (*Solanum lycopersicum*) and its wild relatives. In: Caliskanm, M. (ed.) *Genetic Diversity in Plants*. InTech. Available at: http://www.intechopen.com/books/genetic-diversity-in-plants/genetic-diversity-in-tomato-solanum-lycopersicum-and-its-wild-relatives (accessed 24 November 2016).

Bebeli, P.J. and Mazzucato, A. (2008) The Solanaceae – a review of recent research on genetic resources and advances in the breeding of tomato, pepper and eggplant. *European Journal of Plant Science and Biotechnology* 2 (Special Issue 1), 3–30.

Behera, T.K., Singh, A.K. and Staub, J.E. (2008) Comparative analysis of genetic diversity in Indian bitter gourd (*Momordica charantia* L.) using RAPD and ISSR markers for developing crop improvement strategies. *Scientia Horticulturae* 115, 209–217.

Bekhradi, F., Kashi, A. and Delshad, M. (2011) Effect of three cucurbits rootstocks on vegetative and yield of 'Charleston Gray' watermelon. *International Journal of Plant Production* 5, 105–110.

Bhatt, R.M., Upreti, K.K., Divya, M.H., Bhat, S., Pavithra, C.B. and Sadashiva, A.T. (2015) Interspecific grafting to enhance physiological resilience to flooding stress in tomato (*Solanum lycopersicum* L.) *Scientia Horticulturae* 182, 8–17.

Bhawna Abdin, M.Z., Arya, L., Saha, D., Sureja, A.K., Pandey, C. and Verma, M. (2014) Population structure and genetic diversity in bottle gourd [*Lagenaria siceraria* (Mol.) Standl.] germplasm from India assessed by ISSR markers. *Plant Systematics and Evolution* 300, 767–773.

Black, L.L., Wu, D.L., Wang, J.F., Kalb, T., Abbass, D. and Chen, J.H. (2003) Grafting tomatoes for production in the hot-wet season. Asian Vegetable Research and Development Center, Shanhua, Taiwan. Publication no. 03-551. Available at: https://www.ces.ncsu.edu/fletcher/programs/ncorganic/research/grafting.pdf (accessed 24 November 2016).

Blanca, J., Canizares, J., Lamb, L., Pascual, L., Díez, M.J. and Nuez, F. (2012) Variation revealed by SNP genotyping and morphology provides insight into the origin of the tomato. *PLoS ONE* 7, e48198.

Blanca, J., Montero-Pau, J., Sauvage, C., Bauchet, G., Illa, E., Díez, M.J., Francis, D., Causse, M., van der Knaap, E. and Cañizares, J. (2015) Genomic variation in tomato, from wild ancestors to contemporary breeding accessions. *BMC Genomics* 16, 257.

Bletsos, F.A. and Olympios, C.M. (2008) Rootstocks and grafting of tomatoes, peppers and eggplants for soil-borne disease resistance, improved yield and quality. *European Journal of Plant Science and Biotechnology* 2 (Special issue 1), 62–73.

Bletsos, F., Thanassoulopoulos, C. and Roupakias, D. (2003) Effect of grafting on growth, yield, and *Verticilium* wilt of eggplant. *HortScience* 38, 183–186.

Boiteux, L.S. and Charchar, J.M. (1996) Genetic resistance to root-knot nematode (*Meloidogyne javanica*) in eggplant (*Solanum melongena*). *Plant Breeding* 115, 198–200.

Bulder, H.A.M., den Nijs, A.P.M., Speek, E.J., van Hasselt, P.R. and Kuiper, P.J.C. (1991) The effect of low root temperature on growth and lipid composition of low temperature tolerant rootstock genotypes for cucumber. *Journal of Plant Physiology* 138, 661–666.

Buller, S., Inglis, D. and Miles, C. (2013) Plant growth, fruit yield and quality, and tolerance to Verticillium wilt of grafted watermelon and tomato in field production in the Pacific Northwest. *HortScience* 48, 1003–1009.

Cáceres, A., Ferriol, M., Gisbert, C. and Pico, B. (2016) Use of wild *Cucumis* as potential new rootstocks for melons. In: Kozik, E.U. and Paris, H.S. (eds) *Proceedings of Cucurbitaceae 2016*, Warsaw, Poland, pp. 300–304.

Calvo-Asensio, I., Prohens, J. and Gisbert, C. (2014) Vigor for *in vitro* culture traits in *S. melongena* × *S. aethiopicum* hybrids with potential as rootstocks for eggplant. *Scientific World Journal* 2014, 1–8.

Cantero-Navarro, E., Romero-Aranda, R., Fernández-Muñoz, R., Martínez-Andújar, C., Pérez-Alfocea, F. and Albacete, A. (2016) Improving agronomic water use efficiency in tomato by rootstock-mediated hormonal regulation of leaf biomass. *Plant Science* 251, 90–100.

Chávez-Mendoza, C., Sánchez, E., Carvajal-Millán, E., Munoz-Márquez, E. and Guevara-Aguilar, A. (2013) Characterization of the nutraceutical quality and antioxidant activity in Bell pepper in response to grafting. *Molecules* 18, 15689–15703.

Cheshmehmanesh, A.A., Kashi, A., Moshrefi, M.M. and Khososi, M. (2004) Effect of grafting on growth and yield of two greenhouse cucumber cultivars, Vilmorin and Royal 24189, onto figleaf squash (*Cucurbita ficifolia*) rootstock. *Seed and Plant* 19, 447–456.

Chetelat, R.T. (2016) Overcoming sterility and unilateral incompatibility of *Solanum lycopersicum* × *S. sitiens* hybrids. *Euphytica* 207, 319–330.

Chetelat, R.T. and Petersen, J.P. (2003) Improved maintenance of the tomato-like *Solanum* spp. by grafting. *Report of Tomato Genetics Cooperative* 53, 14–15.

Chomicki, G. and Renner, S.S. (2015) Watermelon origin solved with molecular phylogenetics including Linnaean material: another example of museomics. *New Phytologist* 205, 526–532.

Ciccarese, F., Crinò, P., Raccuia, S.A. and Temperini, A. (2012) Use of resistant cardoons as rootstocks for the control of verticillium wilt in globe artichoke. *Acta Horticulturae* 942, 201–206.

Clain, C., da Silva, D., Fock, I., Vaniet, S., Carmeille, A., Gousset, C., Sihachakr, D., Luisetti, J., Kodja, H. and Besse, P. (2004) RAPD genetic homogeneity and high levels of bacterial wilt tolerance in *Solanum torvum* Sw. (Solanaceae) accessions from Reunion Island. *Plant Science* 166, 1533–1540.

Cohen, R., Omari, N., Porat, A. and Edelstein, M. (2012) Management of *Macrophomina* wilt in melons using grafting or fungicide soil application: pathological, horticultural and economical aspects. *Crop Protection* 35, 58–63.

Cohen, R., Tyutyunik, J., Fallik, E., Oka, Y., Tadmor, Y. and Edelstein, M. (2014) Phytopathological evaluation of exotic watermelon germplasm as a basis for rootstock breeding. *Scientia Horticulturae* 165, 203–210.

Colla, G., Rouphael, Y., Cardarelli, M., Temperini, O., Rea, E., Salerno, A. and Pierandrei, F. (2008) Influence of grafting on yield and fruit quality of pepper (*Capsicum annuum* L.) grown under greenhouse conditions. *Acta Horticulturae* 782, 359–363.

Colla, G., Rouphael, Y., Leonardi, C. and Bie, Z. (2010) Role of grafting in vegetable crops grown under saline conditions. *Scientia Horticulturae* 127, 147–155.

Condurso, C., Verzera, A., Dima, G., Tripodi, G., Crino, P., Paratore, A. and Romano, D. (2012) Effects of different rootstocks on aroma volatile compounds and carotenoid content of melon fruits. *Scientia Horticulturae* 148, 9–16.

Cortez-Madrigal, H. (2012) Grafts of crops on wild relatives as base of an integrated pest management: the tomato *Solanum lycopersicum* as example. In: Larramendy, M.L. and Soloneski, S. (eds) *Integrated Pest Management and Pest Control – Current and Future Tactics*. InTech. Available at: http://www.intechopen.com/books/integrated-pest-management-and-pest-control-current-and-future-tactics/grafts-of-crops-on-wild-relatives-as-base-of-an-integrated-pest-management-the-tomato-solanum-lycope (accessed 24 November 2016).

Cross, H.B., Lira, R. and Motley, T.J. (2006) Origins and diversification of Chayote. In: Motley, T.J., Zerega, N.J.C. and Cross, H.B. (eds) *Darwin's Harvest: New Approaches to the Origins, Evolution, and Conservation of Crops*. Columbia University Press, New York, pp. 171–194.

Dalamu, T.K., Behera, A.B., Gaikwad, Saxena, S. , Bharadwaj, C. and Munshi, A.D. (2012) Morphological and molecular analyses define the genetic diversity of Asian bitter gourd (*Momordica charantia* L.). *Australian Journal of Crop Science* 6, 261–267.

Dane, F. and Liu, J. (2007) Diversity and origin of cultivated and citron type watermelon (*Citrullus lanatus*). *Genetic Resources and Crop Evolution* 54, 1255–1265.

Dasgan, H.Y., Balacheva, E., Yetişir, H., Yarsi, G., Altuntas, O., Akhoundnejad, Y. and Coban, A. (2015) The effectiveness of grafting to improve salt tolerance of sensitive melon when the tolerant melon is use as rootstock. *Procedia Environmental Sciences* 29, 268.

Daunay, M.C. (2008) Eggplant. In: Prohens, J. and Nuez, F. (eds) *Handbook of Plant Breeding. Vegetables II: Fabaceae, Liliaceae, Umbelliferae, and Solanaceae*. Springer, New York, pp. 163–220.

Davis, A.R., Perkins-Veazie, P., Sakata, Y., López-Galarza, S., Maroto, J.V., Lee, S.G., Huh, Y.C., Sun, Z., Miguel, A., King, S.R., Cohen, R. and Lee, J.M. (2008) Cucurbit grafting. *Critical Reviews in Plant Sciences* 27, 50–74.

Decker-Walters, D.S., Wilkins-Ellert, M., Chung, S.M. and Staub, J.E. (2004) Discovery and genetic assessment of wild bottle gourd [*Lagenaria siceraria* (Mol.) Standley; Cucurbitaceae] from Zimbabwe. *Economic Botany* 58, 501–508.

de Oliveira, C.D., Braz, L.T., dos Santos, J.M., Banzatto, D.A. and de Oliveira, P.R. (2009) [Hot peppers resistance to root-knot-nematodes and stump/rootstock compatibility among hot peppers and red pepper hybrids]. *Horticultura Brasileira* 27, 520–526 (in Portuguese with English abstract).

Díez, M.J. and Nuez, F. (2008) Tomato. In: Prohens, J. and Nuez, F. (eds) *Handbook of Plant Breeding. Vegetables II: Fabaceae, Liliaceae, Umbelliferae, and Solanaceae*. Springer, New York, pp. 249–326.

Díez, M.J., van Dooijeweert, W., Maggioni, L. and Lipman, E. (eds) (2008) *First Meeting of the ECPGR Working Group on Cucurbits*, 1–2 September 2005, Plovdiv, Bulgary. Bioversity International, Rome, Italy, p. 67.

Di Gioia, F., Signore, A., Serio, F. and Santamaria, P. (2013) Grafting improves tomato salinity tolerance through sodium partitioning within the shoot. *HortScience* 48, 855–862.

Djidonou, D., Zhao, X., Simonne, E.H., Koch, K.E. and Erickson, J.E. (2013) Yield, water-, and nitrogen-use efficiency in field-grown, grafted tomatoes. *HortScience* 48, 485–492.

Domingos, J.P., Fita, A.M., Picó, M.B., Sifres, A., Daniel, I.H., Salvador, J., Sapalo, F., Mozambique, P. and Díez, M.J. (2016) Angolan vegetable crops have unique genotypes of potential value for future breeding programmes. *South African Journal of Science* 112(3/4), Art. #2015-0285, 12 pages.

Doñas-Uclés, F., Jiménez-Luna, M., Góngora-Corral, J.A., Pérez-Madrid, D., Verde-Fernández, D. and Camacho-Ferre, F. (2014) Influence of three rootstocks on yield and commercial quality of Italian sweet pepper. *Ciência e Agroteccnologia* 38, 538–545.

Ebert, A.W. (2013) *Ex situ* conservation of plant genetic resources of major vegetables. In: Normah, M.N., Chin, H.F. and Reed, B.M. (eds) *Conservation of Tropical Plant Species*. Springer, New York, pp. 373–417.

Ebert, A.W., Astorga, C., Ebert, I.C.M., Mora, A. and Umaña, C. (2007) *Securing Our Future: CATIE's Germplasm Collections*. CATIE, Technical Bulletin, Turrialba, Costa Rica.

Edelstein, M., Ben-Hur, M., Leib, L. and Plaut, Z. (2011) Mechanism responsible for restricted boron concentration in plant shoots grafted on pumpkin rootstocks. *Israel Journal of Plant Sciences* 59, 207–215.

Edelstein, M., Tyutyunik, J., Fallik, E., Meir, A., Tadmor, Y. and Cohen, R. (2014) Horticultural evaluation of exotic watermelon germplasm as potential rootstocks. *Scientia Horticulturae* 165, 196–202.

El-Eslamboly, A.A.S. and Deabes, A.A.A. (2014) Grafting cucumber onto some rootstocks for controlling root-knot nematodes. *Minufiya Journal of Agricultural Research* 39, 1109–1129.

Erickson, D.L., Smith, B.D., Clarke, A.C., Sandweiss, D.H. and Tuross, N. (2005) An Asian origin for a 10,000-year-old domesticated plant in the Americas. *Proceedings of the National Academy of Sciences USA* 102, 18315–18320.

Ferriol, M. and Picó, B. (2008) Pumpkin and winter squash. In: Prohens, J. and Nuez, F. (eds) *Handbook of Plant Breeding. Vegetables I: Asteraceae, Brassicaceae, Chenopoidicaceae, and Cucurbitaceae*. Springer, New York, pp. 317–349.

Fita, A., Pico, B., Roig, C. and Nuez, F. (2007) Performance of *Cucumis melo* ssp. *agrestis* as a rootstock for melon. *Journal of Horticultural Science and Biotechnology* 82, 184–190.

Flores, F.B., Sanchez-Bel, P., Estan, M.T., Martinez-Rodriguez, M.M., Moyano, E., Morales, B., Campos, J.F., Garcia-Abellán, J.O., Egea, M.I., Fernández-Garcia, N., Romojaro, F. and Bolarín, M.C. (2010) The effectiveness of grafting to improve tomato fruit quality. *Scientia Horticulturae* 125, 211–217.

Frary, A., Doganlar, S. and Daunay, M.C. (2007) Eggplant. In: Kole, C. (ed.) *Genome Mapping and Molecular Breeding in Plants*, Vol. 5: *Vegetables*. Springer, Berlin, Germany, pp. 287–313.

Gajc-Wolska, J., Kowalczyk, K., Marcinkowska, M., Radzanowska, J. and Bujalski, D. (2015) Influence of growth conditions and grafting on the yield, chemical composition and sensory quality of tomato fruit in greenhouse cultivation. *Source of the Document Journal of Elementology* 20, 73–81.

Galatti, F., Franco, A., Ito, L., Charlo, H., Gaion, L. and Braz, L. (2013) Rootstocks resistant to *Meloidogyne incognita* and compatibility of grafting in net melon. *Ceres* 60, 432–436.

Garibaldi, A., Baudino, M., Minuto, A. and Gullino, M.L. (2008) Effectiveness of fumigants and grafting against tomato brown root rot caused by *Colletotrichum coccodes*. *Phytoparasitica* 36, 483–488.

Geboloğlu, N., Yanar, Y., Yanar, D., Akyazi, F. and Çakmak, P. (2011) Role of different rootstocks on yield and resistance for *Fusarium oxysporum*, *Verticillium dahliae* and *Meloidogyne incognita* in grafted peppers. *European Journal of Horticultural Science* 76, 41–44.

Gilardi, G., Pugliese, M., Colla, P., Gullino, M.L. and Garibaldi, A. (2014) Management of *Phytophthora capsici* on bell pepper and *Colletotrichum coccodes* on tomato by using grafting and organic amendments. *Acta Horticulturae* 1044, 257–262.

Ginoux, G. and Laterrot, H. (1991) Greffage de l'aubergine: reflexions sur le choix du portegreffe. *PHM Revue Horticole* 321, 49–54.

Gisbert C., Sánchez-Torres P., Raigon M.D. and Nuez F. (2010). *Phytophthora capsici* resistance evaluation in pepper hybrids: agronomic performance and fruit quality of pepper grafted plants. *Journal of Food Agriculture and Environment* 8, 116–121.

Gisbert, C., Prohens, J. and Nuez, F. (2011a) Performance of eggplant grafted onto cultivated, wild, and hybrid materials of eggplant and tomato. *International Journal of Plant Production* 5, 367–380.

Gisbert, C., Prohens, J., Raigon, M.D., Stommel, J.R. and Nuez, F. (2011b) Eggplant relatives as sources of variation for developing new rootstocks: effects of grafting on eggplant yield and fruit apparent quality and composition. *Scientia Horticulturae* 128, 14–22.

Gisbert, C., Trujillo-Moya, C., Sánchez-Torres, P., Sifres, A., Sánchez-Castro, E. and Nuez, F. (2013) Resistance of pepper germplasm to *Meloidogyne incognita*. *Annals of Applied Biology* 162 110–118.

Giuffrida, F., Cassaniti, C. and Leonardi, C. (2014) Tomato and eggplant scions influence the effect of rootstock under Na_2SO_4 salinity. *Acta Agriculturae Scandinavica Section B: Soil and Plant Science* 64, 700–709.

Goreta Ban, S., Zanic, K. Dumicic, G., Raspudic, E., Vuletin Selak, G. and Ban, D. (2014) Growth and yield of grafted cucumbers in soil infested with root-knot nematodes. *Chilean Journal of Agricultural Research* 74, 29–34.

Guan, W., Zhao, X., Dickson, D.W., Mendes, M.L. and Thies, J. (2014) Root-knot nematode resistance, yield, and fruit quality of specialty melons grafted onto *Cucumis metuliferus*. *Hortscience* 49, 1046–1051.

Guler, Z., Candir, E., Yetişir, H., Karaca, F. and Solmaz, I. (2014) Volatile organic compounds in watermelon (*Citrullus lanatus*) grafted onto 21 local and two commercial bottle gourd (*Lagenaria siceraria*) rootstocks. *Journal of Horticultural Science and Biotechnology* 89, 448–452.

Heiser, C.B. (1995) Peppers: *Capsicum* (Solanaceae). In: Smartt, J. and Simmonds, N.W. (eds) *Evolution of Crop Plants*, 2nd edn. Longman Scientific and Technical, Harlow, UK, pp. 449–451.

Hennart, J.W. (2014) Capsicum rootstock. Patent number US20140096289 A1.

Hernández-González, Z., Sahagun-Castellanos, J., Espinosa-Robles, P., Colinas-Leon, M.T. and Rodriguez-Perez, J.E. (2014) Effect of rootstock on yield and fruit size in grafted cucumber. *Revista Fitotecnia Mexicana* 37, 41–47.

Higashide, T., Nakano, A. and Yasuba, K.I. (2014) Yield and dry matter production of a Japanese tomato 'Momotaro York' are improved by grafting onto a Dutch rootstock 'Maxifort'. *Journal of the Japanese Society for Horticultural Science* 83, 235–243.

Hoyos Echebarría, P., Fernández, J.A., Martínez, P.F. and Castilla, N. (2001) Influence of different rootstocks on the yield and quality of greenhouses grown cucumbers. *Acta Horticulturae* 559, 139–143.

Huang, Y., Bie, Z.L., He, S.P., Hua, B., Zhen, A. and Liu, Z.X. (2010) Improving cucumber tolerance to major nutrients induced salinity by grafting onto *Cucurbita ficifolia*. *Environmental and Experimental Botany* 69, 32–38.

Huh, Y.C., Om, Y.H., Woo, Y.H. and Lee, J.M. (2003) growth and characteristics of watermelon grafted onto *Citrullus* rootstocks selected for disease resistance. *Horticulture, Environment, and Biotechnology* 44, 649–654.

Huitrón, M.V., Diaz, M., Diánez, F. and Camacho, F. (2007) The effect of various rootstocks on triploid watermelon yield and quality. *Journal of Food, Agriculture and Environment* 5, 344–348.

Ibiza, V.P., Blanca, J., Cañizares, J. and Nuez, F. (2012) Taxonomy and genetic diversity of domesticated *Capsicum* species in the Andean region. *Genetic Resources and Crop Evolution* 59, 1077–1088.

Ioannou, N. (2001) Integrating soil solarization with grafting on resistant rootstocks for management of soil-borne pathogens of eggplant. *Journal of Horticultural Science and Biotechnology* 76, 396–401.

Ito, L.A., Gaion, L.A., Galatti, F.S., Braz, L.T. and Santos, J.M. (2014) Resistance of cucurbita rootstocks to nematodes and compatibility of grafting with net melon. *Horticultura Brasileira* 32, 297–302.

Iwamoto, Y. and Ezura, H. (2006) Efficient plant regeneration from protoplasts of eggplant rootstock cultivar and its wild relatives. *Plant Biotechnology* 23, 525–529.

Iwamoto, Y., Hirai, M., Ohmido, N., Fukui, K. and Ezura, H. (2007) Fertile somatic hybides between *Solanum integrifolium* and *S. anitwongsei* (syn. *S. kurzii*) as candidates for bacterial wilt-resistant rootstock of eggplant. *Plant Biotechnology* 24, 179–184.

Jang, Y., Moon, J.H., Lee, J.W., Lee, S.G., Kim, S.Y. and Chun, C. (2013) Effects of different rootstocks on fruit quality of grafted pepper (*Capsicum annuum* L.) *Korean Journal of Horticultural Science and Technology* 31, 687–699.

Jang, Y., Huh, Y.C., Park, D.K., Mun, B., Lee, S. and Um, Y. (2014) Greenhouse evaluation of melon rootstock resistance to monosporascus root rot and vine decline as well as of yield and fruit quality in grafted 'Inodorus' melons. *Korean Journal of Horticultural Science and Technology* 32, 614–622.

Janick, J. and Paull, R.E. (2008) *The Encyclopedia of Fruits and Nuts*. British Library, CAB International, London.

Jarret, R.L. (2014) Observations on anatomical aspects of the fruit, leaf and stem tissues of four *Citrullus* spp. *African Journal of Plant Science* 8, 521–527.

Jeffrey, C. (2005) A new system of Cucurbitaceae. *Botanicheskii Zhurnal (St Petersburg)* 90, 332–335.

Ji, Y.F., Pertuze, R. and Chetelat, R.T. (2004) Genome differentiation by GISH in interspecific and intergeneric hybrids of tomato and related nightshades. *Chromosome Research* 12, 107–116.

Jifon, J.L. and Crosby, K.M. (2008) Possible physiological mechanisms for resistance to vine decline diseases in grafted watermelons. *Acta Horticulturae* 782, 329–333.

Johnson, S., Inglis, D. and Miles, C. (2014) Grafting effects on eggplant growth, yield, and Verticillium wilt incidence. *International Journal of Vegetable Science* 20, 3–20.

Kacjan Maršić, N., Mikulič-Petkovšek, M. and Stampar, F. (2014) Grafting influences phenolic profile and carpometric traits of fruits of greenhouse-grown eggplant (*Solanum melongena* L.). *Journal of Agricultural and Food Chemistry* 62, 10504–10514.

Karaagac, O. and Balkaya, A. (2013) Interspecific hybridization and hybrid seed yield of winter squash (*Cucurbita maxima* Duch.) and pumpkin (*Cucurbita moschata* Duch.) lines for rootstock breeding. *Scientia Horticulturae* 149, 9–12.

Karaca, F., Yetışır, H., Solmaz, I., Candir, E., Kurt, S., Sari, N. and Guler, Z. (2012) Rootstock potential of Turkish *Lagenaria siceraria* germplasm for watermelon: plant growth, yield and quality. *Turkish Journal of Agriculture and Forestry* 36, 167–177.

Kashyap, V., Vinod Kumar, S., Collonnier, C., Fusari, F., Haicour, R., Rotino, G.L., Sihachakr, D. and Rajam, M.V. (2003) Biotechnology of eggplant. *Scientia Horticulturae* 97, 1–25.

Keinath, A.P. (2014a) Differential susceptibility of nine cucurbit species to the foliar blight and crown canker phases of gummy stem blight. *Plant Disease* 98, 247–254.

Keinath, A.P. (2014b) Reproduction of *Didymella bryoniae* on nine species of cucurbits under field conditions. *Plant Disease* 98, 1379–1386.

Keinath, A.P. and Hassell, R.L. (2014) Control of Fusarium wilt of watermelon by grafting onto bottlegourd or interspecific hybrid squash despite colonization of rootstocks by *Fusarium*. *Plant Disease* 98, 255–266.

Khah, E. (2011) Effect of grafting on growth, performance and yield of aubergine (*Solanum melongena* L.) in greenhouse and open-field. *International Journal of Plant Production* 5, 359–366.

King, S.R., Davis, A.R., Liu, W. and Levi, A. (2008) Grafting for disease resistance. *HortScience* 43, 1673–1676.

King, S.R., Davis, A.R., Zhang, X. and Crosby, K. (2010) Genetics, breeding and selection of rootstocks for Solanaceae and Cucurbitaceae. *Scientia Horticulturae* 127, 106–111.

Kirkbride, J.H. (1993) *Biosystematic Monograph of the Genus Cucumis (Cucurbitaceae)*. Parkway, Boone, North Carolina.

Kistler, L., Montenegro, A., Smith, B.D., Gifford, J.A., Green, R.E., Newsom, L.A. and Shapiro, B. (2014) Transoceanic drift and the domestication of African bottle gourds in the Americas. *Proceedings of the National Academy of Sciences USA* 25, 111, 2937–2941.

Kivi, S.S., Salehi, R. and Liaghat, A. (2014) Effects of deficit irrigation and grafting on yield and fruit quality of Iranian 'Khatooni' melon. *Acta Horticulturae*, 1038, 573–578.

Knapp, S., Bohs, L., Nee, M. and Spooner, D.M. (2004) Solanaceae – a model for linking genomics with biodiversity. *Comparative and Functional Genomics* 5, 285–291.

Knapp, S., Vorontsova, M.S. and Prohens, J. (2013) Wild relatives of the eggplant (*Solanum melongena* L.: Solanaceae): new understanding of species names in a complex group. *PLoS ONE* 8, e57039.

Kokalis-Burelle, N. and Rosskopf, E.N. (2011) Microplot evaluation of rootstocks for control of *Meloidogyne incognita* on grafted tomato, muskmelon, and watermelon. *Journal of Nematology* 43, 166–171.

Kokalis-Burelle, N., Bausher, M.G. and Rosskopf, E.N. (2009) Greenhouse evaluation of *Capsicum* rootstocks for management of *Meloidogyne incognita* on grafted bell pepper. *Nematropica* 39, 121–132.

Kousik, C.S., Levi, A., Ling, K.S. and Wechter, W.P. (2008) Potential sources of resistance to cucurbit powdery mildew in US plant introductions of bottle gourd. *Hortscience* 43, 1359–1364.

Kousik, C.S., Donahoo, R.S. and Hassell, R. (2012) Resistance in watermelon rootstocks to crown rot caused by *Phytophthora capsici*. *Crop Protection* 39, 18–25.

Krumbein, A. and Schwarz, D. (2013) Grafting: a possibility to enhance health-promoting and flavour compounds in tomato fruits of shaded plants? *Scientia Horticulturae* 149, 97–107.

Kunwar, S., Paret, M.L., Olson, S.M., Ritchie, L., Rich, J.R., Freeman, J. and McAvoy, T. (2015) Grafting using rootstocks with resistance to *Ralstonia solanacearum* against *Meloidogyne incognita* in tomato production. *Plant Disease* 99, 119–124.

Kurokawa, S., Kobayashi, H. and Senda, T. (2009) Genetic diversity of *Sicyos angulatus* in central and north-eastern Japan by inter-simple sequence repeat analysis. *Weed Research* 49, 365–372.

Lee, J.M. and Oda, M. (2003) Grafting of herbaceous vegetable and ornamental crops. *Horticultural Reviews* 28, 61–124.

Lee, H.J., Lee, S.G., Chun, C. and Lee, J.M. (2006) Growth and salt tolerance of grafted hot pepper seedlings. *HortScience* 41, 1072–1073.

Lee, J.M., Kubota, C., Tsao, S.J., Bie, Z., Hoyos Echevarria, P., Morra, L. and Oda, M. (2010) Current status of vegetable grafting: diffusion, grafting techniques, automation. *Scientia Horticulturae* 127, 93–105.

Leonardi, C. and Giuffrida, F. (2006) Variation of plant growth and macronutrient uptake in grafted tomatoes and eggplants on three different rootstocks. *European Journal of Horticultural Science* 71, 97–101.

Lester R.N. (1986). Taxonomy of scarlet eggplants, *Solanum aethiopicum* L. *Acta Horticulturae* 182, 125–132.

Levi, A., Thomas, C.E., Simmons, A.M. and Thies, J.A. (2005) Analysis based on RAPD and ISSR markers reveals closer similarities among *Citrullus* and *Cucumis* species than with *Praecitrullus fistulosus* (stocks) pangalo. *Genetic Resources and Crop Evolution* 52, 465–472.

Levi, A., Thiesa, J., Linga, K., Simmonsa, A.M., Kousik, C. and Hassella, R. (2009) Genetic diversity among *Lagenaria siceraria* accessions containing resistance to root-knot nematodes, whiteflies, ZYMV or powdery mildew. *Plant Genetic Resources* 7, 216–226.

Li, H., Wang, F., Chen, X.J., Shi, K., Xia, X.J., Considine, M.J., Yu, J.Q. and Zhou, Y.H. (2014a) The sub/supra-optimal temperature-induced inhibition of photosynthesis and oxidative damage in cucumber leaves are alleviated by grafting onto figleaf gourd/luffa rootstocks. *Physiologia Plantarum* 152, 571–584.

Li, H., Liu, S.S., Yi, C.Y., Wang, F., Zhou, J., Xia, X.J., Shi, K., Zhou, Y.H. and Yu, J.Q. (2014b) Hydrogen peroxide mediates abscisic acid-induced HSP70 accumulation and heat tolerance in grafted cucumber plants. *Plant, Cell and Environment* 37, 2768–2780.

Lira, R., Eguiarte, L. and Montes, S. (2009) Proyecto de recopilación y análisis de la información existente de las especies de los géneros *Cucurbita* y Sechium que crecen y/o se cultivan en México. Final report. Available at: http://www.biodiversidad.gob.mx/genes/centrosOrigen/Cucurbita%20y%20Sechium/Informe_Final/Informe%20final%20Cucurbita%20y%20Sechium.pdf (accessed 22 November 2016).

Liu, N., Zhou, B., Zhao, X., Lu, B., Li, Y. and Hao, J. (2009) Grafting eggplant onto tomato rootstock to suppress *Verticillium dahliae* infection: the effect of root exudates. *Hortscience* 44, 2058–2062.

Liu, Z.X., Bie, Z.L., Huang, Y., Zhen, A., Lei, B. and Zhang, H.Y. (2012) Grafting onto *Cucurbita moschata* rootstock alleviates salt stress in cucumber plants by delaying photoinhibition. *Photosynthetica* 50, 152–160.

Liu, B., Ren J., Zhang, Y., An, J., Chen, M., Chen, H., Xu, C. and Ren, H.Z. (2015) A new grafted rootstock against root-knot nematode for cucumber, melon, and watermelon. *Agronomy for Sustainable Development* 35, 251–259.

López-Marín, J., González, A., Pérez-Alfocea, F., Egea-Gilabert, C. and Fernández, J.A. (2013) Grafting is an efficient alternative to shading screens to alleviate thermal stress in greenhouse-grown sweet pepper. *Scientia Horticulturae* 149, 39–46.

Louws, F.J., Rivard, C.L. and Kubotac, C. (2010) Grafting fruiting vegetables to manage soilborne pathogens, foliar pathogens, arthropods and weeds. *Scientia Horticulturae* 127, 127–146.

Mahmoud, M.A.A. (2014) Grafting as a tool to improve TYLCV-tolerance in tomato. *Journal of Horticultural Science and Ornamental Plants* 6, 109–115.

Marr, K.L., Bhattarai, N.K. and Xia, Y.M. (2005) Allozymic, morphological, and phenological diversity in cultivated *Luffa acutangula* (Cucurbitaceae) from China, Laos, and Nepal, and Allozyme Divergence between *L. acutangula* and *L. aegyptiaca*. *Economic Botany* 59, 154–165.

Matsumoto, Y., Miyagi, M., Watanabe, N. and Kuboyama, T. (2012) Temperature-dependent enhancement of pollen tube growth observed in interspecific crosses between wild Cucumis spp. and melon (*C. melo* L.) *Scientia Horticulturae*, 138, 144–150.

Matsuzoe, N., Aida, H., Hanada, K., Ali, M., Okubo, H. and Fujieda, K. (1996) Fruit quality of tomato plants grafted on Solanum rootstocks. *Journal of the Japanese Society for Horticultural Science* 65, 73–80.

Meyer, R.S., Karol, K.G., Little, D.P., Nee, M.H. and Litt, A. (2012) Phylogeographic relationships among Asian eggplants and new perspectives on eggplant domestication. *Molecular Phylogenetics and Evolution* 63, 685–701.

Meyer, R.S., Knapp, S., Karol, K.G., Little, D.P., Nee, M.H. and Litt, A. (2013) Reply to J. Samuels: taxonomic notes on several wild relatives of *Solanum melongena* L. *Molecular Phylogenetics and Evolution* 69, 306–307.

Miceli, A., Sabatino, L., Moncada, A., Vetrano, F. and D'Anna, F. (2014) Nursery and field evaluation of eggplant grafted onto unrootedcuttings of *Solanum torvum*. *Scientia Horticulturae* 178, 203–210.

Miguel, A. (2004) Use of grafted plants and IPM methods for the production of tomatoes in the Mediterranean region. In: *Proceedings of Fifth International Conference on Alternatives to Methyl Bromide*. 27–30 September, Lisbon, Portugal, pp. 141–145.

Miguel, A., Marsal, R., Goto, R., San Bautista Primo, A., López Galarza, S.V., Pascual España, B. and Maroto Borrego, J.V. (2011) Improving the affinity of tomato grafted on *Solanum torvum* using an intermediate rootstock. *Acta Horticulturae* 898, 291–295.

Mo, H.S., Jang, K.S., Hwang, J.E., Jeon, S.G. and Kim, B.S. (2015) Horticultural and chemical quality characterization of accessions selected from four species of *Capsicum*. *Horticulture Environment and Biotechnology* 56, 54–66.

Mohsenian, Y. and Roosta, H.R. (2015) Effects of grafting on alkali stress in tomato plants: datura rootstock improve alkalinity tolerance of tomato plants. *Journal of Plant Nutrition* 38, 51–72.

Mohsenian, Y., Roosta, H.R., Karimi, H.R. and Esmaeilizade, M. (2012) Investigation of the ameliorating effects of eggplant, datura, orange nightshade, local Iranian tobacco, and field tomato as rootstocks on alkali stress in tomato plants. *Photosynthetica* 50, 411–421.

Moncada, A., Miceli, A., Vetrano, F., Mineo, V., Planeta, D. and D'Anna, F. (2013) Effect of grafting on yield and quality of eggplant (*Solanum melongena* L.). *Scientia Horticulturae* 149, 108–114.

Morra, L. and Bilotto, M. (2006) Evaluation of new rootstocks for resistance to soil-borne pathogens and productive behaviour of pepper (*Capsicum annuum* L.). *Journal of Horticulturae Science and Biotechnology* 81, 518–524.

Neitzke, R.S., Barbieri, R.L., Vasconcelos, C.S., Priori, D., Fischer, S.Z., Götzke, M.L.D., Romano, C.M. and Heiden, G. (2011) Diversity in capsicum landraces cultivated in Brazil. *Acta Horticulturae* 918, 531–536.

Newstrom, L.E. (1991) Evidence for the origin of chayote, *Sechium Edule* (Cucurbitaceae). *Economic Botany* 45, 410–428.

Nienhuis, J. and Lower, R.L. (1980) Influence of reciprocal donor scions on fruit setting characteristics of recipient scions of *Cucumis sativus* and *C. hardwickii* R. *Cucurbit Genetics Cooperative Report* 3, 17–19.

Nilsen, E.T., Freeman, J., Grene, R. and Tokuhisa, J. (2014) A rootstock provides water conservation for a grafted commercial tomato (*Solanum lycopersicum* L.) line in response to mild-drought conditions: a focus on vegetative growth and photosynthetic parameters. *Plos ONE* 9, e115380.

Nkansah, G.O., Ahwireng, A.K., Amoatey, C. and Ayarna, A.W. (2013) Grafting onto African eggplant enhances growth, yield and fruit quality of tomatoes in tropical Forest ecozones. *Journal of Applied Horticulture* 15, 16–20.

Nunes, G.H.S., Nunes, W.W.L.P., Ferriol, M., Alfaro-Fernández, A., Font, I., Gisbert, C. and Picó, B. (2013) Potential new *Cucurbita* spp. rootstocks with viral and fungal resistance. In: First Annual Conference COST ACTION FA 1204. November 2013, Murcia, Spain, Oral presentation, p. 16.

Oda, M. (2006) Use of grafted seedlings for vegetable production in Japan. *Acta Horticulturae* 770, 15–20.

Oda, M., Nagata, M., Tsuji, K. and Sasaki, H. (1996) Effects of scarlet eggplant rootstock on growth, yield, and sugar content of grafted tomato fruits. *Journal of the Japanese Society for Horticultural Science* 65, 531–536.

Oka, Y., Offenbach, R. and Pivonia, S. (2004) Pepper rootstock graft compatibility and response to *Meloidogyne javanica* and *M. incognita. Journal of Nematology* 36, 137–141.

Orsini, F., Sanoubar, R., Oztekin, G.B., Kappel, N., Tepecik, M., Quacquarelli, C., Tuzel, Y., Bona, S. and Gianquinto, G. (2013) Improved stomatal regulation and ion partitioning boosts salt tolerance in grafted melon. *Functional Plant Biology* 40, 628–636.

Ozarslandan, A., Sogut, M.A., Yetişir, H. and Elekcioglu, I.H. (2011) Screening of bottle gourds (*Lagenaria siceraria* (Molina) Standley) genotypes with rootstock potential for watermelon production for resistance against *Meloidogyne incognita* (Kofoid and White, 1919) Chitwood and *Meloidogyne javanica* (Treub, 1885) Chitwood. *Turkish Journal of Entomology* 35, 687–697.

Palada, M.C. and Wu, D.L. (2008) Evaluation of chili rootstocks for grafted sweet pepper production during the hot-wet and hot-dry seasons in Taiwan. *Acta Horticulturae* 767, 167–174.

Pandey, S., Kumar, S., Mishra, U., Rai, A., Singh, M. and Rai, M. (2008) Genetic diversity in Indian ash gourd (*Benincasa hispida*) accessions as revealed by quantitative traits and RAPD markers. *Scientia Horticulturae* 118, 80–86.

Pandravada, S.R., Sivaraj, N., Jairam, R., Sunil, N., Begum, H., Reddy, T.M., Chakrabarty, S.K., Bisht, I.S. and Bansal, K.C. (2014) *Luffa hermaphrodita*: first report of its distribution and cultivation in Adilabad, Andhra Pradesh, South India. *Asian Agri-History* 18, 123–132.
Paran, I. and Falik, E. (2011) Breeding for fruit quality in pepper (*Capsicum spp.*). In: Jenks, M.A. and Bebeli, P.J. (eds) *Breeding for Fruit Quality*. Wiley, Hoboken, New Jersey, pp. 307–322.
Paris, H.S. (2015) Origin and emergence of the sweet dessert watermelon, *Citrullus lanatus*. *Annals of Botany* 116, 133–148.
Paris, H.S. and Kabelka, E. (2009) Gene list for *Cucurbita species*. *Cucurbit Genetics Cooperative Report* 31–32, 44–69.
Passam, H.C., Stylianoy, M. and Kotsiras, A. (2005) Performance of eggplant grafted on tomato and eggplant rootstocks. *European Journal of Horticultural Science* 70, 130–134.
Penella, C., Nebauer, S.G., Lopéz-Galarza, S., SanBautista, A., Gorbe, E. and Calatayud, A. (2013) Evaluation for salt stress tolerance of pepper genotypes to be used as rootstocks. *Journal of Food, Agriculture and Environment* 11, 1101–1107.
Penella, C., Nebauer, S.G., Bautista, A.S., López-Galarza, S. and Calatayud, T. (2014) Rootstock alleviates PEG-induced water stress in grafted pepper seedlings: physiological responses. *Journal of Plant Physiology* 171, 842–851.
Penella, C., Nebauer, S.G., Quiñones, A., Bautista, A.S., López-Galarza, S. and Calatayud, A. (2015) Some rootstocks improve pepper tolerance to mild salinity through ionic regulation. *Plant Science* 230, 12–22.
Peralta, I.E., Knapp, S. and Spooner, D.M. (2005) New species of wild tomatoes (*Solanum* section *Lycopersicon:* Solanaceae) from Northern Peru. *Systematic Botany* 30, 424–434.
Peralta, I.E., Knapp, S. and Spooner, D.M. (2006) Nomenclature for wild and cultivated tomatoes. *Tomato Genetics Cooperative Report* 56, 6–12.
Peralta, I.E., Spooner, D.M. and Knapp, S. (2008) Taxonomy of wild tomatoes and their relatives (*Solanum sect. Lycopersicoides, sect. Juglandifolia, sect. Lycopersicon;* Solanaceae). *Systematic Botany Monographs* 84, 1–186.
Petran, A. and Hoover, E. (2014) *Solanum torvum* as a compatible rootstock in interspecific tomato grafting. *Journal of Horticulture* 1, 103.
Petropoulos, S.A., Khah, E.M. and Passam, H.C. (2012) Evaluation of rootstocks for watermelon grafting with reference to plant development, yield and fruit quality. *International Journal of Plant Production* 6, 481–491.
Petropoulos, S.A., Olympios, C., Ropokis, A., Vlachou, G., Ntatsi, G., Paraskevopoulos, A. and Passam, H.C. (2014) Fruit volatiles, quality, and yield of watermelon as affected by grafting. *Journal of Agricultural Science and Technology* 16, 873–885.
Pickersgill, B. (1997) Genetic resources and breeding of *Capsicum* spp. *Euphytica* 96, 129–133.
Pinheiro, J.B., Boiteux, L.S., Almeida, M.R.A., Pereira, R.B., Galhardo, L.C.S. and Carneiro, R.M.D.G. (2015) First report of *Meloidogyne enterolobii* in *Capsicum* rootstocks carrying the *Me1* and *Me3/Me7* genes in Central Brazil. *Nematropica* 45, 184–188.
Pitrat, M. (2008) Melon. In: Prohens, J. and Nuez, F. (eds) *Handbook of Plant Breeding. Vegetables I. Asteraceae, Brassicaceae, Chenopoidicaceae, and Cucurbitaceae*. Springer, New York, pp. 283–315.
Pofu, K.M., Mashela, P.W. and Mphosi, M.S. (2011) Management of *Meloidogyne incognita* in nematode susceptible watermelon cultivars using nematode resistant *Cucumis africanus* and *Cucumis myriocarpus* rootstocks. *African Journal of Biotechnology* 10, 8790–8793.
Pofu, K.M., Mashela, P.W. and Mafeo, T.P. (2013) Optimising stem diameters of watermelon cultivars and indigenous *Cucumis species* for improving compatibility of inter-generic grafts. *Acta Horticulturae* 1007, 807–812.
Prohens, J., Plazas, M., Raigon, M.D., Seguı-Simarro, J.M., Stommel, J.R. and Vilanova, S. (2012) Characterization of interspecific hybrids and first backcross generations from crosses

between two cultivated eggplants (*Solanum melongena* and *S. aethiopicum* Kumba group) and implications for eggplant breeding. *Euphytica* 186, 517–538.

Punithaveni, V., Jansirani, P. and Sivakumar, M. (2015) Screening of cucurbitaceous rootstocks and cucumber scions for root knot nematode resistance (*Meloidogyne incognita* Kofoid and White). *Electronic Journal of Plant Breeding* 6, 486–492.

Ranil, R.H.G., Niran, H.M.L., Plazas, M., Fonseka, R.M., Fonseka, H.H., Vilanova, S., Andujar, I., Gramazio, P., Fita, A. and Prohens, J. (2015) Improving seed germination of the eggplant rootstock *Solanum torvum* by testing multiple factors using an orthogonal array design. *Scientia Horticulturae* 193, 174–181.

Renner, S.S. and Pandey, A.K. (2013) The Cucurbitaceae of India: accepted names, synonyms, geographic distribution, and information on images and DNA sequences. *PhytoKeys* 20, 53–118.

Rivard, C.L. and Louws, F.J. (2008) Grafting to manage soilborne diseases in Heirloom tomato production. *HortScience* 43, 2104–2111.

Rivard, C.L., O'Connell, S., Peet, M.M. and Louws, F.J. (2010) Grafting tomato with interspecific rootstock to manage diseases caused by *Sclerotium rolfsii* and southern root-knot nematode. *Plant Disease* 94, 1015–1021.

Rivero, R.M., Ruiz, J.M., Sanchez, E. and Romero, L. (2003) Does grafting provide tomato plants an advantage against H_2O_2 production under conditions of thermal shock? *Physiologia Plantarium* 117, 44–50.

Robertson, L.D. and Labate, J.A. (2007) Genetic resources of tomato (*Lycopersicon esculentum* Mill.) and wild relatives. In: Razdan, M.K. and Mattoo, A.K. (eds) *Genetic Improvement of Solanaceous crops*, Vol. 2: *Tomato*. Science Publishers, Enfield, UK, pp. 25–75.

Rodriquez, M.M. and Bosland, P.W. (2010) Grafting capsicum to tomato rootstocks. *Journal of Young Investigators* 20, 2–6.

Roosta, H.R. and Karimi, H.R. (2012) Effects of alkali-stress on ungrafted and grafted cucumber plants: using two types of local squash as rootstocks. *Journal of Plant Nutrition* 35, 1843–1852.

Ros-Ibáñez, C., Robertson, L., Martínez-Lluch, M.C., Cano-García, A. and Lacasa-Plasencia, A. (2014) Development of virulence to *Meloidogyne incognita* on resistant pepper rootstocks. *Spanish Journal of Agricultural Research* 12, 225–232.

Rotino, G.L., Sala, T. and Toppino, L. (2014) Eggplant. In: Pratap, A. and Kumar, J. (eds) *Alien Gene Transfer in Crop Plants*, Vol. 2: *Achievements and Impacts*. Springer, New York, pp. 381–409.

Rouphael, Y., Schwarz, D., Krumbein, A. and Colla, G. (2010) Impact of grafting on product quality of fruit vegetables. *Scientia Horticulturae* 127, 172–179.

Rouphael, Y., Cardarelli, M., Rea, E. and Colla, G. (2012) Improving melon and cucumber photosynthetic activity, mineral composition, and growth performance under salinity stress by grafting onto *Cucurbita* hybrid rootstocks. *Photosynthetica* 50, 180–188.

Saadoun, M. and Allagui, M.B. (2013) Management of chili pepper root rot and wilt (caused by *Phytophthora nicotianae*) by grafting onto resistant rootstock. *Phytopathologia Mediterranea* 52, 141–147.

Sabatino, L., Palazzolo, E. and D'Anna, F. (2013) Grafting suitability of Sicilian eggplant ecotypes onto *Solanum torvum*: fruit composition, production and phenology. *Journal of Food, Agriculture and Environment* 11, 1195–1200.

Salar, N., Salehi, R. and Delshad, M. (2015) Effect of grafting and nitrogen application on yield and fruit quality of grafted and non-grafted melon. *Acta Horticulturae* 1086, 225–230.

Samuels, J. (2013) Taxonomic notes on several wild relatives of *Solanum melongena* L. (Solanaceae): comments on Meyer *et al.* (2012). *Molecular Phylogenetics and Evolution* 67, 297–299.

Sanjur, O.I., Piperno, D.R., Andres, T.C. and Wessel-Beaver, L. (2002) Phylogenetic relationships among domesticated and wild species of *Cucurbita* (Cucurbitaceae) inferred from a mitochondrial gene: implications for crop plant evolution and areas of origin. *Proceedings of the National Academy of Sciences USA* 99, 535–540.

Savvas, D., Papastavrou, D., Ntatsi, G., Ropokis, A., Olympios, C., Hartmann, H. and Schwarz, D. (2009) Interactive effects of grafting and Mn-supply level on growth, yield and nutrient uptake by tomato. *HortScience* 44, 1978–1982.

Schaefer, H. and Renner, S. (2011) Phylogenetic relationships in the order Cucurbitales and a new classification of the gourd family (Cucurbitaceae). *Taxon* 60, 122–138.

Schaefer, H., Heibl, C. and Renner, S.S. (2009) Gourds afloat: a dated phylogeny reveals an Asian origin of the gourd family (Cucurbitaceae) and numerous oversea dispersal events. *Proceedings of the Royal Society B* 276, 843–851.

Schwarz, D., Beuch, U., Bandte, M., Fakhro, A., Buttner, C. and Obermeier, C. (2010a) Spread and interaction of *Pepino mosaic virus* (PepMV) and *Pythium aphanidermatum* in a closed nutrient solution recirculation system: effects on tomato growth and yield. *Plant Pathology* 59, 443–452.

Schwarz, D., Rouphael, Y., Colla, G. and Venema, J.H. (2010b) Grafting as a tool to improve tolerance of vegetables to abiotic stresses: thermal stress, water stress and organic pollutants. *Scientia Horticulturae* 127, 162–171.

Sebastian, P., Schaefer, H., Telford, I.R. and Renner, S.S. (2010) Cucumber (*Cucumis sativus*) and melon (*C. melo*) have numerous wild relatives in Asia and Australia, and the sister species of melon is from Australia. *Proceedings of the National Academy of Sciences USA* 107, 14269–14273.

Selvi, N.A.T., Pugalendhi, L. and Sivakumar, M. (2013) Screening of cucurbitaceous rootstocks against root knot nematode *Meloidogyne incognita* Kofoid and White. *Asian Journal of Horticulture* 8, 720–725.

Sigüenza, C., Schochow, M., Turini, T. and Ploeg, A. (2005) Use of *Cucumis metuliferus* as a rootstock for melon to manage *Meloidogyne incognita*. *Journal of Nematology* 37, 276–280.

Sikora, B. and Nowaczyk, P. (2014) Application of RAPD technique for identification of interspecific hybrids from genus *Capsicum*. *Acta Scientiarum Polonorum, Hortorum Cultus* 13, 155–166.

Silva, E.de.S.da., Palangana, F.C., Goto, R., Furtado, E.L., Fernandes, D.M. (2012) Net melon resistance to *Didymella bryoniae* according to grafting and potassium levels. *Summa Phytopathologica* 38, 139–143.

Singh, S., Singh, D.R., Kumar, K. and Birah, A. (2014) Eco-friendly management modules for bacterial wilt (*Ralstonia solanacearum*) of tomato for protected cultivation in a tropical island ecosystem. *Biological Agriculture and Horticulture* 30, 219–227.

Sirohi, P.S., Munshi, A.D., Kumar, G. and Behera, T.K. (2005) Cucurbits. In: Dhillon, B.S., Tyagi, R.K., Saxena, S. and Randhawa, G.J. (eds) *Plant Genetic Resources: Horticultural Crops*. Narosa Publishing House, New Delhi, India, pp. 34–58.

Sonnante, G., Pignone, D. and Hammer, K. (2007) The domestication of artichoke and cardoon: from Roman times to the genomic age. *Annals of Botany* 100, 1095–1100.

Soteriou, G.A., Kyriacou, M.C., Siomos, A.S. and Gerasopoulos, D. (2014) Evolution of watermelon fruit physicochemical and phytochemical composition during ripening as affected by grafting. *Food Chemistry* 165, 282–289.

Soteriou, G.A., Papayiannis, L.C. and Kyriacou, M.C. (2016) Indexing melon physiological decline to fruit quality and vine morphometric parameters. *Scientia Horticulturae* 203, 207–215.

Stevens, M.A. and Rick, C.M. (1986) Genetics and breeding. In: Atherton, J.G. and Rudich, J. (eds) *The Tomato Crop: a Scientific Basis for Improvement*. Chapman & Hall, London, pp. 35–109.

Sureja, A.K., Sirohi, P.S., Behera, T.K. and Mohapatra, T. (2006) Molecular diversity and its relationship with hybrid performance and heterosis in ash gourd [*Benincasa hispida* (Thunb.) Cogn]. *Journal of Horticultural Science and Biotechnology* 81, 33–38.

Temperini, O., Calabrese, N., Temperini, A., Rouphael, Y., Tesi, R., Lenzi, A., Carito, A. and Colla, G. (2013) Grafting artichoke onto cardoon rootstocks: graft compatibility, yield and Verticillium wilt incidence. *Scientia Horticulturae* 149, 4, 22–27.

Thies, J.A. and Levi, A. (2007) Characterization of watermelon (*Citrullus lanatus* var.*citroides*) germplasm for resistance to rootknot nematodes. *Hortscience* 42, 1530–1533.

Thies, J.A., Ariss, J., Kousik, C.S. and Hassell, R. (2008) Is grafting useful for managing root-knot nematodes in watermelon. *Hortscience* 43, 623.

Thies, J.A., Ariss, J.J., Hassell, R.L., Olson,S., Kousik, C.S. and Levi., A. (2010) Grafting for management of southern root-knot nematode, *Meloidogyne incognita*, in watermelon. *Plant Disease* 94, 1195–1199.

Thies, J.A., Buckner, S., Horry, M., Hassell, R. and Levi, A. (2015a) Influence of *Citrullus lanatus* var. *citroides* rootstocks and their F_1 hybrids on yield and response to root-knot nematode, *Meloidogyne incognita*, in grafted watermelon. *HortScience* 50, 9–12.

Thies, J.A. Levi, A. and Ariss, J.J. (2015b) RKVL-318, a root-knot nematode resistant watermelon line as rootstock for grafted watermelon. *HortScience* 50, 141–142.

Toppino, L., Valè, G. and Rotino, G.L. (2008) Inheritance of Fusarium wilt resistance introgressed from *Solanum aethiopicum gilo and aculeatum* groups into cultivated eggplant (*S. melongena*) and development of associated PCR-based markers. *Molecular Breeding* 22, 237–250.

Traka-Mavrona, E., Koutsika-Sotiriou, M. and Pritsa, T. (2000) Response of squash (*Cucurbita* spp.) as rootstock for melon (*Cucumis melo* L.). *Scientia Horticulturae* 83, 353–362.

Trinchera, A., Pandozy, G., Rinaldi, S., Crinò, P, Temperini, O. and Rea, E. (2013a) Graft union formation in artichoke grafting onto wild and cultivated cardoon: an anatomical study. *Journal of Plant Physiology* 170, 1569–1578.

Trinchera, A., Pandozy, G., Temperini, A., Caprio, F., Rea, E. and Rinaldi, S. (2013b) Grafting of globe artichoke plants onto wild and cultivated cardoon: agronomical and physiological aspects. *Acta Horticulturae* 983, 231–235

Trionfetti Nisini, P.T., Colla, G., Granati, E., Temperini, E., Crino, P. and Saccardo, F. (2002) Rootstock resistance to Fusarium wilt and effect on fruit yield and quality of two muskmelon cultivars. *Scientia Horticulture* 93, 281–288.

Tsaballa, A., Athanasiadis, C., Pasentsis, K., Ganopoulos, I., Nianiou-Obeidat, I. and Tsaftaris, A. (2013) Molecular studies of inheritable grafting induced changes in pepper (*Capsicum annuum*) fruit shape. *Scientia Horticulturae* 149, 2–8.

Turhan, A., Ozmen, N., Serbeci, M.S. and Seniz, V. (2011) Effects of grafting on different rootstocks on tomato fruit yield and quality. *Horticultural Science* 38, 142–149.

Tzortzakakis, E.A., Bletsos, F.A. and Avgelis, A.D. (2006) Evaluation of *Solanum* rootstock accessions for control of root-knot nematodes and tobamoviruses. *Journal of Plant Diseases and Protection* 113, 188–189.

USDA APHIS (2012) Federal noxious weed list (1 February 2012). Plant Protection and Quarantine (PPQ), US Department of Agriculture (USDA), Animal and Plant Health Inspection Service (APHIS), Washington, DC.

Venema, J.H., Posthumus, F. and van Hasselt, P.R. (1999) Impact of suboptimal temperature on growth, photosynthesis, leaf pigments and carbohydrates of domestic and high-altitude wild *Lycopersicon* species. *Journal of Plant Physiology* 155, 711–718.

Venema, J.H., Linger, P., van Heusden, A.W., van Hasselt, P.R. and Bruggemann, W. (2005) The inheritance of chilling tolerance in tomato (*Lycopersicon* spp.). *Plant Biology* 7, 118–130.

Venema, J.H., Dijk, B.E., Bax, J.M., van Hasselt, P.R. and Elzenga, J.T.M. (2008) Grafting tomato (*Solanum lycopersicum*) onto the rootstock of a high-altitude accession of *Solanum habrochaites* improves suboptimal-temperature tolerance. *Environmental and Experimental Botany* 63, 359–367.

Verzera, A., Dima, G., Tripodi, G., Condurso, C., Crinò, P., Romano, D., Mazzaglia, A., Lanza, C.M., Restuccia, C. and Paratore, A. (2014) Aroma and sensory quality of honeydew melon fruits

(*Cucumis melo* L. subsp. Melo var. *inodorus* H. Jacq.) in relation to different rootstocks. *Scientia Horticulturae* 169, 118–124.

Weng, Y. (2010) Genetic diversity among *Cucumis metuliferus* populations revealed by cucumber microsatellites. *HortScience* 45, 214–219.

Wimer, J.A., Miles, C.A. and Inglis, D.A. (2014) Evaluation of watermelon rootstocks for resistance to Verticillium wilt in northwestern Washington. *Acta Horticulturae* 1085, 363–364.

Yamaguchi, N., Mori, S., Baba, K., Kaburagi-Yada, S., Arao, T., Kitajima, N., Hokura, A. and Terada, Y. (2011) Cadmium distribution in the root tissues of solanaceous plants with contrasting root-to-shoot Cd translocation efficiencies. *Environmental and Experimental Botany* 71, 198–206.

Yang, R.Y. and Ojiewo, C. (2013a) African nightshades and African eggplants: taxonomy, crop management, utilization, and phytonutrients. In: Juliani, H., Simon, J.E. and Ho, C.-T. *African Natural Plant Products*, Vol. II: *Discoveries and Challenges in Chemistry, Health, and Nutrition*. ACS Symposium Series. American Chemical Society, Washington, DC, pp. 137–165.

Yang, Y.J., Lu, X.M., Yan, B., Li, B., Sun, J., Guo, S.R. and Tezuka, T. (2013b) Bottle gourd rootstock-grafting affects nitrogen metabolism in NaCl-stressed watermelon leaves and enhances short-term salt tolerance. *Journal of Plant Physiology* 170, 653–661.

Yetişir, H. and Sari, N. (2004) Effect of hypocotyl morphology on survival rate and growth of watermelon seedlings grafted on rootstocks with different emergence performance at various temperatures. *Turkish Journal of Agriculture and Forestry* 28, 231–237.

Yetişir, H. and Uygur, V. (2009) Plant growth and mineral element content of different gourd species and watermelon under salinity stress. *Turkish Journal of Agriculture and Forestry* 33, 65–77.

Yetişir, H., Caliskan, M.E.C., Soylu, S. and Saka, M. (2006) Some physiological and growth responses of watermelon [*Citrullus lanatus* (Thunb.) Matsum. and Nakai] grafted on *Lagenaria siceraria* to flooding. *Environmental and Experimental Botany* 58, 1–8.

Yetişir, H., Kurt, S., Sari, N. and Tok, F.M. (2007) Rootstock potential of Turkish *Lagenaria siceraria* germplasm for watermelon: plant growth, graft compatibility, and resistance to Fusarium. *Turkish Journal of Agriculture and Forestry* 31, 381–388.

Zhang, S.P., Gu, X.F. and Wang, Y. (2008) Effect of low temperature stress on the physiological and biochemical indexes in cucumber seedling grafted on bur cucumber (*Sicyos angulatus* L.). *Acta Horticulturae* 771, 243–247.

Zhang, X.Y., Fu, Q., Zhu, H., Wang, H. (2014) Effects of grafting on melon fruit growth and quality. *China Vegetables* 6, 13–19.

Zheng, Y., Alverson, A.J., Wang, Q. and Palmer, J.D. (2013) Chloroplast phylogeny of *Cucurbita*: evolution of the domesticated and wild species. *Journal of Systematics and Evolution* 51, 326–334.

Zhou, Y.H., Zhou, J., Huang, L.F., Ding, X.T., Shi, K. and Yu, J.Q. (2009) Grafting of *Cucumis sativus* onto *Cucurbita ficifolia* leads to improved plant growth, increased light utilization and reduced accumulation of reactive oxygen species in chilled plants. *Journal of Plant Research* 122, 529–540.

Zhou, B.L., Meng, Z.H., Li, J. and Chen, Z.X. (2012) Effects of grafting on the growth, physiology, and biochemistry of eggplants under water stress. *Chinese Journal of Ecology* 31, 2804–2809.

Zhou, X.Z., Wu, Y.F., Chen, S., Chen, Y., Zhang, W.G., Sun, X.T. and Zhao, Y.J. (2014) Using *Cucurbita* rootstocks to reduce Fusarium wilt incidence and increase fruit yield and carotenoid content in oriental melons. *Hortscience* 49, 1365–1369.

3 Rootstock Breeding: Current Practices and Future Technologies

Andrew J. Thompson,[1]* Maria Belen Pico,[2] Halit Yetişir,[3] Roni Cohen[4] and Penelope J. Bebeli[5]

[1]*Cranfield University, Cranfield, Bedfordshire, UK;* [2]*Universitat Politècnica de València, Valencia, Spain;* [3]*University of Erciyes, Kayseri, Turkey;* [4]*Newe Ya'ar Research Center, Agricultural Research Organization, Ramat Yishay, Israel;* [5]*Agricultural University of Athens, Athens, Greece*

3.1 Introduction

The aim of plant breeding is to 'maximize the probability of creating, and identifying, superior genotypes which will make successful new cultivars. In other words, they will contain all the desirable characteristics/traits necessary for use in a production system' (Brown and Caligari, 2011). Prior to commercial release of a new cultivar, the breeding process requires: (i) identification of variable germplasm; (ii) hybridization to combine genetic materials from different sources into a single entity; (iii) selection of superior genotypes with a favourable combination of characteristics; and (iv) multiplication of stable cultivars.

The available germplasm resources for vegetable rootstock breeding are described in detail in Chapter 2 (this volume). Here, we will consider how to combine different sources of genetic variation, for example by overcoming species barriers, and how to select useful rootstocks. The many traits that can be associated with the root system or conferred by a rootstock to the scion are outlined in Plate 6, and are the topic of more detailed discussion elsewhere in this book. First, we will consider the impact of the practice of grafting on breeding strategies.

3.2 Stacking Traits: Meiosis or Grafting or Both?

Combining characteristics from different sources into a single seed line, known as 'trait stacking' or 'pyramiding', is challenging. The first reason is genetic linkage:

* a.j.thompson@cranfield.ac.uk

positive and negative alleles may occur at closely linked loci, and recombining them in the most favourable way may require screening of huge numbers of individual plants to find rare meiotic recombination events, a process that becomes increasingly difficult as the number of genetic loci involved increases. Secondly, there may be negative consequences from pleiotropic effects of a single allele on many traits; for example, a gene that increases root growth may also be expressed in a fruit and have a negative effect on fruit quality, but the two effects cannot be separated in a genetically homogenous non-grafted plant. Thirdly, there may be complex epistatic interactions whereby the effect of an allele at one locus depends on the presence of a specific allele at another locus; such interactions can be described, but their complexity makes breeding more difficult. Finally, we must recognize the difference between qualitative and quantitative traits; the former are determined by a single locus that can produce a large effect on the trait, for example a disease resistance gene, which is relatively easy to track. In contrast, quantitative traits are controlled by a large number of quantitative trait loci (QTLs), each of which has a small effect on the trait, leading to a continuous distribution of trait values. Selecting for improvement of a quantitative trait clearly is far more challenging because of the need to combine alleles at a large number of genetic loci.

Historically, the primary goals of most breeders of graftable vegetable crops were enhanced yield, fruit quality and disease resistance. This has led to the release of commercial elite scion cultivars that are high yielding and have fruit characteristics suitable for postharvest handling, and that are attractive to consumers, usually combined with a range of key qualitative disease resistance loci. Adding further rootzone-expressed traits to such cultivars requires crossing and reselection of all favourable traits, a long and expensive process that can take up to 10 years. In addition, desirable rootzone-expressed traits tend to be highly dependent on the environment due to local soil conditions and the associated abiotic and biotic pressures, and therefore differ among growing systems. Trait stacking by grafting has many advantages: the breeding goals for the rootstock and scion are different, and they can be bred independently, breaking the breeding scheme into two more easily addressed challenges. The rootstock can also include a much larger proportion of wild-species DNA than the scion without impacting on domestication traits, such as fruit size and quality. The impact of the traits from each are then combined through the graft, and scion and rootstock combinations can be chosen by the grower or plant nursery to address specific market or environmental requirements. In comparison, a non-grafted system requires a larger number of whole-plant cultivars to provide the same degree of flexibility and choice to the grower, and the deployment of wild-species germplasm is severely restricted due to impacts on marketable yield.

For growers, the only disadvantage of the grafting strategy is the additional expense of purchasing grafted transplants, but it is clear from the rapid increase in the industry take-up of grafting that the economic advantages outweigh the extra costs. For commercial seed companies, the situation is more complex: grafted crops tend to have lower planting densities, often with two main stems grown from a single rootstock, and therefore seed sale volumes are lower in grafted crops, even if supplying both rootstock and scion seeds; rootstock × scion compatibility is an additional complication to the selection criterion when breeding rootstocks, and selection protocols must discover successful rootstock × scion × environment combinations (Cohen *et al.*, 2007). Seed companies must also pay great attention to rootstock seed quality to provide the very high germination uniformity and disease-free status demanded by grafting nurseries. Finally, if not policed, the possibility of illegal propagation of commercial F_1 hybrids by grafting side shoots on to rootstocks becomes a threat to seed company sales when grafting is widely practised. However, seed companies have been reacting strongly to the increasing trend towards grafting, with most vegetable seed companies actively involved in rootstock breeding, and there are significant commercial opportunities in some markets to develop effective rootstocks where the benefits of grafting are still marginal and uptake is low (e.g. in pepper), or to increase market share where a few well known rootstock cultivars dominate (e.g. in tomato). Vegetable seed companies, including those that breed rootstocks (Table 3.1) have been undergoing considerable consolidation, resulting in a handful of large multinational corporations (Howard, 2009), a process that has accelerated in recent years, and inevitably this reduces the tendency to breed for specific local environments. As a result, there is also an important role for breeding of vegetable rootstocks for 'public good', especially where there can be rapid gains in crop production by overcoming biotic and abiotic constraints in developing countries (Palada and Wu, 2008; Keatinge *et al.*, 2014).

Perhaps the ultimate goal of the vegetable breeder is to combine all the rootstock and scion traits into one non-grafted cultivar, a challenge that could be met eventually if a greater understanding of the genetic loci is achieved. However, some alleles have positive and negative effects in the root and shoot, respectively, which might give a neutral or negative overall effect on yield in a non-grafted

Table 3.1. Major companies that breed and supply vegetable rootstock seeds. This is a non-exhaustive list in alphabetical order. All websites accessed 16 April 2016.

Company/group	Website
Asahi Industries	https://www.asahi-kg.co.jp/en/products/agriculture/seed/index.html
Bejo Zaden	http://www.bejo.com/
BHN Seed	http://www.bhnseed.com/
Capgen Seeds	http://www.capgenseeds.com/en/
De Ruiter/Monsanto	http://www.monsanto.com/products/pages/deruiter-seeds.aspx
DP Seeds	http://www.dpseeds.com/rootstock
Enza Zaden	http://www.enzazaden.us/products/fruitvegets/
Gautier Semences	http://www.gautiersemences.com/
Hazera/Limagrain	http://hazerainc.com/
Vilmorin/Limagrain	http://www.vilmorin.com/
Nunhams/Bayer	http://www.nunhems.com/
Origene	http://www.origeneseeds.com/
Rijk Zwaan	http://www.rijkzwaan.com/
Sakata	http://www.sakata.com/
Seminis/Monsanto	http://www.monsanto.com/products/pages/seminis.aspx
Syngenta	http://www.syngenta.com/global/corporate/en/Pages/home.aspx
Takii Seed	http://www.takiiseed.com/
Zeraim Gedera/Syngenta	http://www.zeraim.com/

plant but a positive effect when deployed only in the rootstock of a grafted crop, as reported for IL8-3 in tomato (Gur *et al.*, 2011). In addition, grafting will always provide the most rapid strategy to deploy new traits, such as a new soilborne disease resistance allele in the rootstock or a new fruit-quality gene in the scion, without the need to recombine and reselect all traits into a single non-grafted cultivar: breaking the complex breeding task into two simpler parts, later joined by grafting, speeds up the process.

3.3 Developing Stable Core Collections of Germplasm for Breeding

Breeding programmes, including for rootstocks, rely on the genetic diversity present in germplasm collections; in Chapter 2 (this volume), the wealth of plant genetic resources for the Solanaceae and Cucurbitaceae was presented. Typically, germplasm collections include landraces (local varieties), locally and internationally bred cultivars, wild species and wild relatives, each of which is registered in a database and conserved by seed multiplication. Core collections are established after morphological and genetic diversity analysis (Yetişir *et al.*, 2008) to efficiently represent the full range of allele diversity and wealth, without unnecessary duplication. Where possible, the open-pollinated members of the core collection are often converted to inbred lines by five or more rounds of self-fertilization and selection to gradually fix segregating loci to homozygosity; this allows the collection to be easily conserved and multiplied by avoiding further genetic segregation, but the process can take several years (Fig. 3.1).

Alternatively, homozygous lines can be generated by forming doubled haploid (DH) plants. In this process, haploid plants are produced from gametophytic cells by anther, microspore or ovary culture. The resulting haploid plants are sterile, but their chromosome numbers can be doubled through natural processes or chemically (using colchicine) to produce fertile DH plants in a much shorter time than classical self-pollinating methods (Germanà, 2011). However, the success of this process varies among species.

In pepper (Irikova *et al.*, 2011; Ochoa-Alejo, 2012; Cheng *et al.*, 2013; Kim *et al.*, 2013) and aubergine (Başay *et al.*, 2011; Salas *et al.*, 2011; Başay and Ellialtioğlu, 2013), anther culture is used effectively to produce DH lines by seed companies, but this does not work in tomato (Bal and Abak, 2007; Seguí-Simarro *et al.*, 2011; Moreno *et al.*, 2012). Anther culture has also been used for the production of dihaploids by halving the chromosome number from the tetraploids that arise from somatic hybridization in aubergine (Rizza *et al.*, 2002).

Regenerating haploid plants through ovule and ovary culture in the Cucurbitaceae family is possible although often challenging (Li *et al.*, 2013). However, haploid embryos may be induced by pollination with irradiated pollen or with pollen of a different species (Sari *et al.*, 1994; Kiełkowska *et al.*, 2014), and this has been investigated in melon (Abak *et al.*, 1996), watermelon (Gursoz *et al.*, 1991; Sari *et al.*, 1994), cucumber (Caglar and Abak, 1999) and pumpkins (Kurtar, 2009; Kurtar *et al.*, 2002); these techniques are now used routinely by breeders and researchers.

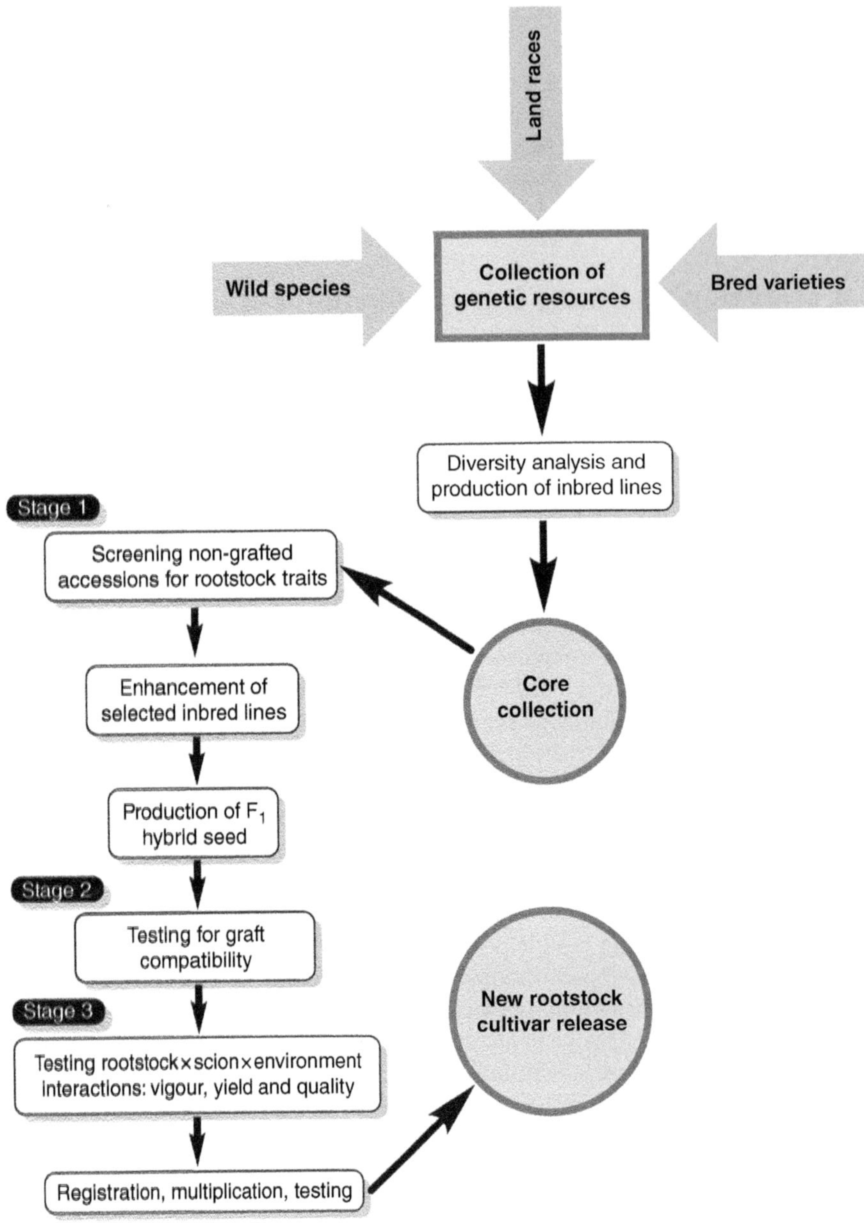

Fig. 3.1. Example of the process of breeding a new rootstock cultivar. The different stages are described in more detail in the main text.

3.4 Deploying Genetic Diversity for Rootstocks

3.4.1 General principles

Useful traits, particularly disease resistances, exist in wild relatives of the key grafted crops tomato, aubergine, pepper, cucumber, melon and watermelon, but these wild species often have poor germination, low vigour or poor graft compatibility. For example, germination rate and uniformity can be a problem with the wild species *Solanum torvum* Sw. (Hernández-Verdugo *et al.*, 2010). Development and selection of heterotic F_1 hybrids has been a successful approach for creating rootstock cultivars that can overcome some of the issues associated with using pure wild-species rootstocks while retaining useful traits. These hybrid rootstocks are most often grafted to F_1 hybrid scions to create a grafted transplant with four interacting genomes; hence, there are many opportunities for interactions and compatibilities controlled by the molecular factors that influence heterosis (Ryder *et al.*, 2014), including the bidirectional movement of small RNAs and mRNAs across graft junctions (Haroldsen *et al.*, 2012; Tsaballa *et al.*, 2013; Goldschmidt, 2014; Avramidou *et al.*, 2015; Yang *et al.*, 2015; Lewsey *et al.*, 2016). However, it must be recognized that the breeding of commercial hybrids is done in secrecy; the origins of the most successful hybrids are trade secrets, and rootstock cultivars are protected by plant breeders' rights. Equally, whether there has been a scientific basis to the selection of the most successful hybrids, or whether it has been through the 'art' of skilled and experienced breeders, is not publically known; and the degree to which marketing versus crop performance influences how growers and nurseries choose commercial elite rootstock cultivars is unclear.

3.4.2 Use of *Cucurbita* F_1 hybrids

Historically, *Cucurbita* rootstock breeding has been dominated by work in Korea, China and Japan, and as a result of this breeding effort, commercial cucurbit scions, including cucumber, watermelon and melon, are most often grafted on to interspecific *Cucurbita maxima* Duch. (pumpkin) × *Cucurbita moschata* Duch. (squash) F_1 hybrid rootstocks. This interspecific hybrid has become the accepted paradigm for vigorous rootstocks that improve yield and disease resistance in cucurbits; this paradigm appears to have arisen largely from the worldwide historical popularity of one such hybrid rootstock, 'Tetsakabuto', developed by Japanese breeders, which is believed to be exceptionally vigorous due to heterosis. In addition to being used as a rootstock, 'Tetsukabuto' is marketed as a winter squash cultivar and is a *C. maxima* 'Delicious' × *C. moschata* 'Kurokawa No. 2' interspecific cross; the apparently good seed production and combining ability are probably the factors that led to its popularity (Robinson, 1999) and to the widespread use of interspecific hybrid rootstocks in general. *C. maxima* × *C. moschata* crosses generally give rather poor fruit set and seed yield, and usually produce a sterile hybrid plant; a recent study in Turkey tested many combinations, and, although some were more favourable than others, further breeding work is required to reach acceptable seed yields (Karaagˆaç and Balkaya, 2013). The latter report underlines the remarkable commercial success of the Asian breeding programmes that created 'Tetsukabuto'; however, there is a lack of published scientific evidence to show that this particular hybrid type is superior to other hybrid or non-hybrid lines. Indeed, a recent study has shown that the performance of *C. maxima* × *C. moschata* F_1 hybrid rootstocks is not conclusively better than the parental lines, depending on the specific cross or the key traits examined (M. Edelstein *et al.*, unpublished data), and, as discussed in Chapter 2 (this volume), there is a huge amount of genetic diversity still not used or tested in hybrid rootstocks; for example, highly diverse *C. moschata* inbreds are available (Kong *et al.*, 2014) and, although *Lagenaria* are used extensively in the Far East as rootstocks, landraces in Europe and Turkey have untapped potential (Yetişir *et al.*, 2008; Karaca *et al.*, 2012).

3.4.3 Use of *Solanum* F_1 hybrids

In the Solanaceae, utilizing a hybrid with at least one parent that retains a large complement of wild-species DNA is typical, and *Solanum lycopersicum* L. (tomato) × *Solanum habrochaites* S. Knapp & D.M. Spooner (wild species) is the most widely used commercial hybrid rootstock for tomato. However, as with *Cucurbita*, there is a lack of published evidence for the superiority of *S. lycopersicum* × *S. habrochaites* F_1 hybrid rootstocks in comparison with the many other hybrid combinations or non-hybrids that are possible; further public research to explore wider genetic diversity, to understand the role of heterosis and to link phenotypes to genetic loci is required.

Despite the understandable lack of public information about the origins and reasons for the predominance of specific commercial rootstocks, there is a considerable scientific literature, described in the following sections, that considers *how* to deploy genetic diversity, that is, from studies on the sexual compatibility between different wild species and the methodologies available for breaking species barriers, including embryo rescue and somatic hybridization. These are the techniques that enable alleles to be moved between species in breeding materials, and that underpin the first stages of rootstock breeding.

3.4.4 Interspecific hybrids and hybridization barriers

Chapter 2 (this volume) details the sexual compatibility within the various groups of Solanaceae and Cucurbitaceae. In breeding programmes, the cultivated Solanaceae parent is usually used as the female and the wild species is the pollen donor (Bletsos *et al.*, 1998, 2004; Premabati Devi *et al.*, 2015), and the success of the cross is determined by the percentage of fruit set, number of seeds per fruit and percentage germination of the F_1 seed. The hybridization barriers can act before fertilization, for example by inhibition of pollen-tube growth, as for *Capsicum annuum* L. × *Capsicum pubescens* Ruiz & Pav. (da Silva Monteiro *et al.*, 2011), and this can be overcome by somatic hybridization. Where barriers occur after fertilization, embryo rescue has been used extensively to produce viable F_1 hybrids. The success of producing interspecific hybrids is easily tested using molecular markers (Reddy *et al.*, 2015).

In somatic hybridization, also called protoplast fusion, cells from two different species of plants are fused together to create a new plant with the characteristics of both species. Protoplasts are first isolated through mechanical or enzymatic procedures, and osmotic agents or electricity are then used to cause fusion. One of the earliest examples was the pomato, the potato–tomato fusion product (Melchers *et al.*, 1978). In aubergine, the first interspecific somatic hybrid was obtained between *Solanum melongena* L. and *Solanum sisymbriifolium* Lam. to provide new sources of disease resistance (Collonnier *et al.*, 2003; Daunay, 2008).

The cultivated and wild relatives of aubergine are an important source of tolerance to drought (*Solanum macrocarpon* L.), salinity (*Solanum linnaeanum* Hepper & P.-M.L. Jaeger) and frost (*Solanum grandiflorum* Ruiz & Pav., *Solanum mammosum* L. and *Solanum khasianum* C.B. Clarke) (Rotino *et al.*, 2014), and efforts have been made to develop somatic hybrids between *S. melongena* and wild or cultivated relatives for the improvement of aubergine rootstocks. Somatic hybrids have been developed for *S. melongena* × *Solanum integrifolium*, *S. melongena* × *Solanum sanitwongsei* W.G. Craib (Asao *et al.*, 2001) and *S. integrifolium* × *S. sanitwongsei*, producing fertile hybrids with increased resistance to bacterial wilt (Iwamoto *et al.*, 2007), and for *S. melongena* × *Solanum marginatum* L.f. (an arborescent species) (Borgato *et al.*, 2007).

Embryo rescue was one of the earliest successful applications of *in vitro* culture used to assist the development of plant embryos that otherwise would be non-viable because of incompatibilities between genomes in a particular hybrid (Cisneros and Tel-Zur, 2010). Embryos are removed from the ovary after fertilization but before abortion, and cultured aseptically on suitable medium to generate viable plants. This has been applied, for example, to interspecific hybrids: *S. melongena* × *S. torvum* (Kumchai *et al.*, 2013), *S. melongena* × *S. khasianum* (Rattan *et al.*, 2015) and *Capsicum annuum* × *Capsicum baccatum* (Jae *et al.* 2006; Eggink *et al.*, 2014), although further development of embryo-rescue techniques in *Capsicum* peppers is needed (Manzur *et al.*, 2013). In the case of *S. lycopersicum* × *Solanum peruvianum* L., the development of lines carrying resistance to tomato yellow leaf curl virus, tomato spotted wilt virus and tomato chlorosis virus was possible (Picó *et al.*, 2002; Encina *et al.*, 2012; Julián *et al.*, 2013). Embryo rescue has been reported to be successful from the following crosses in the cucurbits: *Cucurbita ficifolia* Bonché × *Cucurbita pepo* L. and *Cucurbita martinezii* L.H. Bailey × *C. pepo* (Rakha *et al.*, 2012); *C. maxima* × *C. pepo*, *C. pepo* × *C. moschata*, *C. ficifolia* × *C. maxima*, (*C. maxima* × *C. moschata*) × *C. pepo* and *Cucurbita argyrosperma* C. Huber × *C. moschata* (de Oliveira *et al.*, 2003; Šiško *et al.*, 2003).

The capture of wild-species DNA in a somatic or embryo-rescued hybrid necessarily requires substantial further germplasm enhancement before this novel germplasm can be deployed as true breeding parental lines for the creation of F_1 hybrid rootstock cultivars. Further crossing and then inbreeding, or production of DH lines to produce stable parental lines, must be completed and then many F_1 hybrids must be tested phenotypically for rootstock traits (Fig. 3.1).

3.5 Grafting as a Tool for Genetic Hybridization and Chimera Production

The agronomic practice of vegetable grafting demands that breeders continually generate and select new rootstock genotypes, which is the main subject of this chapter. However, the formation of the graft union itself also opens up new opportunities, described below, for: (i) combining genetic material within cells (genetic hybridization); and (ii) creating novel cellular chimeras. Grafting can therefore be regarded as an extra tool for the researcher and breeder to address fundamental questions or pursue genetic improvement in graftable crops.

3.5.1 Genetic hybridization: transfer of nuclear and organellar DNA between cells of the graft union

Grafting of two genetically different plants together to form a graft union brings cells of distinct genotypes into intimate proximity. This proximity offers the opportunity for the exchange of genetic material, either through cell fusion, in a process akin to protoplast fusion, or by movement of nuclei, chromatin or organelles through widened plasmodesmata (Fuentes *et al.*, 2014). This genetic exchange could occur naturally when plants fuse together, as is commonly observed in appressed stems of woody perennial species (Warschefsky *et al.*, 2016); if the resulting cells then underwent shoot regeneration, for example through the wound-induced formation of adventitious shoots, this could lead to individual plants with unique combinations of genetic material that could be inherited in subsequent generations. This mechanism for genetic exchange can be exploited by researchers and plant breeders to generate novel genetic resources. Grafting can therefore be considered a potential tool for genetic hybridization that can overcome sexual compatibility barriers.

Early evidence of the transfer of plastid genomes came from experiments where two transgenic tobacco lines were grafted together, one with a nuclear genome marker and one with a plastid marker. Cells that contained both markers were detected at the graft unions at high frequency (Stegemann and Bock, 2009). Later, transfer of whole plastid genomes in the absence of nuclear genome introgression (Stegemann *et al.*, 2012) was observed to occur across graft junctions, and this was proposed as a mechanism for the natural horizontal transfer of DNA between sexually incompatible species that had been predicted previously from phylogenetic analysis. Similarly, cell-to-cell movement of mitochondria was observed in tobacco graft junctions: restoration of *Nicotiana tabacum* L. fertile flower anatomy was observed in plants regenerated from an *N. tabacum*/*Nicotiana sylvestris* Speg. & S. Comes graft due to replacement of a defective mitochondrial genome (Gurdon *et al.*, 2016). More dramatically, the transfer of the entire nuclear genome was observed in the junction of a *N. tabacum*/*Nicotiana glauca* Graham graft; in this case, the *N. tabacum* parent ($2n = 4x = 48$ chromosomes) carried a hygromycin resistance gene, and the *N. glauca* parent ($2n = 2x = 24$ chromosomes) carried a kanamycin resistance gene. The graft junction was excised and cultured in the presence of kanamycin and hygromycin to produce callus and then shoots; some of the regenerated plants were stable and fertile and carried a full complement of parental chromosomes ($2n = 6x = 72$) that were successfully transmitted to the subsequent generations on meiosis by disomic inheritance segregating as an amphidiploid (Fuentes *et al.*, 2014). These plants therefore represented a new allopolyploid plant species (named *Nicotiana tabauca*) created by a grafting-induced asexual process.

Allopolyploidization is highly advantageous and common in crop species, and is believed to have occurred through sexual hybridization and chromosome doubling; grafting-induced allopolyploidization provides a new method for crop improvement that can combine genomes without the need for sexual compatibility between parents, and may be a mechanism that has played its part in genome evolution and crop domestication. Although grafting-induced genetic hybridization is conceptually similar to protoplast fusion in its outcome, it may be technically easier to achieve and could give a viable alternative in graft-compatible species where protoplast fusion has failed.

3.5.2 Use of grafting to generate chimeras

While genetic hybridization allows new combinations of DNA to occur in a single cell, chimeric plants are defined as those that contain cells arising from more than one genetic origin that are propagated by cell division within an individual plant. Interchange of DNA between cells is not a requirement for chimeric plant formation, and chimerism cannot therefore be inherited through the gametes. Here, chimeras are of interest because they can arise from the graft union, and their potential utility is discussed.

According to the Tunica-Corpus model (Reeve, 1948), shoot apical meristems of dicotyledonous plants contain three cell layer classes: L1, L2 and L3. Of these, L1 is the outermost epidermal layer, L2 is the next inner mesophyll/palisade layer and L3 comprises the remaining internal cells including vascular cell types. As a meristem develops, the genetic lineage of cells in each layer is preserved in the developing shoot because the cells of L1 and L2 divide only in the anticlinal plane, while the innermost cells divide in both anticlinal and periclinal directions. If all the cells of one layer arise from a single genotype that differs from another layer, then a periclinal chimera (i.e. differing between layers) can be produced that is stable when vegetatively propagated (Filippis *et al.*, 2013); this is the basis of many variegated ornamental plants. In the case of adventitious shoot formation from a graft union, a meristem can be generated from a group of cells, and if this group contains cells from the rootstock and scion, it can develop into a chimera, and sometimes will form a stable periclinal chimera if one cell layer becomes homogeneous (Zhu *et al.*, 2007). Although *in vitro* protoplast fusion and protoplast co culture can also be used to create chimeras, these techniques are technically challenging, and many researchers prefer to use grafting followed by adventitious shoot formation to generate chimeras for studying multicellular development and its molecular and genetic control; typically, parental lines with visible markers such as colour and surface traits are chosen to allow visible detection of chimeras. Some examples are described below.

Splice grafts made between *N. glauca* and *N. tabacum* were decapitated just above the graft union and treated with auxin/lanolin pastes to stimulate callus formation. Of 209 adventitious shoots that were formed from such graft unions, three were found to be interspecific mericlinal chimeras, and these could later be stabilized as periclinal chimeras (Marcotrigiano and Gouin, 1984). In this study, no chimeras could be produced from alternative *in vitro* methodologies such as mixed callus cultures, so grafting was essential to recover chimeric shoots in this case.

Approach grafting between red and green cabbage varieties, followed by sectioning and culturing of the graft junction, produced up to 53% of resulting adventitious shoots showing a visible red/green chimeric structure, and microscopic examination of the presence of anthocyanins arising from the red cabbage cells showed chimeric structures classified as 'complex sectorial-peripheral' (Noguchi *et al.*, 1992). These authors considered that symmetrical vertical heterografting through the centre of the shoot apex should give the optimal production of chimeric plants. In another *Brassica* study, the shoot apical meristems of *Brassica juncea* L. Czern. (tuber mustard) and *Brassica oleracea* L. (cabbage, red cultivar) seedlings were grafted vertically in the presence of different concentrations of auxin and cytokinin; after graft formation, the fused shoot tips were cut and cultured *in vitro* for adventitious shoot formation, and the most successful treatment resulted in 6% of adventitious shoots being visibly chimeric (Chen *et al.*, 2006).

In *Solanum*, stable periclinal chimeras were produced between tomato (*S. lycopersicum*) and the nightshade *Solanum nigrum* from graft unions (Lindsay *et al.*, 1995), and in a more recent study in the post-genomic era, tomato 'Heinz 1706' was grafted to *Solanum pennellii* Correll LA716, and one periclinal chimera was observed in the adventitious shoots derived from 50 grafts (Filippis *et al.*, 2013). This chimera had an L1 layer derived from *S. pennellii*, while the L2 and L3 cell layers were from the cultivar. Due to the known sequence polymorphisms between the two species, both of which now have reference genomes (Bolger *et al.*, 2014), the chimera could be used to study the gene expression differences between L1 and L2/L3 using RNA sequencing.

The above examples show that chimeras are readily generated, but are they useful? In clonally propagated fruit trees, the production of stable chimeric plants via grafting could be used to combine different traits that are controlled by a specific cell layer for agronomic advantage (Zhou *et al.*, 2002), but, as commercial grafted vegetable crops are invariably raised from seedlings, and stable propagation of chimeric plants occurs only through vegetative (clonal) means, chimeric individuals generated in vegetable crop species are unlikely to have direct commercial applications. However, formation of chimeric plants may create greater opportunities for DNA transfer (i.e. between genetically distinct cell layers of the whole shoot in contrast to the rather limited graft union), and could allow further possibilities for generating novel stable genetic combinations.

3.6 Selection of Improved Rootstocks

3.6.1 Phenotypic selection

Before defining breeding aims, information on the performance of the currently used commercially available rootstocks in different locally conducted experiments should be gathered, where it exists (Kubota *et al.*, 2008). Breeding aims can then be shaped by the market needs, and suitable selection schemes developed.

Each breeder will have their own unique views and strategies for breeding an improved rootstock, but a convenient framework for discussion divides the process into three stages (Fig. 3.1). The first stage is the development and selection of promising inbred lines in the non-grafted state, and the use of these as parental lines to make F_1 hybrid seed. Generally, non-grafted accessions and breeding lines are selected for resistance to soilborne pests and diseases, abiotic stresses, root vigour, germination and seedling uniformity. The second stage is to select lines with the desired grafting affinity and scion compatibility, and the third stage is to evaluate the effects of the selected rootstock × scion combinations on crop performance, focusing on fruit yield and quality under targeted agroclimatic conditions and specific production systems.

Stage 1: screening of non-grafted breeding lines

Inbred lines from core collections of germplasm can be evaluated for key traits before moving to more complex evaluations. Seed traits are one such area for early selection: seed evaluation can be based on seed yield, seed mass, germination rate and uniformity of emergence. These seed traits tend to have high heritability, that is, a high influence of genetic factors relative to environmental factors, and germination of rootstock seeds takes place in protected grafting nursery conditions where the environment is controlled and has less impact on the outcomes for seed traits (Huarachi Morejon, 2013; Premabati Devi *et al.*, 2015).

The non-grafted genetic material can easily be screened phenotypically for soilborne diseases at this stage in glasshouse pot experiments if molecular markers are not available for a particular disease; for example, aubergine and pepper cultivars and breeding lines were pre-screened for root rot resistance to *Phytophthora capsici* isolates (Foster *et al.*, 2013), tomato germplasm was screened for nematode resistance (Cervantes-Moreno *et al.*, 2014) and exotic watermelon accessions were screened for multiple pests and diseases (Cohen *et al.*, 2014). Particularly for disease resistance in the Solanaceae, there are a range of molecular markers available, so phenotypic screening is only necessary if resistances are overcome or new diseases appear. However, for resistance to abiotic stresses, to the best of our knowledge, there are no molecular markers that are currently in commercial use, so phenotypic screening is the only option. For example, screens for resistance to low temperature or high salinity are used by many breeding companies, some of whom have a special emphasis on breeding for such root traits (Rootility, 2016).

The ability to maintain growth at lower rootzone temperatures can be taken as a general indicator of vigour under optimum conditions, although there is little literature on this type of germplasm screening because it is usually carried out as part of proprietary breeding programmes and therefore the scientific basis is unknown. Although rapid growth of non-grafted germplasm can be taken as an indication of vigour, this is likely to be a highly complex trait, in which heterosis and graft compatibility play major roles, so it is not clear if the screening of non-grafted plants for vigour is a good indicator of later performance of grafted F_1 hybrids. However, it is common practice in rootstock breeding to select for accessions with large, well developed root systems after excavating or pulling up soil-grown plants, an approach that can be quantified for QTL studies, at least in the non-grafted crop maize, by 'shovelomics' (Colombi *et al.*, 2015). There is considerable current research to discover QTLs and gene alleles that can improve abiotic stress traits and vigour in rootstocks so that molecular markers can be developed to replace some aspects of phenotypic selection (*see* Marker-assisted selection, below). A greater understanding of the genetic loci that control heterosis, where only a few examples have been described so far (Krieger *et al.*, 2010), is also needed to improve the selection of vigour.

Typically, selected inbred accessions would go through further rounds of genetic enhancement to combine traits. This is achieved by backcrossing and recurrent selection to create parental lines suitable for the production of F_1 hybrid seeds.

Stage 2: testing for graft compatibility

F_1 hybrid rootstocks are next tested for their ability to form graft junctions and to support the growth and development of the grafted plant. This 'graftability' depends on having suitable hypocotyl diameters to match the scion, and the developmental cellular capacity for making vascular connections. Graft compatibility refers to the success of specific rootstock × scion combinations in terms of both graft formation and the subsequent growth and development of the successfully grafted plant. Compatibility screens can only be achieved by grafting each rootstock genotype to a range of scion genotypes and then evaluating the formation of the graft junction and the continued development of the grafted transplant. Such trials are expensive, and therefore the skill of the breeder is to bring forward a relatively small number of promising hybrids to this second step.

Knowledge of the genetic basis of grafting compatibility is limited, but information on molecular and cellular aspects of graft formation, including heterografts, is beginning to emerge (Milien *et al.*, 2012; Cookson *et al.*, 2014; Moreno *et al.*, 2014; Melnyk *et al.*, 2015) (*see* Chapter 5, this volume). There are no known molecular genetic markers for rootstock × scion compatibility, but this could be addressed by using suitable grafted mapping populations to find QTLs.

Stage 3: evaluation of the performance of rootstock × scion combinations

Breeding for rootstocks is a complicated process because it involves not only the rootstock genotype but also its effects on the scion, which can vary in different environments and cultivating systems (King *et al.*, 2010). Rootstock cultivars that will cover large market areas are desirable, and thus the compatible rootstock × scion combinations should be tested across many locations and over multiple years to evaluate the genotype (rootstock) × genotype (scion) × environment interactions against existing elite rootstocks and self-grafted scions in the presence of the most important biotic and abiotic stresses. Typical variables for assessment are: (i) vegetative versus generative crop development; (ii) marketable fruit yield in the early and late season; (ii) fruit quality; and (iv) resistance to pest and diseases.

Alongside the activities of commercial breeders, scientific researchers often collect existing commercial rootstock cultivars to evaluate their effectiveness at improving scion performance under different environmental conditions, and to understand the physiological mechanisms involved (Leonardi and Giuffrida, 2006; Rivard and Louws, 2008; Rivard *et al.*, 2010; Zhang *et al.*, 2016). Such scientific understanding could lead to the development of phenotypic screens that have greater predictive power for selecting improved rootstocks.

3.6.2 Marker-assisted Selection

In breeding populations, seedlings with a specific complement of chromosome segments of known origin can be selected by marker-assisted selection (MAS) using DNA-based genetic markers. If a causal relationship is already established between a genetic locus and a phenotype, then expensive phenotypic selection of plants at more advanced stages of development, such as late-season fruit yield due to rootstock genotype, can be greatly reduced. MAS is applied routinely in breeding companies using high-throughput genotyping platforms; some of these have the capacity to run more than 200,000 polymerase chain reaction (PCR)-based genotyping assays per day (Douglas, 2016). Technologies for genotyping have developed rapidly over the last decade, with single-nucleotide polymorphisms (SNPs) now dominating; SNPs are currently assayed by technologies such as the PCR-based KASP™ (He *et al.*, 2014), and genotype-by-sequencing using next-generation sequencing is emerging as an alternative approach (Kim *et al.*, 2016). Molecular markers for well characterized pest and disease resistance alleles are commonly used commercially and in publicly funded breeding programmes (Hanson *et al.*, 2016) but are more available in the Solanaceae than in the Cucurbitaceae. The challenge remains to find molecular markers for large-effect, robust QTLs for selection for resistance to abiotic stress, vigour, yield and fruit quality.

Identification of genetic markers for rootstock traits

Classical QTL approaches can be used to identify chromosomal locations that control rootstock traits, with the eventual aim of developing markers for MAS in rootstock breeding, and there are a few examples reported in the scientific literature.

Field trials with an introgression line (IL) mapping population, where single introgressions of *S. pennellii* LA716 were present in the background of processing tomato 'M82', were conducted using the ILs as rootstocks; this showed that IL8-3 contained a recessive QTL that conferred approximately 20–75% greater yield on the scion compared with non-grafted plants over three seasons (Gur *et al.*, 2011). The detection of an effect in IL8-3 but not in IL8-3-1 or IL8-2, implied that the rootstock effect was within a 1.4 Mb region of chromosome 8 containing approximately 180 genes (Chitwood *et al.*, 2013).

Using two recombinant inbred line populations from crosses between *S. lycopersicum* var. *cerasiforme* E9 and *Solanum pimpinellifolium* L. L5 or *Solanum cheesmaniae* (Riley) Fosberg L2, in which each member of the population was grafted to a common scion, several QTLs were discovered that affected scion traits under salinity stress. The most robust QTL with the highest log of odds (LOD) scores from this work, *gTW3.1* (total fruit weight per plant, LOD=4.29) and *gFN3.1* (fruit number per plant, LOD=4.59), were located on chromosome 3 at approximately 50 cM; the two QTLs at this locus had additive effects of 2.3 fruits per plant and 130 g fruit per plant, compared with mean trait values of around 37 fruits per plant and 1750 g fruit per plant, respectively, in the self-grafted controls (Estañ *et al.*, 2009). However, when the same population was grown at a different level of salinity, different QTLs for grams of fruit per plant were found and their LOD scores were marginal (Asins *et al.*, 2015). This illustrates the difficulty of defining complex quantitative traits controlled by many small-effect QTLs with environmental interactions. In the same study, a robust QTL for the effect of rootstock genotype on scion Na^+ concentration was found when plants were grown under moderate salinity, but this did not co-locate to QTLs for fruit yield, so its utility for breeding is unclear. QTLs for mineral content were also described (boron, potassium, magnesium and molybdenum), and these could be useful to breed rootstocks that enhance dietary nutritional value of fruit, or that mitigate plant mineral deficiencies (Asins *et al.*, 2015). To the best of our knowledge, there are no other published studies on vegetable rootstock QTLs, but the literature on perennial rootstock

QTLs, such as apple, pear and citrus, is more expansive, despite the problems with perennial genetics; for example, two rootstock loci, *Dw1* and *Dw2*, have been well characterized for their ability to confer dwarfing, vigour and precocity (Rusholme Pilcher *et al.*, 2008; Fazio *et al.*, 2014; Knäbel *et al.*, 2015), and are beginning to be used in MAS (Bassett *et al.*, 2015).

Genomic resources for vegetable rootstock breeding

QTL studies are underpinned by the genomic resources now available in all grafted vegetable crops. Next-generation sequencing technologies have facilitated the development of reference genomes for all of the major grafted crops: cucumber (Huang *et al.*, 2009), tomato (Tomato Genome Consortium, 2012), melon (Garcia-Mas *et al.*, 2012), watermelon (Xu *et al.*, 2013), hot pepper (Kim *et al.*, 2014) and aubergine (Hirakawa *et al.*, 2014). We are now in the post-genomic era, with large numbers of accessions being resequenced to show their DNA sequence differences (polymorphisms) compared with reference sequences. For example, in tomato, more than 500 accessions have been resequenced and the data are publically available (Aflitos *et al.*, 2014; Lin *et al.*, 2014); in addition, wild species related to tomato are being sequenced by *de novo* assembly to create species-specific reference genomes, with the *S. pennellii* LA716 genome completed to a high standard (Bolger *et al.*, 2014). The parents of a tomato population that has been used extensively to study rootstock QTLs (Estañ *et al.*, 2009; Asins *et al.*, 2010, 2015) and physiological mechanisms (Albacete *et al.*, 2015a,b) have also been resequenced (Kevei *et al.*, 2015). These genomic sequences, and genotyping platforms such as Illumina BeadChip (Asins *et al.*, 2015), greatly accelerate the process of identifying genetic markers associated with traits, and also in finding the causative polymorphisms; these huge advances in the description and assay of genetic variation leave phenotyping as the bottleneck in the discovery of useful markers for rootstock MAS.

3.7 Transgenic Rootstocks

Transgenic rootstocks have been created and tested as potential means to enhance scion performance or to understand signalling mechanisms. For example, overexpression of the cytokinin biosynthesis gene *IPT* in tomato rootstocks led to increased cytokinin in xylem sap and a 30% increase in yield under a moderate salinity stress of 75 mM NaCl (Ghanem *et al.*, 2011a,b); overexpression of *S*-adenosyl-1-methionine led to the accumulation of polyamines in the root system, and as a rootstock this genotype increased yield under alkali stress (Gong *et al.*, 2014); and overexpression of a H^+-pyrophosphatase in *Lagenaria* rootstocks increased root system size and the salinity tolerance of watermelon scions to which they were grafted (Han *et al.*, 2015). Engineering virus resistance in rootstocks has also been attempted in watermelon by overexpression of coat proteins from the cucumber green mottle mosaic virus (Park *et al.*, 2005). Numerous further ideas can be tested for their ability to transgenically enhance rootstocks based on current knowledge, either to alter root system vigour and architecture or to generate hormonal or RNA-based signals, and this can be done using genome editing, which may or may not be considered a gene-modification technology from a regulatory perspective (Cyranoski, 2015).

It has been suggested that transgenic rootstocks could be used to produce non-genetically modified fruit from a non-transgenic scion, to avoid some regulatory issues, but recent descriptions of the movement of nucleic acids across the graft junction (Zhang *et al.*, 2008; Haroldsen *et al.*, 2012; Tsaballa *et al.*, 2013; Avramidou *et al.*, 2015), and the broad nature of the regulatory framework currently make this an unworkable proposition.

3.8 Rootstock Registration and Commercialization

The International Union for the Protection of New Varieties of Plants (UPOV) has established test guidelines for tomato rootstocks, including hybrids of *S. lycopersicum* with *S. habrochaites*, *S. peruvianum* or *S. cheesmaniae*, which were adopted in March 2014 (CPVO, 2014). UPOV has also established test guidelines for a *C. maxima* × *C. moschata* interspecific hybrid (UPOV, 2014b) and for bottle gourd (UPOV, 2014a). A database of commercial rootstocks for cucurbits, aubergine, pepper and tomato is updated annually (USDA, 2016). A database of National Listings of plant varieties is also available (CPVO, 2016). A list of major companies that breed and supply vegetable rootstock seeds is given in Table 3.1.

Acknowledgements

M.B.P. thanks the MINECO project AGL2014-53398-C2-2-R, co-funded with FEDER funds. A.J.T. acknowledges the support of BBSRC grant BB/L01954X/1.

References

Abak, K., Sarı, N., Paksoy, M., Yilmaz, H., Aktaş, H. and Tunali, C. (1996) Genotype response to haploid embryo induction with pollination by irradiated pollens in melon, obtaining of dihaploid lines, determination of haploid and diploid plants by different techniques. *Turkish Journal of Agriculture and Forestry* 20, 425–430.

Aflitos, S., Schijlen, E., de Jong, H., de Ridder, D., Smit, S., Finkers, R., Wang, J., Zhang, G., Li, N., Mao, L., Bakker, F., Dirks, R., Breit, T., Gravendeel, B., Huits, H., Struss, D., Swanson-Wagner, R., van Leeuwen, H., van Ham, R.C., Fito, L., Guignier, L., Sevilla, M., Ellul, P., Ganko, E., Kapur, A., Reclus, E., de Geus, B., van de Geest, H., te Lintel Hekkert, B., van Haarst, J., Smits, L., Koops, A., Sanchez-Perez, G., van Heusden, A.W., Visser, R., Quan, Z., Min, J., Liao, L., Wang, X., Wang, G., Yue, Z., Yang, X., Xu, N., Schranz, E., Smets, E., Vos, R., Rauwerda, J., Ursem, R., Schuit, C., Kerns, M., van den Berg, J., Vriezen, W., Janssen, A., Datema, E., Jahrman, T., Moquet, F., Bonnet, J. and Peters, S. (2014) Exploring genetic variation in the tomato (*Solanum* section Lycopersicon) clade by whole-genome sequencing. *The Plant Journal* 80, 136–148.

Albacete, A., Andújar, C., Dodd, I., Giuffrida, F., Hichri, I., Lutts, S., Thompson, A. and Asins, M. (2015a) Rootstock-mediated variation in tomato vegetative growth under drought, salinity and soil impedance stresses. *Acta Horticulturae* 1086, 141–146.

Albacete, A., Andújar, C., Pérez-Alfocea, F., Lozano, J. and Asins, M. (2015b) Rootstock-mediated variation in tomato vegetative growth under low potassium or phosphorous supplies. *Acta Horticulturae* 1086, 147–152.

Asao, H., Arai, S. and Hirai, M. (2001) Somatic hybridization between *Solanum melongena* L. (Eggplant) and *Solanum sanitwongsei* Craib. In: Nagata, T. and Bajaj, Y.P.S. (eds) *Somatic Hybridization in Crop Improvement II*. Springer, New York, pp. 233–244.

Asins, M.J., Bolarin, M.C., Perez-Alfocea, F., Estan, M.T., Martinez-Andujar, C., Albacete, A., Villalta, I., Bernet, G.P., Dodd, I.C. and Carbonell, E.A. (2010) Genetic analysis of physiological components of salt tolerance conferred by *Solanum* rootstocks. What is the rootstock doing for the scion? *Theoretical and Applied Genetics* 121, 105–115.

Asins, M.J., Raga, V., Roca, D., Belver, A. and Carbonell, E.A. (2015) Genetic dissection of tomato rootstock effects on scion traits under moderate salinity. *Theoretical and Applied Genetics* 128, 667–679.

Avramidou, E., Kapazoglou, A., Aravanopoulos, F.A., Xanthopoulou, A., Ganopoulos, I., Tsaballa, A., Madesis, P., Doulis, A.G. and Tsaftaris, A. (2015) Global DNA methylation changes in Cucurbitaceae inter-species grafting. *Crop Breeding and Applied Biotechnology* 15, 112–116.

Bal, U. and Abak, K. (2007) Haploidy in tomato (*Lycopersicon esculentum* Mill.): a critical review. *Euphytica* 158, 1–9.

Başay, S. and Ellialtioğlu, S.S. (2013) Effect of genotypical factors on the effectiveness of anther culture in eggplant (*Solanum melongena* L.). *Turkish Journal of Biology* 37, 499–505.

Başay, S., Şeniz, V. and Ellialtioğlu, S. (2011) Obtaining dihaploid lines by using anther culture in the different eggplant cultivars. *Journal of Food, Agriculture and Environment* 9, 188–190.

Bassett, H., Malone, M., Ward, S., Foster, T., Chagné, D. and Bus, V. (2015) Marker assisted selection in an apple rootstock breeding family. *Acta Horticulturae* 1100, 25–28.

Bletsos, F.A., Rolpakias, D.G., Tsaktsira, M.L., Scaltsoyjannes, A.B. and Thanassoulopoulos, C.C. (1998) Interspecific hybrids between three eggplant (*Solanum melongena* L.) cultivars and two wild species (*Solanum torvum* Sw. and *Solanum sisymbriifolium* Lam.). *Plant Breeding* 117, 159–164.

Bletsos, F., Roupakias, D., Tsaktsira, M. and Scaltsoyjannes, A. (2004) Production and characterization of interspecific hybrids between three eggplant (*Solanum melongena* L.) cultivars and *Solanum macrocarpon* L. *Scientia Horticulturae* 101, 11–21.

Bolger, A., Scossa, F., Bolger, M.E., Lanz, C., Maumus, F., Tohge, T., Quesneville, H., Alseekh, S., Sorensen, I., Lichtenstein, G., Fich, E.A., Conte, M., Keller, H., Schneeberger, K., Schwacke, R., Ofner, I., Vrebalov, J., Xu, Y., Osorio, S., Aflitos, S.A., Schijlen, E., Jiménez-Goméz, J.M., Ryngajllo, M., Kimura, S., Kumar, R., Koenig, D., Headland, L.R., Maloof, J.N., Sinha, N., van Ham, R.C., Lankhorst, R.K., Mao, L., Vogel, A., Arsova, B., Panstruga, R., Fei, Z., Rose, J.K., Zamir, D., Carrari, F., Giovannoni, J.J., Weigel, D., Usadel, B. and Fernie, A.R.(2014) The genome of the stress-tolerant wild tomato species *Solanum pennellii*. *Nature Genetics* 46, 1034–1038.

Borgato, L., Conicella, C., Pisani, F. and Furini, A. (2007) Production and characterization of arboreous and fertile *Solanum melongena* + *Solanum marginatum* somatic hybrid plants. *Planta* 226, 961–969.

Brown, J. and Caligari, P. (2011) *An Introduction to Plant Breeding*. Wiley-Blackwell, Oxford, UK.

Caglar, G. and Abak, K. (1999) Progress in the production of haploid embryos, plants and doubled haploids in cucumber (*C. sativus* L.) by gamma irradiated pollen, in Turkey. *Acta Horticulturae* 492, 317–322.

Cervantes-Moreno, R., Rodríguez-Pérez, J.E., Fonseca, C.C., Sahagún-Castellanos, J. and Rodríguez-Guzmán, E. (2014) Tolerance of 26 native tomato collections from Mexico to nematode *Meloidogyne incognita* (Kofoid and White) Chitwood. *Revista Chapingo, Serie Horticultura* 20, 5–18.

Chen, L.P., Ge, Y.M. and Zhu, X.Y. (2006) Artificial synthesis of interspecific chimeras between tuber mustard (*Brassica juncea*) and cabbage (*Brassica oleracea*) and cytological analysis. *Plant Cell Reports* 25, 907–913.

Cheng, Y., Ma, R.L., Jiao, Y.S., Qiao, N. and Li, T.T. (2013) Impact of genotype, plant growth regulators and activated charcoal on embryogenesis induction in microspore culture of pepper (*Capsicum annuum* L.). *South African Journal of Botany* 88, 306–309.

Chitwood, D.H., Kumar, R., Headland, L.R., Ranjan, A., Covington, M.F., Ichihashi, Y., Fulop, D., Jiménez-Gómez, J.M., Peng, J., Maloof, J.N. and Sinha, N.R. (2013) A quantitative genetic basis for leaf morphology in a set of precisely defined tomato introgression lines. *Plant Cell* 25, 2465–2481.

Cisneros, A. and Tel-Zur, N. (2010) Embryo rescue and plant regeneration follow interspecific crosses in the genus *Hylocereus* (Cactaceae) *Euphytica* 174, 73–82.

Cohen, R., Burger, Y., Horev, C., Koren, A. and Edelstein, M. (2007) Introducing grafted cucurbits to modern agriculture: the Israeli experience. *Plant Disease* 91, 916–923.

Cohen, R., Tyutyunik, J., Fallik, E., Oka, Y., Tadmor, Y. and Edelstein, M. (2014) Phytopathological evaluation of exotic watermelon germplasm as a basis for rootstock breeding. *Scientia Horticulturae* 165, 203–210.

Collonnier, C., Fock, I., Daunay, M.C., Servaes, A., Vedel, F., Siljak-Yakovlev, S., Souvannavong, V. and Sihachakr, D. (2003) Somatic hybrids between *Solanum melongena* and *S. sisymbrifolium*, as a useful source of resistance against bacterial and fungal wilts. *Plant Science* 164, 849–861.

Colombi, T., Kirchgessner, N., Le Marié, C.A., York, L.M., Lynch, J.P. and Hund, A. (2015) Next generation shovelomics: set up a tent and REST. *Plant and Soil* 338, 1–20.

Cookson, S.J., Clemente Moreno, M.J., Hevin, C., Nyamba Mendome, L.Z., Delrot, S., Magnin, N., Trossat-Magnin, C. and Ollat, N. (2014) Heterografting with nonself rootstocks induces genes involved in stress responses at the graft interface when compared with autografted controls. *Journal of Experimental Botany* 65, 2473–2481.

CPVO (2014) Protocol for tests on distinctiveness, uniformity and stability: *Solanum lycopersicum L.* × *Solanum habrochaites* S. Knapp and D.M. Spooner; *Solanum lycopersicum L.* × *Solanum peruvianum* (L.) Mill.; *Solanum lycopersicum* L. × *Solanum cheesmaniae* (L. Ridley) Fosberg. Community Plant Variety Office, document code: CPVO-TP/294/1.

CPVO (2016) Variety Finder Database. Community Plant Variety Office. Available at: http://www.cpvo.europa.eu/main/en/home/databases/cpvo-variety-finder (accessed 10 April 2016).

Cyranoski, D. (2015) CRISPR tweak may help gene-edited crops bypass biosafety regulation. *Nature News*. Available at: http://www.nature.com/news/crispr-tweak-may-help-gene-edited-crops-bypass-biosafety-regulation-1.18590.

da Silva Monteiro, C.E., Pereira, T.N.S. and de Campos, K.P. (2011) Reproductive characterization of interspecific hybrids among *Capsicum* species. *Crop Breeding and Applied Biotechnology* 11, 241–249.

Daunay, M.C. (2008) Eggplant. In: Prohens-Tomás J. and Nuez F. (eds) *Handbook of Plant Breeding. Vegetables II: Fabaceae, Liliaceae, Solanacea, Umbelliferae*. Springer, New York, pp. 163–220.

de Oliveira, A.C.B., Maluf, W.R., Pinto, J.E.B.P., and Azevedo, S.M. (2003) Resistance to papaya ringspot virus in summer squash *Cucurbita pepo* L. introgressed from an interspecific *C. pepo* × *C. moschata* cross. *Euphytica* 132, 211–215.

Douglas (2016) LGC acquires Douglas Scientific to expand its high throughput pcr platform offering. Available at: http://www.douglasscientific.com/NewsEvents/News/2016-04-06-LGC-Acquires-Douglas-Scientific.pdf (accessed 16 April 2016).

Eggink, P.M., Tikunov, Y., Maliepaard, C., Haanstra, J.P.W., de Rooij, H., Vogelaar, A., Gutteling, E.W., Freymark, G., Bovy, A.G. and Visser, R.G. (2014) Capturing flavors from *Capsicum baccatum* by introgression in sweet pepper. *Theoretical Applied Genetics* 127, 373–390.

Encina C.L., Carmona-Martin E., Fernandez-Munoz R. (2012) Embryo rescue of *Solanum lycopersicum* × *S. peruvianum* tomato cross for TOCV resistance. *Acta Horticulturae* 935, 91–93.

Estañ, M.T., Villalta, I., Bolarín, M.C., Carbonell, E.A. and Asins, M.J. (2009) Identification of fruit yield loci controlling the salt tolerance conferred by solanum rootstocks. *Theoretical and Applied Genetics* 118, 305–312.

Fazio, G., Wan, Y., Kviklys, D., Romero, L., Adams, R., Strickland, D. and Robinson, T. (2014) *Dw2*, a new dwarfing locus in apple rootstocks and its relationship to induction of early bearing in apple scions. *Journal of the American Society for Horticultural Science* 139, 87–98.

Filippis, I., Lopez-Cobollo, R., Abbott, J., Butcher, S. and Bishop, G.J. (2013) Using a periclinal chimera to unravel layer-specific gene expression in plants. *The Plant Journal* 75, 1039–1049.

Foster, J.M., Naegele, R.P. and Hausbeck, M.K. (2013) Evaluation of eggplant rootstocks and pepper varieties for potential resistance to isolates of *Phytophthora capsici* from Michigan and New York. *Plant Disease* 97, 1037–1041.

Fuentes, I., Stegemann, S., Golczyk, H., Karcher, D. and Bock, R. (2014) Horizontal genome transfer as an asexual path to the formation of new species. *Nature* 511, 232–235.

Garcia-Mas, J., Benjak, A., Sanseverino, W., Bourgeois, M., Mir, G., Gonźalez, V.M., Heńaff, E., Cam̂ara, F., Cozzuto, L., Lowy, E., Alioto, T., Capella-Gutiérrez, S., Blanca, J., Cañizares, J., Ziarsolo, P., Gonzalez-Ibeas, D., Rodríguez-Moreno, L., Droege, M., Du, L., Alvarez-Tejado, M., Lorente-Galdos, B., Melé, M., Yang, L., Weng, Y., Navarro, A., Marques-Bonet, T., Aranda, M.A., Nuez, F., Picó, B., Gabaldón, T., Roma, G., Guigó, R., Casacuberta, J.M., Arús, P. and Puigdomènech, P. (2012) The genome of melon (*Cucumis melo* L.). *Proceedings of the National Academy of Sciences USA* 109, 11872–11877.

Germanà, M.A. (2011) Gametic embryogenesis and haploid technology as valuable support to plant breeding. *Plant Cell Reports* 30, 839–857.

Ghanem, M.E., Albacete, A., Smigocki, A.C., Frébort, I., Pospíilová, H., Martínez-Andújar, C., Acosta, M., Sánchez-Bravo, J., Lutts, S., Dodd, I.C. and Pérez-Alfocea, F. (2011a) Root-synthesized cytokinins improve shoot growth and fruit yield in salinized tomato (*Solanum lycopersicum* L.) plants. *Journal of Experimental Botany* 62, 125–140.

Ghanem, M.E., Hichri, I., Smigocki, A.C., Albacete, A., Fauconnier, M.L., Diatloff, E., Martinez-Andujar, C., Lutts, S., Dodd, I.C. and Pérez-Alfocea, F. (2011b) Root-targeted biotechnology to mediate hormonal signalling and improve crop stress tolerance. *Plant Cell Reports* 30, 807–823.

Goldschmidt, E.E. (2014) Plant grafting: new mechanisms, evolutionary implications. *Frontiers in Plant Science* 5, 527.

Gong, B., Li, X., Vandenlangenberg, K.M., Wen, D., Sun, S., Wei, M., Li, Y., Yang, F., Shi, Q. and Wang, X. (2014) Overexpression of *S*-adenosyl-1-methionine synthetase increased tomato tolerance to alkali stress through polyamine metabolism. *Plant Biotechnology Journal* 12, 694–708.

Gur, A., Semel, Y., Osorio, S., Friedmann, M., Seekh, S., Ghareeb, B., Mohammad, A., Pleban, T., Gera, G., Fernie, A.R. and Zamir, D. (2011) Yield quantitative trait loci from wild tomato are predominately expressed by the shoot. *Theoretical and Applied Genetics* 122, 405–420.

Gurdon, C., Svab, Z., Feng, Y., Kumar, D. and Maliga, P. (2016) Cell-to-cell movement of mitochondria in plants. *Proceedings of the National Academy of Sciences USA* 113, 3395–3400.

Gursoz, N., Abak, K., Pitrat, M., Rode, J.C. and Dumas de Vaulx, R. (1991) Obtention of haploid plants induced by irradiated pollen in watermelon (*Citrullus lanatus*). *Cucurbit Genetics Cooperative* 14, 109–110.

Han, J.S., Park, K.I., Jeon, S.M., Park, S., Naing, A.H. and Kim, C.K. (2015) Assessments of salt tolerance in a bottle gourd line expressing the *Arabidopsis* H^+-pyrophosphatase *AVP1* gene and in a watermelon plant grafted onto a transgenic bottle gourd rootstock. *Plant Breeding* 134, 233–238.

Hanson, P., Lu, S.F., Wang, J.F., Chen, W., Kenyon, L., Tan, C.W., Tee, K.L., Wang, Y.Y., Hsu, Y.C., Schafleitner, R., Ledesmaa, D. and Yanga, R.Y. (2016) Conventional and molecular marker-assisted selection and pyramiding of genes for multiple disease resistance in tomato. *Scientia Horticulturae* 201, 346–354.

Haroldsen, V.M., Szczerba, M.W., Aktas, H., Lopez-Baltazar, J., Odias, M.J., Chi-Ham, C.L., Labavitch, J.M., Bennett, A.B. and Powell, A.L.T. (2012) Mobility of transgenic nucleic acids and proteins within grafted rootstocks for agricultural improvement. *Frontiers in Plant Science* 3, 39.

He, C., Holme, J. and Anthony, J. (2014) SNP genotyping: the KASP assay. *Methods in Molecular Biology* 1145, 75–86.

Hernández-Verdugo, S., López-España, R.G., Porras, F., Parra-Terraza, S., Villarreal-Romero, M. and Osuna-Enciso, T. (2010) Variation in germination among populations and plants of wild chili pepper. *Agrociencia* 44, 667–677.

Hirakawa, H., Shirasawa, K., Miyatake, K., Nunome, T., Negoro, S., Ohyama, A., Yamaguchi, H., Sato, S., Isobe, S., Tabata, S. and Fukuoka, H. (2014) Draft genome sequence of eggplant (*Solanum melongena* L.): the representative *Solanum* species indigenous to the Old World. *DNA Research* 21, 649–660.

Howard, P.H. (2009) Visualizing consolidation in the global seed industry: 1996–2008. *Sustainability* 1, 1266–1287.

Huang, S., Li, R., Zhang, Z., Li, L., Gu, X., Fan, W., Lucas, W.J., Wang, X., Xie, B., Ni, P., Ren, Y., Zhu, H., Li, J., Lin, K., Jin, W., Fei, Z., Li, G., Staub J., Kilian A., van der Vossen, E.A., Wu, Y., Guo, J., He, J., Jia, Z., Ren, Y., Tian, G., Lu, Y., Ruan, J., Qian, W., Wang, M., Huang, Q., Li, B., Xuan, Z., Cao, J., Asan Wu, Z., Zhang, J., Cai, Q., Bai, Y., Zhao, B., Han, Y., Li, Y., Li, X., Wang, S., Shi, Q., Liu, S., Cho, W.K., Kim, J.Y., Xu, Y., Heller-Uszynska, K., Miao, H., Cheng, Z., Zhang, S., Wu, J., Yang, Y., Kang, H., Li, M., Liang, H., Ren, X., Shi, Z., Wen, M., Jian, M., Yang, H., Zhang, G., Yang, Z., Chen, R., Liu, S., Li, J., Ma, L., Liu, H., Zhou, Y., Zhao, J., Fang, X., Li, G., Fang, L., Li, Y., Liu, D., Zheng, H., Zhang, Y., Qin, N., Li, Z., Yang, G., Yang, S., Bolund, L., Kristiansen, K., Zheng, H., Li, S., Zhang, X., Yang, H., Wang, J., Sun, R., Zhang, B., Jiang, S., Wang, J., Du, Y. and Li, S. (2009) The genome of the cucumber, *Cucumis sativus* L. *Nature Genetics* 41, 1275–1281.

Huarachi Morejon, N. (2013) Genetic and environmental factors affecting improvement of rootstocks for tomato. MSc thesis, Ohio State University, Ohio, USA.

Irikova, T., Grozeva, S. and Rodeva, V. (2011) Anther culture in pepper (*Capsicum annuum* L.) *in vitro*. *Acta Physiologiae Plantarum* 33, 1559–1570.

Iwamoto, Y., Hirai, M., Ohmido, N., Fukui, K. and Ezura, H. (2007) Fertile somatic hybrids between *Solanum integrifolium* and *S. sanitwongsei* (syn. *S. kurzii*) as candidates for bacterial wilt-resistant rootstock of eggplant. *Plant Biotechnology* 24, 179–184.

Jae BY, Dong CY, Jae WD, Hyo GP. (2006) Overcoming two post-fertilization genetic barriers in interspecific hybridization between *Capsicum annuum* and *C. baccatum* for introgression of anthracnose resistance. Breeding Science 56, 31–38.

Julián, O., Herráiz, J., Corella, S., di-Lolli, I., Soler, S., Díez, M.J. and Pérez-de-Castro, A. (2013) Initial development of a set of introgression lines from *Solanum peruvianum* PI 126944 into tomato: exploitation of resistance to viruses. *Euphytica* 193, 183–196.

Karaağaç, O. and Balkaya, A. (2013) Interspecific hybridization and hybrid seed yield of winter squash (*Cucurbita maxima* Duch.) and pumpkin (*Cucurbita moschata* Duch.) lines for rootstock breeding. *Scientia Horticulturae* 149, 9–12.

Karaca, F., Yetişir, H., Solmaz, I., Çandir, E., Kurt, S., Sari, N. and Güler, Z. (2012) Rootstock potential of Turkish *Lagenaria siceraria* germplasm for watermelon: plant growth, yield and quality. *Turkish Journal of Agriculture and Forestry* 36, 167–177.

Keatinge, J.D.H., Lin, L.J., Ebert, A.W., Chen, W.Y., Hughes, J.D.A., Luther, G.C., Wang, J.F. and Ravishankar, M. (2014) Overcoming biotic and abiotic stresses in the Solanaceae through grafting: current status and future perspectives. *Biological Agriculture and Horticulture* 30, 272–287.

Kevei, Z., King, R.C., Mohareb, F., Sergeant, M.J., Awan, S.Z. and Thompson, A.J. (2015) Resequencing at ≥ 40-fold depth of the parental genomes of a *Solanum lycopersicum* × *S. pimpinellifolium* recombinant inbred line population and characterization of frame-shift indels that are highly likely to perturb protein function. *G3: Genes, Genomes, Genetics* 5, 971–981.

Kiełkowska, A., Adamus, A. and Baranski, R. (2014) An improved protocol for carrot haploid and doubled haploid plant production using induced parthenogenesis and ovule excision in vitro. *In Vitro Cellular and Developmental Biology – Plant* 50, 376–383.

Kim, M., Park, E.J., An, D. and Lee, Y. (2013) High-quality embryo production and plant regeneration using a two-step culture system in isolated microspore cultures of hot pepper (*Capsicum annuum* L.). *Plant Cell, Tissue and Organ Culture* 112, 191–201.

Kim, S., Park, M., Yeom, S.I., Kim, Y.M., Lee, J.M., Lee, H.A., Seo, E., Choi, J., Cheong, K., Kim, K.T., Jung, K., Lee, G.W., Oh, S.K., Bae, C., Kim, S.B., Lee, H.Y., Kim, S.Y., Kim, M.S., Kang, B.C., Jo, Y.D., Yang, H.B., Jeong, H.J., Kang, W.H., Kwon, J.K., Shin, C., Lim, J.Y., Park, J.H., Huh, J.H., Kim, J.S., Kim, B.D., Cohen, O., Paran, I., Suh, M.C., Lee, S.B., Kim, Y.K., Shin, Y., Noh, S.J., Park, J., Seo, Y.S., Kwon, S.Y., Kim, H.A., Park, J.M., Kim, H.J., Choi, S.B., Bosland, P.W., Reeves, G., Jo, S.H., Lee, B.W., Cho, H.T., Choi, H.S., Lee, M.S., Yu, Y., Do, Choi, Y., Park, B.S., van, Deynze, A., Ashrafi, H., Hill, T., Kim, W.T., Pai, H.S., Ahn, H.K., Yeam, I., Giovannoni, J.J., Rose, J.K., Sørensen, I., Lee, S.J., Kim, R.W., Choi, I.Y., Choi, B.S., Lim, J.S., Lee, Y.H. and Choi D. (2014) Genome sequence of the hot pepper provides insights into the evolution of pungency in *Capsicum* species. *Nature Genetics* 46, 270–278.

Kim, C., Guo, H., Kong, W., Chandnani, R., Shuang, L.S. and Paterson, A.H. (2016) Application of genotyping by sequencing technology to a variety of crop breeding programs. *Plant Science* 242, 14–22.

Knäbel, M., Friend, A.P., Palmer, J.W., Diack, R., Wiedow, C., Alspach, P., Deng, C., Gardiner, S.E., Tustin, D.S., Schaffer, R., Foster, T., and Chagné, D (2015) Genetic control of pear rootstock-induced dwarfing and precocity is linked to a chromosomal region syntenic to the apple *Dw1* loci. *BMC Plant Biology* 15, 230.

Kong, Q., Chen, J., Liu, Y., Ma, Y., Liu, P., Wu, S., Huang, Y. and Bie, Z. (2014) Genetic diversity of *Cucurbita* rootstock germplasm as assessed using simple sequence repeat markers. *Scientia Horticulturae* 175, 150–155.

Krieger, U., Lippman, Z.B. and Zamir, D. (2010) The flowering gene *SINGLE FLOWER TRUSS* drives heterosis for yield in tomato. *Nature Genetics* 42, 459–463.

Kubota, C., McClure, M.A., Kokalis-Burelle, N., Bausher, M.G. and Rosskopf, E.N. (2008) Vegetable grafting: history, use, and current technology status in North America. *HortScience* 43, 1664–1669.

Kumchai, J., Wei, Y.C., Lee, C.Y., Chen, F.C. and Chin, S.W. (2013) Production of interspecific hybrids between commercial cultivars of the eggplant (*Solanum melongena* L.) and its wild relative *S. torvum*. *Genetics and Molecular Research* 12, 755–764.

Kurtar, E.S. (2009) Influence of gamma irradiation on pollen viability, germination ability, and fruit and seed-set of pumpkin and winter squash. *African Journal of Biotechnology* 8, 6918–6926.

Kurtar, E.S., Sarl, N. and Abak, K. (2002) Obtention of haploid embryos and plants through irradiated pollen technique in squash (*Cucurbita pepo* L.). *Euphytica* 127, 335–344.

Leonardi, C. and Giuffrida, F. (2006) Variation of plant growth and macronutrient uptake in grafted tomatoes and eggplants on three different rootstocks. *European Journal of Horticultural Science* 71, 97–101.

Lewsey, M.G., Hardcastle, T.J., Melnyk, C.W., Molnar, A., Valli, A., Urich, M.A., Nery, J.R., Baulcombe, D.C. and Ecker, J.R. (2016) Mobile small RNAs regulate genome-wide DNA methylation. *Proceedings of the National Academy of Sciences USA* 113, E801–E810.

Li, J.W., Si, S.W., Cheng, J.Y., Li, J.X. and Liu, J.Q. (2013) Thidiazuron and silver nitrate enhanced gynogenesis of unfertilized ovule cultures of *Cucumis sativus*. *Biologia Plantarum* 57, 164–168.

Lin, T., Zhu, G., Zhang, J., Xu, X., Yu, Q., Zheng, Z., Zhang, Z., Lun, Y., Li, S., Wang, X., Huang, Z., Li, J., Zhang, C., Wang, T., Zhang, Y., Wang, A., Zhang, Y., Lin, K., Li, C., Xiong, G., Xue, Y., Mazzucato, A., Causse, M., Fei, Z., Giovannoni, J.J., Chetelat, R.T., Zamir, D., Städler, T., Li, J., Ye, Z., Du, Y. and Huang, S. (2014) Genomic analyses provide insights into the history of tomato breeding. *Nature Genetics* 46, 1220–1226.

Lindsay, G.C., Hopping, M.E., Binding, H. and Burge, G.K. (1995) Graft chimeras and somatic hybrids for new cultivars. *New Zealand Journal of Botany* 33, 79–92.

Manzur, J.P., Penella, C. and Rodríguez-Burruezo, A. (2013) Effect of the genotype, developmental stage and medium composition on the in vitro culture efficiency of immature zygotic embryos from genus Capsicum. *Scientia Horticulturae* 161, 181–187.

Marcotrigiano, M. and Gouin, F.R. (1984) Experimentally synthesized plant chimeras 2. A comparison of *in vitro* and *in vivo* techniques for the production of interspecific *Nicotiana* chimeras. *Annals of Botany* 54, 513–521.

Melchers, G., Sacristán, M.D. and Holder, A.A. (1978) Somatic hybrid plants of potato and tomato regenerated from fused protoplasts. *Carlsberg Research Communications* 43, 203–218.

Melnyk, C.W., Schuster, C., Leyser, O. and Meyerowitz, E.M. (2015) A developmental framework for graft formation and vascular reconnection in *Arabidopsis thaliana*. *Current Biology* 25, 1306–1318.

Milien, M., Renault-Spilmont, A.S., Cookson, S.J., Sarrazin, A. and Verdeil, J.L. (2012) Visualization of the 3D structure of the graft union of grapevine using X-ray tomography. *Scientia Horticulturae* 144, 130–140.

Moreno, Á., Claveria, E., Pujol, M. and Dolcet-Sanjuan, R. (2012) Development of a methodology for the production of doubled haploid lines in *Solanum lycopersicum* L. *Acta Horticulturae* 935, 95–100.

Moreno, M.J.C., Hevin, C., Ollat, N. and Cookson, S.J. (2014) Developments at the graft interface in homo- and hetero-grafts gene expression and histological changes during the first month after grafting. *Plant Signaling and Behavior* 9, e28852.

Noguchi, T., Hirata, Y. and Yagishita, N. (1992) Intervarietal and interspecific chimera formation by *in vitro* graft-culture method in *Brassica*. *Theoretical and Applied Genetics* 83, 727–732.

Ochoa-Alejo, N. (2012) Anther culture of chili pepper (*Capsicum* spp.). *Methods in Molecular Biology* 877, 227–231.

Palada, M.C. and Wu, D.L. (2008) Evaluation of chili rootstocks for grafted sweet pepper production during the hot-wet and hot-dry seasons in Taiwan. *Acta Horticulturae* 767, 151–158.

Park, S.M., Lee, J.S., Jegal, S., Jeon, B.Y., Jung, M., Park, Y.S., Han, S.L., Shin, Y.S., Her, N.H., Lee, J.H., Lee, M.Y., Ryu, K.H., Yang, S.G. and Harn, C.H. (2005) Transgenic watermelon rootstock resistant to CGMMV (cucumber green mottle mosaic virus) infection. *Plant Cell Reports* 24, 350–356.

Picó, B., Herraiz, J., Ruiz, J.J. and Nuez, F. Widening the genetic basis of virus resistance in tomato. (2002) *Science Horticulturae* 94, 73–89.

Premabati Devi, C., Munshi, A.D., Behera, T.K., Choudhary, H., Vinod, Gurung, B. and Saha, P. (2015) Cross compatibility in interspecific hybridization of eggplant, *Solanum melongena*, with its wild relatives. *Scientia Horticulturae* 193, 353–358.

Rakha, M.T., Metwally, E.I., Moustafa, S.A., Etman, A.A. and Dewir, Y.H. (2012) Production of Cucurbita interspecific hybrids through cross pollination and embryo rescue technique. *World Applied Sciences Journal* 20, 1366–1370.

Rattan, P., Kumar, S., Salgotra, R. K., Samnotra, R. K. and Sharma, F. (2015) Development of interspecific F_1 hybrids (*Solanum melongena* × *Solanum khasianum*) in eggplant through embryo rescue technique. *Plant Cell, Tissue and Organ Culture* 120, 379–386.

Reddy, D.C.L., Khandagale, K.S., Victorathisayam, T., Aswath, C. and Singh, T.H. (2015) Genetic purity of interspecific hybrids of *Solanum melongena* and *S. macrocarpon* assessed using SSR markers. *International Journal of Vegetable Science* 21, 111–117.

Reeve, R.M. (1948) The "tunica-corpus" concept and development of shoot apicies in certain dicotyledons. *American Journal of Botany* 35, 65–75.

Rivard, C.L. and Louws, F.J. (2008) Grafting to manage soilborne diseases in heirloom tomato production. *Hortscience* 43, 2104–2111.

Rivard, C.L., O'Connell, S., Peet, M.M. and Louws, F.J. (2010) Grafting tomato with interspecific rootstock to manage diseases caused by *Sclerotium rolfsii* and southern root-knot nematode. *Plant Disease* 94, 1015–1021.

Rizza, F., Mennella, G., Collonnier, C., Sihachakr, D., Kashyap, V., Rajam, M., Presterà, M. and Rotino, G. (2002) Androgenic dihaploids from somatic hybrids between *Solanum melongena* and *S. aethiopicum* group gilo as a source of resistance to *Fusarium oxysporum* f. sp. *melongenae*. *Plant Cell Reports* 20, 1022–1032.

Robinson, R.W. (1999) Rationale and methods for producing hybrid cucurbit seed. *Journal of New Seeds* 1, 1–47.

Rootility (2016) Company profile. Available at: http://www.rootility.com/about (accessed April 2016).

Rotino, G.L., Sala, T. and Toppino, L. (2014) Eggplant. In: Pratap, A. and Kumar, J. (eds) *Alien Gene Transfer in Crop Plants*, Vol. 2: *Achievements and Impacts*. Springer, New York, pp. 381–409.

Rusholme Pilcher, R.L., Celton, J.M., Gardiner, S.E. and Tustin, D.S. (2008) Genetic markers linked to the dwarfing trait of apple rootstock 'Malling 9'. *Journal of the American Society for Horticultural Science* 133, 100–106.

Ryder, P., McKeown, P.C., Fort, A. and Spillane, C. (2014) Epigenetics and heterosis in crop plants. In: Alvarez-Venegas, R., de la Pena, C. and Casas-Mollano, J.A. (eds) *Epigenetics in Plants of Agronomic Importance: Fundamentals and Applications: Transcriptional Regulation and Chromatin Remodelling in Plants*. Springer, New York, pp. 13–31.

Salas, P., Prohens, J. and Seguí-Simarro, J.M. (2011) Evaluation of androgenic competence through anther culture in common eggplant and related species. *Euphytica* 182, 261–274.

Sari, N., Abak, K., Pitrat, M., Rode, J.C. and Dumas de Vaulx, R. (1994) Induction of parthenogenetic haploid embryos after pollination by irradiated pollen in watermelon. *HortScience* 29, 1189–1190.

Seguí-Simarro, J.M., Corral-Martínez, P., Parra-Vega, V. and González-García, B. (2011) Androgenesis in recalcitrant solanaceous crops. *Plant Cell Reports* 30, 765–778.

Šiško, M., Ivančič, A., Bohanec, B. (2003) Genome size analysis in the genus *Cucurbita* and its use for determination of interspecific hybrids obtained using the embryo-rescue technique. *Plant Science* 165, 663–669.

Stegemann, S. and Bock, R. (2009) Exchange of genetic material between cells in plant tissue grafts. *Science* 324, 649–651.

Stegemann, S., Keuthe, M., Greiner, S. and Bock, R. (2012) Horizontal transfer of chloroplast genomes between plant species. *Proceedings of the National Academy of Sciences USA* 109, 2434–2438.

Tomato Genome Consortium (2012) The tomato genome sequence provides insights into fleshy fruit evolution. *Nature* 485, 635–641.

Tsaballa, A., Athanasiadis, C., Pasentsis, K., Ganopoulos, I., Nianiou-Obeidat, I. and Tsaftaris, A. (2013) Molecular studies of inheritable grafting induced changes in pepper (*Capsicum annuum*) fruit shape. *Scientia Horticulturae* 149, 2–8.

UPOV (2014a) Guidelines for the conduct of tests for distinctness, uniformity and stability: bottle gourd, Calabash. International Union for the Protection of New Varieties of Plants, document code: TG/313/1.

UPOV (2014b) Guidelines for the conduct of tests for distinctness, uniformity and stability: *Cucurbita maxima* × *Cucurbita moschata* interspecific hybrids. Draft Report of the International Union for the Protection of New Varieties of Plants, document code: CUCUR_MMO.

USDA (2016) Vegetable grafting research based information portal. Available at: http://www.vegetablegrafting.org/rootstock-tables/ (accessed 10 April 2016).

Warschefsky, E.J., Klein, L.L., Frank, M.H., Chitwood, D.H., Londo, J.P., von Wettberg, E.J.B. and Miller, A.J. (2016) Rootstocks: diversity, domestication, and impacts on shoot phenotypes. *Trends in Plant Science* 21, 418–437.

Xu, Y., Wang, J., Guo, S., Zhang, J., Sun, H., Salse, J., Lucas, W.J., Zhang, H., Zheng, Y., Mao, L., Ren, Y., Wang, Z., Min, J., Guo, X., Murat, F., Ham, B.K, Zhang, Z., Gao, S., Huang, M., Xu, Y., Zhong, S., Bombarely, A., Mueller LA, Zhao, H., He, H., Zhang, Y., Zhang, Z., Huang, S., Tan, T., Pang, E., Lin, K., Hu, Q., Kuang, H., Ni, P., Wang, B., Liu, J., Kou, Q., Hou, W., Zou, X., Jiang, J., Gong, G., Klee, K., Schoof, H., Huang, Y., Hu, X., Dong, S., Liang, D., Wang, J., Wu, K., Xia, Y., Zhao, X., Zheng, Z., Xing, M., Liang, X., Huang, B., Lv, T., Wang, J., Yin, Y., Yi, H., Li, R., Wu, M., Levi, A., Zhang, X., Giovannoni, J.J., Wang, J., Li, Y., Fei, Z. and Xu, Y. (2013) The draft genome of watermelon (*Citrullus lanatus*) and resequencing of 20 diverse accessions. *Nature Genetics* 45, 51–58.

Yang, Y., Mao, L., Jittayasothorn, Y., Kang, Y., Jiao, C., Fei, Z. and Zhong, G.Y. (2015) Messenger RNA exchange between scions and rootstocks in grafted grapevines. *BMC Plant Biology* 15, 251.

Yetişir, H., Şakar, M. and Serçe, S. (2008) Collection and morphological characterization of *Lagenaria siceraria* germplasm from the Mediterranean region of Turkey. *Genetic Resources and Crop Evolution* 55, 1257–1266.

Zhang, Z.J., Wang, Y.M., Long, L.K., Lin, Y., Pang, J.S. and Liu, B. (2008) Tomato rootstock effects on gene expression patterns in eggplant scions. *Russian Journal of Plant Physiology* 55, 93–100.

Zhang, Z.H., Han, M., Zhang, Y., Wang, Y. and Xu, K. (2016) Identification and evaluation of tomato rootstock seedlings for drought tolerance. *Chinese Journal of Ecology* 35, 719–725.

Zhou, J., Hirata, Y., Nou, I.S., Shiotani, H. and Ito, T. (2002) Interactions between different genotypic tissues in citrus graft chimeras. *Euphytica* 126, 355–364.

Zhu, X.Y., Zhao, M., Ma, S., Ge, Y.M., Zhang, M.F. and Chen, L.P. (2007) Induction and origin of adventitious shoots from chimeras of *Brassica juncea* and *Brassica oleracea*. *Plant Cell Reports* 26, 1727–1732.

4 Rootstock–Scion Signalling: Key Factors Mediating Scion Performance

Jan Henk Venema,[1]* Francesco Giuffrida,[2] Ivan Paponov,[3] Alfonso Albacete,[4] Francisco Pérez-Alfocea[4] and Ian C. Dodd[5]

[1]*University of Groningen, Groningen, The Netherlands;* [2]*Università degli Studi di Catania, Catania, Italy;* [3]*Norwegian Institute of Bioeconomy Research, Klepp Station, Norway;* [4]*Department of Plant Nutrition, CEBAS-CSIC, Murcia, Spain;* [5]*Lancaster Environment Centre, Lancaster University, Lancaster, UK*

4.1 Introduction

Although the scion genotype often has a dominant effect on most agronomic traits, the rootstock introduces new sources of genotypic and phenotypic variability in the crop, due to rootstock × scion × environment interactions. These interactions are essentially based on hydraulic and chemical signals through the xylem (root to shoot) and the phloem (shoot to root). Although maximizing crop yield depends on the leaves receiving an optimal supply of water and mineral nutrients from the root system via the xylem, integrated plasticity involves long-distance communication (Sachs, 2005; Albacete *et al.*, 2008), where small organic molecules play key roles. Except for some chlorophyll- and photosynthesis-related compounds, roots are believed to produce every type of compound that exists in plants (Gregory, 2006). Although many recent reviews have focused on specific chemical compounds in root-to-shoot communication, mainly phytohormones (Ghanem *et al.*, 2011b; Pérez-Alfocea *et al.*, 2010, 2011), other small molecules in xylem sap can modulate whole-plant processes and/or stress responses. Progress in identifying the roles of various signals in regulating root-to-shoot communication may be slow because at least four kinds of signals allow roots to influence shoots (Jackson, 1993; Gregory, 2006): (i) increased root export of an existing signalling compound or generating a new one; (ii) decreased root export of an

* j.h.venema@rug.nl

existing signalling compound; (iii) accumulation of signal/message in the shoot (source) due to reduced demand in the root; and (iv) a reduction in signalling compounds in the shoot due to increased root demand.

While this conceptual framework conveniently discriminates root-to-shoot (xylem sap composition: (i) and (ii)) and shoot-to-root (phloem sap composition: (iii) and (iv)), many signalling molecules can be synthesized in both root and shoot tissues and can travel in both xylem and phloem saps. This is recognized in empirical flow models (e.g. Jeschke and Pate, 1991), which have been applied to determine uptake, synthesis, transport and metabolism of signalling molecules. This chapter examines ionic, phytohormonal, proteomic and RNA signalling *in planta*, highlighting specific examples of signalling between the scion and rootstock. We draw attention to examples of grafting between specific combinations of scions and rootstocks that are deficient in (or overexpress) these signals, thus highlighting their importance (or otherwise) in regulating plant responses.

4.2 Current Knowledge of Ionic and Chemical Signalling Between Rootstock and Scion

4.2.1 Ionic signalling

While the ionic composition of transport fluids is sometimes assayed to indicate momentary plant nutrient status (as an adjunct to tissue measurements), there is increasing recognition that ion concentrations in both the xylem and phloem can specifically regulate processes such as leaf growth and water use, as well as root architecture. Augmenting the transport fluids of isolated leaves (e.g. using a detached leaf transpiration assay; Zhang and Davies, 1991) or intact plants (Reid *et al.*, 2012) with specific compounds can assess their physiological significance. Alternatively, reciprocal grafting provides another important method to establish the importance of ionic signalling in controlling shoot responses, most commonly when plants are exposed to salt stress (e.g. Albacete *et al.*, 2009; Ursini *et al.*, 2013).

Xylem sap pH

Many plant species show significant changes in both xylem sap and apoplastic pH in response to altered environmental conditions around the roots (e.g. drought, low temperature, flooding, soil nutrient status; Wilkinson, 1999) or leaves (e.g. fungal infection; Felle *et al.*, 2005). Xylem sap alkalization can be an important component of root-to-shoot signalling and may act synergistically with abscisic acid (ABA) to reduce stomatal conductance and leaf growth (Schachtman and Goodger, 2008; Shabala *et al.*, 2016). An alkaline xylem sap pH promotes stomatal closure by keeping ABA in the deprotonated state, thereby preventing ABA uptake by the mesophyll and epidermal symplast, and thus enhancing apoplastic ABA concentrations (Wilkinson, 1999). In transit from the roots to the leaves, xylem sap becomes significantly more alkaline, presumably through the action of xylem parenchyma-associated ATPases (Jia and Davies, 2007). The transpiration rate itself affects xylem and apoplastic pH, and limited transit time between the root and leaf (as in rapidly transpiring plants) will limit the impact of ATPases on sap pH. Nevertheless, the precise mechanism(s) causing increases in xylem sap pH is not completely clear and may include changes in: (i) ATPase activity of cells; (ii) nitrate uptake; (iii) the ratio of cations and anions; and (iv) the assimilation and production of organic acids (e.g. malate) (Wilkinson and Davies, 2002; Korovetska *et al.*, 2014). Moreover, xylem sap and apoplastic pH signalling is far from universal among plant species, with some species showing xylem sap alkalization in response to soil drying, while others show no effect or even xylem sap acidification (Jia and Davies, 2007; Sharp and Davies, 2009; Gloser *et al.*, 2016). Interestingly, species that did not show changes in xylem sap pH following soil drying still showed stomatal closure after spraying the leaves with alkaline buffers, which presumably caused the apoplast to become alkaline (Sharp and Davies, 2009). As xylem sap can become alkaline in transit from the root to the shoot (Sharp and Davies, 2009), the relative importance of the root and shoot in mediating long-distance pH signalling could be investigated by reciprocal grafting with closely related species that show opposing changes in xylem sap pH following soil drying.

Nitrate and ammonium ions

Plants use either ammonium (NH_4^+) or nitrate (NO_3^-) as their main nitrogen source (Glass and Siddiqi, 1995). In most plants, NO_3^- uptake from culture medium into the root is mediated by the plasma membrane NO_3^-/H^+ co-transporter (Mistrik and Ullrich, 1996). NO_3^- is actively transported across the root plasma membrane by both high (between 1 μM and 1 mM external NO_3^- concentration) and low (above 1 mM NO_3^-) affinity proton symporters (Crawford and Glass, 1998). NO_3^- can be effluxed from the cell, stored in the vacuole or loaded into the xylem and transported to the shoot. This xylem transport is stimulated by potassium ions (K^+), which are co-transported to maintain the electrical charge balance (Ivashikina and Feyziev, 1998). Following xylem transport, NO_3^- must leave the xylem and reach the mesophyll cells, where it is either stored in the vacuole or reduced (Crawford and Glass, 1998). The high NO_3^- concentrations of the xylem sap (7.16 ± 9.18 mM) and the low concentrations (0.59 ± 0.97 mM) in the phloem (Peuke, 2010) indicate that NO_3^- is absorbed efficiently by leaf cells. Once in the cell, NO_3^- can be reduced to NH_4^+ via the action of NAD(P)H-dependent cytosolic nitrate reductase and nitrite reductase. These conversions can occur in both the roots and shoots.

NH_4^+ converted from NO_3^- or absorbed from nutrient solution can be further converted to amino acids. Alternatively, NH_4^+ can be translocated from the root to the shoot and may constitute a significant part (about 10%) of xylem nitrogen transport (Schjoerring *et al.*, 2002). Xylem NH_4^+ concentrations reflect NO_3^- or NH_4^+ uptake, NO_3^- reduction and NH_4^+ assimilation. Root NH_4^+ assimilation is mediated by cytosolic glutamine synthetase and glutamate synthase. However, plant species differ in their capacity to assimilate NH_4^+ in the root, and the glutamine synthetase activity can also be repressed at high (>3 mM) nitrogen application levels, independent of the nitrogen form (Schjoerring *et al.*, 2002). In plants exposed to high NO_3^- levels with high uptake rates, the nitrogen in their roots cannot be fully reduced and assimilated, and a significant proportion is transported to the shoot for reduction (NO_3^-) and/or assimilation (NH_4^+) (Lexa and Cheeseman, 1997). Thus the ratio between NO_3^- and NH_4^+ in the xylem can change according to root and shoot nitrate reductase activity (plants can be root or shoot reducers; Lexa and Cheeseman, 1997).

NO_3^- not only is a major source of nitrogen for plants but also acts as a signalling molecule that regulates many aspects of nutrient uptake, metabolism and gene expression (Krapp *et al.*, 2014). Xylem NO_3^- concentrations change diurnally due to changes in transpiration rate, showing higher concentrations at night (Siebrecht *et al.*, 2003). NO_3^- efflux into the xylem is maintained during high xylem NO_3^- concentrations (as occurs at night) due to positive feedback on NO_3^- loading (Miller *et al.*, 2007). However, root NO_3^- assimilation (nitrate reductase activity) can increase when the shoot is provided with an excess of NO_3^-, leading to small changes in the overall capacity of the plant to reduce nitrogen (Lexa and Cheeseman, 1997).

It is now well established that roots are able to sense external NO_3^- availability (Krapp *et al.*, 2014). Leaf growth inhibition following NO_3^- deprivation is preceded by declining xylem concentrations of the mobile forms of cytokinin, zeatin and zeatin riboside. Conversely, the rapid stimulation of leaf growth by NO_3^- resupply is associated with a corresponding increase in root cytokinin biosynthesis and xylem translocation of zeatin and zeatin riboside (Rahayu *et al.*, 2005). In this context, a model for cytokinin as a long-distance signal to modulate shoot responses to root NO_3^- perception was proposed: following NO_3^- application, genes are upregulated in the roots to synthesize xylem-mobile cytokinins that are translocated to the shoot, where signal transduction pathways that regulate leaf expansion are activated (Sakakibara *et al.*, 2006). In addition, it has been hypothesized that the NO_3^- signal perceived in the shoot may block the shoot-to-root transport of indole-3-acetic acid, reducing lateral root growth (Forde, 2002a,b).

Several experiments have described changes in xylem NO_3 concentrations in response to environmental cues, suggesting that nitrate can play a long-distance signalling role in altering shoot responses. However, the physiological effects of these changes may be indirect, as NO_3^- fed through the transpiration stream can cause the leaf apoplast to become alkaline, which may then influence apoplastic ABA concentrations (Jia and Davies, 2007). However, soil drying either increased (*Commelina communis*) or decreased (*Solanum lycopersicum*) xylem NO_3^- concentration, concurrent with apoplastic alkalization (Jia and Davies, 2007), demonstrating that apoplastic alkalization could occur independently of changes in xylem NO_3^- concentration. Moreover, the negative relationship between NO_3^- concentration and xylem sap alkalization that preceded stomatal closure of flooded tomato (*S. lycopersicum*) plants suggested that NO_3^- was not regulating pH (Jackson *et al.*, 2003).

Rootstock characteristics can increase root NO_3^- absorption and translocation, and scion NO_3^- accumulation, thereby stimulating nitrate reductase and NO_3^- assimilation (Martínez-Ballesta *et al.*, 2010). For example, xylem sap NO_3^- concentrations were lower in tomato plants grafted on to commercial tomato rootstocks compared with ungrafted plants when grown hydroponically in aerated Hoagland nutrient solution under both control and salt-stress conditions (Fernández-García *et al.*, 2002).

Sulfate

Sulfate has been suggested to act as an ionic signal during drought, as its xylem sap concentration increased with progressive soil drying (Goodger *et al.*, 2005; Ernst *et al.*, 2010; Korovetska *et al.*, 2014). Moreover, sulfate-enhanced ABA-mediated decreases in stomatal aperture of bean (*Vicia faba*) epidermal peels (Ernst *et al.*, 2010) and the transpiration rate of detached maize (*Zea mays*; Ernst *et al.*, 2010) and hop (*Humulus lupulus*; Korovetska *et al.*, 2014) leaves. An increased xylem sulfate concentration of water-stressed plants was correlated with a threefold increase in the expression of a root sulfate transporter gene in maize roots, prior to any increase in expression of ABA biosynthesis genes (Ernst *et al.*, 2010). Thus, sulfate was the first detected chemical change in the xylem sap of maize roots after water stress and triggered stomatal closure in the presence of ABA (Ernst *et al.*, 2010). However, in hop plants, although increased xylem sulfate concentrations occurred at the same soil water contents as xylem sap alkalization, changes in xylem sap pH were sufficient to elicit stomatal closure in detached leaves, while changes in xylem sulfate concentration were not (Korovetska *et al.*, 2014). Thus, the temporal sequence of changes in xylem sap composition may vary between species, although this should be confirmed by withholding water from both species grown in the same substrate.

Sulfate may also be an ionic signal of salinity, and was transported from the roots to the leaves in greater amounts than chloride (Cl^-) when *Prosopis strombulifera* was watered with equimolar concentrations of either Na_2SO_4 or NaCl (Reinoso *et al.*, 2005). Although foliar ABA concentrations were higher in sulfate-treated plants, stomatal conductance remained high, and it was suggested that sulfate antagonized the ABA-induced stomatal closure (Llanes *et al.*, 2014), in contrast to the effects observed in the isolated systems outlined above. Further work is needed to understand the role of xylem sulfate as a potential anti-transpirant in salinized plants.

Rootstocks can substantially influence sulfate transport to the shoot. Although salt stress increased xylem sulfate concentration fourfold in ungrafted tomato plants, specific rootstocks could eliminate this increase entirely, although the effect on plant water use was not determined (Fernández-García *et al.*, 2002). Furthermore, compared with ungrafted plants, grafting pepper (*Capsicum annuum*) on to F_1 pepper hybrids reduced sulfate uptake and its transport to the shoot under both NaCl and sulfate stresses (Giuffrida *et al.*, 2013). In aubergine (*Solanum melongena*) the sulfate uptake increased when grafted on to tomato and interspecific hybrid rootstocks (Leonardi and Giuffrida, 2006; Giuffrida *et al.*, 2015) and was decreased when grafted on to *Solanum torvum* rootstocks (Cassaniti *et al.*, 2011). *S. torvum* rootstocks maintained sulfate transport to aubergine leaves at the same level as non-salinized plants under moderate (20.5 mM) Na_2SO_4 stress, unlike *S. lycopersicum* × *Solanum habrochaites* (Beaufort and He-Man) rootstocks, which are commonly used commercially (Giuffrida *et al.*, 2015). In contrast, grafting did not influence sulfate uptake under optimal conditions in tomato (Leonardi and Giuffrida, 2006) and melon (*Cucumis melo*; Ruiz *et al.*, 1997), nor under Na_2SO_4 stress in tomato (Giuffrida *et al.*, 2014) and cucumber (*Cucumis sativus*; Colla *et al.*, 2012). Clearly, rootstocks can significantly modify sulfate uptake and xylem transport, although the physiological effects of these changes may be difficult to separate from other ionic changes in xylem sap composition.

Role of K^+ and Na^+/Cl^- export in salt stress

The plant's response to salinity depends to some extent on the ability to maintain (via root exclusion, phloem recirculation, ion compartmentalization in vacuoles and/or dilution by growth) the rate of accumulation of toxic ions (Na^+/Cl^-) as close as possible to a physiological homeostatic capacity (Sacher *et al.*, 1982). These properties can also be transferred to a more saline-sensitive cultivar by using tolerant genotypes as rootstocks (Martínez-Rodríguez *et al.*, 2008). Although maintenance of ionic homeostasis is necessary for plant survival, generally tissue ion concentration is not a universal indicator of salt tolerance (Villalta *et al.*, 2008; Rajendran *et al.*, 2009).

While much research has focused on tissue ion accumulation, several studies have tested whether xylem sodium ion (Na^+) concentrations can explain rootstock-mediated variation in salt tolerance. Leaf xylem Na^+ concentration in grafted tomato does not seem to be a good indicator of salt tolerance, because it was not correlated with plant production (Albacete *et al.*, 2009). Nevertheless, rootstock effects on leaf growth were evident, while the xylem Na^+ concentration was five times less than that in the external solution (Albacete *et al.*, 2009), suggesting that a root Na^+ exclusion mechanism was still functional (Pérez-Alfocea *et al.*, 1996). Similarly, the lower xylem Na^+ concentrations of cucumber plants grafted on to pumpkin (*Cucurbita pepo*) rootstocks compared with self-grafted cucumber was attributed to the lower Na^+ absorption capacity of pumpkin roots than cucumber. This species difference is probably due to altered Na^+ influx into the roots and Na^+ efflux into the soil solution (Huang *et al.*, 2013). As rootstock mediation of salinity responses has also been correlated with phytohormonal effects (Albacete *et al.*, 2009; Ghanem *et al.*, 2011a), comparing xylem phytohormone concentrations at the same xylem Na^+ concentrations will be instructive to distinguish the mechanism(s) by which rootstocks affect salt tolerance.

Instantaneous measurements of xylem sap composition may not always be related to the capacity of plants to maintain Na^+ exclusion over time, irrespective of their initial capacity (Pérez-Alfocea *et al.*, 1993; Martínez-Rodríguez *et al.*, 2008). In tomato, less productive rootstocks elicited the highest increases in xylem Na^+ concentrations after 100 days of salt treatment, while the xylem Na^+ concentration remained unchanged in the tolerant combinations. In contrast, both xylem K^+ concentration and the K^+/Na^+ ratio were positively correlated with leaf biomass across all rootstocks (Albacete *et al.*, 2009). However, xylem K^+ and Na^+ concentrations were not correlated, suggesting that these ions use different transport mechanisms (Malagoli *et al.*, 2008) and that the root capacity to maintain K^+ transport under salt stress is a better indicator of salt tolerance than tissue concentrations.

Calcium

Calcium ion (Ca^+) uptake into the root is primarily via the mass flow of soil water driven by transpiration (McLaughlin and Wimmer, 1999). Apoplastic movement of Ca^{2+} through the root was classically thought to be impeded by the presence of the Casparian band (Clarkson, 1984) where solutes would be forced to move symplastically to bypass this impermeable barrier, thereby regulating delivery of Ca^{2+} to the xylem. However, there is strong evidence for the existence of both apoplastic (regulated by transpirational flux) and symplastic (regulated by active transport) pathways regulating calcium transport to the shoot (White, 2001).

An elevated rhizospheric Ca^{2+} concentration can increase the xylem sap Ca^{2+} concentration. Increasing the nutrient solution Ca^{2+} concentration from 1 to 8 mM increased the root xylem sap Ca^{2+} concentration (from 0.75 to 2.32 mM) in *Commelina* (Atkinson *et al.*, 1990). If delivered to the apoplast in the vicinity of the stomata, this elevated Ca^{2+} level might interfere with normal stomatal function. Indeed, *Commelina* grown in a high-calcium (15 mM) nutrient solution exhibited a 15% reduction in stomatal conductance after 13 days when compared with plants grown at 4 mM calcium (Ruiz *et al.*, 1993). While elevated rhizospheric calcium concentration can decrease stomatal conductance, few studies have attempted to establish a role for increased xylem Ca^{2+} concentrations in altering physiological responses under other abiotic stresses.

Nevertheless, it has been hypothesized that root-to-shoot Ca^{2+} delivery (and its regulation by the rootstock) under various environmental stresses can influence shoot responses and fruit quality since xylem Ca^{2+} supply has been associated with blossom end rot (Guichard *et al.*, 2001). In grafted tomato grown at 75 mM NaCl, the rootstock genotype resulted in a threefold variation in xylem Ca^{2+} concentrations (Fernández-García *et al.*, 2011). Although rootstock genotype can profoundly affect the incidence of blossom end rot (Donas-Ucles *et al.*, 2014), rootstock-mediated effects on blossom end rot have not been linked unequivocally to the xylem Ca^{2+} supply (Ho *et al.*, 1993).

4.2.2 Plant hormone signalling

ABA

For many years, it was thought that soil drying decreased stomatal conductance (g_s) by directly decreasing guard-cell turgor, thereby closing the stomata (Pierce and Raschke, 1981). However, more detailed observations, especially of plants grown in the field where soil water deficits slowly develop over weeks, demonstrated that plants exposed to drying soils had both a higher leaf water potential (Ψ_{leaf}) and lower g_s than well watered plants. Alternatively, it was suggested that enhanced root ABA biosynthesis and/or increased xylem loading of ABA in the root system altered root-to-shoot chemical signalling that caused stomatal closure (reviewed by Dodd, 2005). Several lines of evidence have pointed to long-distance ABA transport being instrumental in regulating g_s. Some experiments have demonstrated that increased root and xylem ABA concentration is temporally associated with stomatal closure and occurs prior to any increase in leaf ABA concentration or decrease in Ψ_{leaf}. Moreover, the correlation between g_s and xylem ABA concentration in plants grown in drying soil was identical to the relationship generated in well watered plants that were fed synthetic ABA via the stem, or in detached leaves supplied with ABA via the transpiration stream. Removing the ABA from maize xylem sap using an immunoaffinity column eliminated its anti-transpirant activity in some species (Zhang and Davies, 1991). Taken together, these experiments mount a persuasive case for ABA initiating stomatal closure of plants exposed to drying soil and other abiotic stresses (Rothwell *et al.*, 2015).

Several lines of evidence have now questioned the theory that ABA-mediated root-to-shoot signalling closes the stomata. First, imposing water stress around the roots (by bathing the roots in an osmoticum) decreases leaf turgor and causes an initial ABA accumulation in the leaves, a response that can be reversed by spraying the leaves with water (Christmann *et al.*, 2007). Secondly, guard-cell ABA biosynthesis can be necessary and sufficient to induce stomatal closure in response to atmospheric water deficits (Bauer *et al.*, 2013) and perhaps also in response to soil water deficits. Lastly, grafted tomato plants comprising a wild-type (WT; ABA-sufficient) shoot and an ABA-deficient rootstock show the same stomatal response to drying soil as self-grafted WT plants (Holbrook *et al.*, 2002), indicating that stomatal closure in response to drying soil is independent of root-supplied ABA.

Nevertheless, there are experiments in which rootstock ABA status has been associated with altered shoot physiological responses. Constitutively elevated ABA production in the autotetraploid rootstock clone of Rangpur lime (*Citrus limonia*) increased leaf ABA concentrations and tolerance to water stress in sweet orange (*Citrus sinensis*) scions by limiting both shoot gas exchange and growth under control conditions, and by inducing expression of drought-responsive genes, compared with the diploid rootstocks (Allario *et al.*, 2013). Augmenting root-supplied ABA by grafting WT scions on to the ABA overexpressing tomato transgenics sp5 and sp12 (Thompson *et al.*, 2007) substantially increased the fruit yield of plants supplied with saline (3 dS m^{-1}) irrigation water compared with self-grafted WT plants (Albacete *et al.*, 2015). However, it is uncertain whether these physiological responses can be attributed directly to altered root-to-shoot ABA signalling (as this was not measured) or to changes in shoot/fruit ABA biosynthesis. The latter explanation is possible because concentration-dependent effects of ABA status on root hydraulic conductivity (Dodd, 2013) may alter shoot water relations and hence ABA biosynthesis (McAdam *et al.*, 2016).

1-Aminocyclopropane-1-carboxylic acid

1-Aminocyclopropane-1-carboxylic acid (ACC), the soluble precursor of the plant hormone ethylene, may act as a root-to-shoot signal of rhizosphere conditions, assuming that enhanced ACC delivery from the roots stimulates foliar ethylene biosynthesis, thereby affecting shoot growth and development. However, ACC may be conjugated in the shoot (van de Poel and van der Straeten, 2014) and deaminated to form α-ketobutyrate (McDonnell *et al.*, 2009), and its conversion to ethylene can also be modulated by ACC oxidase activity (English *et al.*, 1996). While there is increasing evidence that ACC can be biologically active independent of its conversion to ethylene (van de Poel and van der Straeten, 2014), the various metabolic pathways for ACC *in planta* means that there may be no simple relationship between root-to-shoot ACC delivery and foliar ethylene biosynthesis. Furthermore, shoot expression of ACC synthase genes (Rodriguespousada *et al.*, 1993) makes the relative contribution of the rootstock to shoot ethylene relations difficult to distinguish.

Following soil flooding, changes in rhizosphere conditions (e.g. soil oxygen status) can be correlated with different measurements of plant ethylene status and physiological responses. Although early work showed that the concentration of the ethylene precursor ACC increased in the xylem sap of flooded tomato plants (Bradford and Yang, 1980), petiole ethylene evolution lagged behind enhanced root-to-shoot ACC delivery and leaf epinasty (Else and Jackson, 1998). Additional studies are required to determine why flooding stimulates foliar ethylene biosynthesis. Prolonged over-irrigation of tomato induced significant foliar nitrogen deficiency and increased ethylene production, but daily supplementation of small volumes of 10 mM $Ca(NO_3)_2$ to over-irrigated soil restored foliar nitrogen concentrations and ethylene emission to the levels of well-drained plants (Fiebig and Dodd, 2016). Thus, plant nitrogen status (and not soil oxygen status) seemed to be the principal cause of enhanced ethylene evolution. However, identifying the source of the additional ethylene (localized shoot biosynthesis and/or increased root export of ACC) in over-irrigated plants requires further work involving reciprocal grafting experiments with WT plants and those transformed with an antisense gene for ACC synthase (Oeller *et al.*, 1991), along with measurements of xylem ACC concentration.

More recent work has aimed to understand the regulation of xylem ACC transport in response to other abiotic stresses, such as soil drying and salinity. Transplanting bare-rooted mandarin seedlings to dry sand for 24 h decreased Ψ_{leaf} to −3.0 MPa and increased xylem ACC concentration (Tudela and Primo-Millo, 1992). Imposing a more moderate (soil matric potential did not decline below −0.25 MPa), relatively stable soil water status on pea (*Pisum sativum*) roughly doubled the xylem ACC concentration (Belimov *et al.*, 2009). Whether rootstocks can modulate ACC transport (and hence leaf ethylene evolution) as the soil dries has not yet been established, but the rootstock caused significant (twofold) variation in xylem ACC concentration when tomato plants were exposed to 75 mM NaCl (Albacete *et al.*, 2009).

Jasmonic acid

Jasmonic acid (JA) and its derivatives, referred to as jasmonates (JAs), belong to the family of oxygenated fatty acid derivatives collectively termed oxylipins. JAs act as key signalling compounds in wounding and defence against herbivores and necrotrophic pathogens (Howe *et al.*, 1996; Ballaré, 2011; Campos *et al.*, 2014; Larrieu *et al.*, 2015) and several growth and developmental processes (reviewed by Creelman and Mullet, 1995; Wasternack and Hause, 2013; Wasternack, 2014). Their role in signalling of several abiotic stresses has only recently been recognized (Dar *et al.*, 2015; Kazan, 2015; de Ollas and Dodd, 2016). Moreover, JAs are involved in regulating beneficial root–microbe interactions such as arbuscular mycorrhizal fungi and plant-growth-promoting rhizobacteria (Mabood and Smith, 2005; Wasternack, 2014) and re-allocation of primary metabolites between roots and shoots (Gómez *et al.*, 2010).

JAs are synthesized in plastids and peroxisomes in leaves and flowers by means of the octadecanoid pathway (Fig. 4.1). JA itself can be further metabolized into biologically active or inactive derivatives. Methyl-jasmonate (MeJA) is a volatile compound that is potentially responsible for interplant communication.

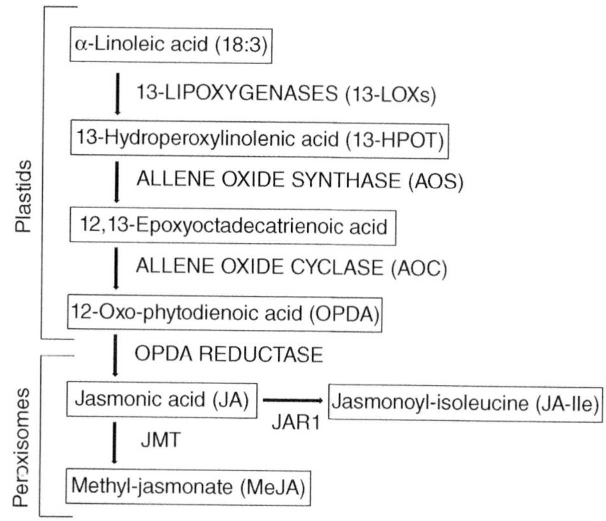

Fig. 4.1. Simplified overview of the biosynthesis of JAs in the octadecanoic pathway. α-Linolenic acid is oxygenated by lipoxygenases (LOXs) to form 13-hydroperoxylinolenic acid (13-HPOT), which is subsequently converted by allene oxide synthase (AOS) and allene oxide cyclase (AOC) to 12-oxo-phytodienoic acid (OPDA). Jasmonic acid (JA) is synthesized from OPDA through reduction and three steps of β-oxidation, and may then be further converted to, for example, methyl jasmonate (MeJA) by jasmonic carboxyl methyl transferase (JMT) or to jasmonoyl-isoleucine (JA-Ile) by the adenylate-forming enzyme Jasmonate resistant1 (JAR1) (Adapted from Wasternack, 2007).

JA conjugated with the amino acid isoleucine (JA-Ile) is currently the only known JA derivative needed for JA signalling due to its interaction with COI1 (JA-dependent receptor). 12 Oxo phytodienoic acid (OPDA) is also biologically active as a regulator of stomatal closure (Savchenko and Dehesh, 2014; Savchenko *et al.*, 2014), although further research is needed to untangle a possible COI1-independent signalling branch controlled by OPDA (Wasternack *et al.*, 2012).

In contrast to leaves and flowers, little is known about the biosynthesis of JAs in roots. As the expression of some of the plastid-localized enzymes of JA synthesis is very low (Zimmermann *et al.*, 2004), the capacity of roots to synthesize JAs was questioned. To answer this question, Grebner *et al.* (2013) conducted reciprocal grafting experiments with the JA-deficient *Arabidopsis* mutant *dde2*, with low expression of the JA biosynthesis enzyme allene oxide synthase. The results demonstrated that roots can produce JAs independently of leaves, without requiring a shoot with JA-biosynthetic competence.

Levels of JAs in leaves and roots may increase rapidly (within 30 min) and sometimes transiently in response to wounding, herbivores and pathogens (Koo *et al.*, 2009; Hasegawa *et al.*, 2011; Chauvin *et al.*, 2013; Grebner *et al.*, 2013; Wasternack and Hause, 2013; Bosch *et al.*, 2014; Fragoso *et al.*, 2014; Savchenko *et al.*, 2014) and several abiotic stress stimuli (de Ollas *et al.*, 2013; Du *et al.*, 2013; Hu *et al.*, 2013; Zhang and Huang, 2013; Savchenko *et al.*, 2014; de Ollas *et al.*, 2015). Moreover, exogenously applied MeJA enhanced resistance to biotic (Turner *et al.*, 2002) and abiotic (Lee *et al.*, 1996; Li *et al.*, 2012; Hu *et al.*, 2013) stresses, further demonstrating the role of JAs in plant signalling responses.

However, there is still little known about JA functions in roots and the role of root-derived JAs in orchestrating above-ground processes (Erb *et al.*, 2009; Hasegawa *et al.*, 2011; Grebner *et al.*, 2013; Fragoso *et al.*, 2014). The following sections review the involvement of JAs in the systemic signalling of wounding, drought and salt stress.

INVOLVEMENT OF JAS IN LONG-DISTANCE WOUND SIGNALLING Wounding of roots or leaves, induced by either mechanical damage or herbivore attack, elicits both local (Grebner *et al.*, 2013) and systemic (Koo *et al.*, 2009; Chauvin *et al.*, 2013) production of JAs. Wounding *Arabidopsis* roots elicited a JA burst in the leaves within 30 min (Hasegawa *et al.*, 2011), and wounding tobacco leaves resulted in a transient increase of JA and JA-Ile within 3 h in the roots (Fragoso *et al.*, 2014). An important question is whether this systemic JA accumulation results from *de novo* synthesis in undamaged distal tissue or from JAs transport from the wounded tissue (Erb *et al.*, 2009; Tytgat *et al.*, 2013).

Reciprocal grafting experiments with tomato mutants defective in JA biosynthesis (*spr2*) or JA perception (*jai1*) revealed that systemic signalling requires JA synthesis at the site of wounding, as well as JA perception in the distal tissues (Li *et al.*, 2002). Thus, it was proposed that the graft-transmissible signal might be a member of the JAs family of oxylipins, or a downstream JA-activated signalling component (Stratmann, 2003). To test the mobility of JAs, experiments with deuterium-labelled and radiolabelled JAs were conducted (Zhang and Baldwin, 1997; Matsuura *et al.*, 2012; Tamogami *et al.*, 2012). They indicated that JA, MeJA and JA-Ile could be transported via the vasculature. Recently, the use of a JA-inducible reporter (JAZ10p:GUS) in combination with reciprocal grafting experiments (in which WT scions were grafted on to *aos* JA biosynthesis mutant rootstocks) demonstrated that JA and/or its precursors might be translocated from wounded aerial tissues to the unwounded roots where they activated *JAZ10* expression (Gasperini *et al.*, 2015). This finding supports experiments in which a fluorescent JA perception biosensor with the Jas motif of the *AtJAZ9* gene (termed Jas9–VENUS) was developed in *Arabidopsis* (Larrieu *et al.*, 2015). These plants revealed that a two-step JA wound-signalling response is initiated in the root after wounding of cotyledons: a rapid minor phase and a slower major phase. However, which endogenous members of the JAs family of oxylipins act as long-distance mobile signals remains to be elucidated.

Four 13-lipoxygenases (13-LOXs) are involved in catalysing the first step in lipid oxidation leading to JA synthesis. To analyse the contribution of the different 13-LOXs on oxylipin production in wounded leaves and roots, *Arabidopsis* mutants defective in the expression of 13-LOXs were used. LOX6 was required for the fast increase in JA and JA-Ile in the distal leaves after wounding (Chauvin *et al.*, 2013) and for the synthesis of JAs in wounded roots (Grebner *et al.*, 2013). Thus, oxylipin products of LOX6 seem to be good candidates to be involved in root-to-shoot communication of wounding (Hasegawa *et al.*, 2011; Grebner *et al.*, 2013).

In contrast to the results of Grebner *et al.* (2013), Hasegawa *et al.* (2011) reported no increase in JA and OPDA levels in wounded *Arabidopsis* roots. It was suggested that root wounding induced alternative forms of systemic communication, such as hydraulic signals, which elicited systemic production of JA and OPDA in the shoot. Hydraulic signals consist of hydrostatic pressure changes in the xylem and/or changes in the transmembrane ion flux in the phloem vasculature, followed by slower secondary signals that finally lead to JA synthesis and JA-mediated responses (Stahlberg and Cosgrove, 1995; Herde *et al.*, 1996; Hasegawa *et al.*, 2011; Tytgat *et al.*, 2013; Salvador-Recatalà *et al.*, 2014). A model proposed that wounding rapidly propagated axial pressure changes along xylem vessels (hydraulic signalling), which are then converted to slower radial pressure changes that squeeze xylem contact cells (Farmer *et al.*, 2014). This cellular squeezing subsequently modulates the activity of glutamine receptor-like proteins, leading to ion fluxes (Ca^{2+}) and to JA accumulation in and beyond contact cells, by activating LOX and OPDA release (*see* Plate 7). These xylem contact cells probably correspond to the niche where LOX is necessary for rapid JA accumulation (e.g. through *LOX6* expression) at sites distal to wounds (Chauvin *et al.*, 2013). The model presented in Plate 7 shares similarities with hydraulic signalling of water stress, the systemic signalling in pathogenesis and the rapid minor wound-signalling response initiated in the root after damaging of cotyledons (Farmer *et al.*, 2014; Larrieu *et al.*, 2015).

INVOLVEMENT OF JAS IN DROUGHT-STRESS SIGNALLING In *Arabidopsis* leaves, shoot desiccation (rosettes were air dried) decreased JA, JA-Ile and OPDA concentrations within 6 h (de Ollas *et al.*, 2015). Withholding water for 5 days resulted in a fourfold accumulation of the JA precursor OPDA in *Arabidopsis* leaves, whereas the conversion of OPDA to JA seemed to be uncoupled from drought-stress signals (Savchenko *et al.*, 2014). While these biochemical differences in JA accumulation are probably due to differences in tissue water relations, the latter study indicated that OPDA induced stomatal closure independently and cooperatively with ABA.

In roots dehydrated in air or in dried perlite, drought stress transiently increased JA, JA-Ile and OPDA levels within 12 h (de Ollas *et al.*, 2013, 2015; Grebner *et al.*, 2013). Lack of the enzyme LOX6, one of the four 13-LOXs required for OPDA synthesis, enhanced drought sensitivity in *Arabidopsis* (Grebner *et al.*, 2013). By contrast, the *Arabidopsis dde2* mutant, unable to produce JAs due to a defect in allene oxidase synthase, did not show increased sensitivity to drought in the same study. This result suggested that oxylipins different from JAs are important in drought tolerance. Reciprocal grafting experiments using the *lox6* mutant of *Arabidopsis* would be useful to elucidate the possible role of LOX6-derived oxylipins (such as OPDA) as root-to-shoot signals that promote stomatal closure (Savchenk and Dehesh, 2014; Kazan, 2015).

To test the hypothesis that JAs were involved in the biosynthesis of ABA, *Arabidopsis* mutants impaired in JAs synthesis (*opr3*, *lox6* and *jar1-1*) and in JA-dependent signalling (*coi1*) were analysed for their root hormone profiles under drought stress (de Ollas *et al.*, 2015). It was found that early JA accumulation leading to JA-Ile build-up and JA signalling were both necessary for full induction of ABA biosynthesis in dehydrated roots. JAs thus interact with ABA, which may also act as a root-to-shoot signal under drought stress (as described above).

Apart from the role JAs play in stomatal behaviour, they may also increase root hydraulic conductivity (L) (Lee *et al.*, 1996). Thus, exogenous application of 100 μM MeJA to bean, tomato and *Arabidopsis* roots increased L by 27–50% after 24 h (Sánchez-Romera *et al.*, 2014). The modulation of L by MeJA could be an important mechanism to improve the water status of plants under limited water availability.

INVOLVEMENT OF JAS IN SALT-STRESS SIGNALLING Salt stress (200 mM NaCl for 15 days) increased root and decreased leaf JA concentrations, and increased the MeJA/JA ratio in tomato (Garcia-Abellan *et al.*, 2015). Intraspecific differences in salt tolerance in tomato correlated positively with the basal JA concentration in both root and shoot (Pedranzani *et al.*, 2003). Moreover, MeJA levels increased in the roots of the salt-sensitive cultivar but not in the salt-tolerant one following salinization. Many studies have reported that the exogenous application of JAs to either leaves or roots alleviated the effects of salt stress (Dar *et al.*, 2015), which was correlated with increased antioxidant enzyme activity, polyamine and lignin accumulation, and changes in cell membrane and cell wall composition (Lee *et al.*, 1996; Shi *et al.*, 2011; Zhao *et al.*, 2013, and references therein).

Recently, a tomato mutant (*res*: restored cell structure by salinity) was shown to accumulate JA specifically in the root but not in the leaf under non-saline conditions (Garcia-Abellan *et al.*, 2015). This mutant exhibited a dwarf phenotype with very few and short roots, in which pericycle and stele cells were disorganized. Remarkably, also in the non-JA-accumulating chlorotic *res* leaves, leaf epidermal and palisade parenchyma cells, and their chloroplast ultrastructure, were disorganized. Reciprocal grafting between the *res* mutant and WT tomatoes showed that the *res* rootstock inhibited WT scion growth, while growth of the *res* scion was stimulated by the WT rootstock. These results suggest that the growth inhibition and cellular alterations observed in the *res* shoot were root based. The observation of lower leaf JA and MeJA concentrations in *res* than in WT (non-grafted) plants under non-saline conditions implies that the *res* root inhibits shoot growth via mechanisms other than increased root-to-shoot transport of JA or MeJA. Moreover, these findings suggest that roots synthesize JA independently of the shoot, as in *Arabidopsis* (Grebner *et al.*, 2013) and *Brassica oleracea* L. (Tytgat *et al.*, 2013). Interestingly, growing the *res* mutant under saline conditions restored root and shoot growth (Garcia-Abellan *et al.*, 2015). Salt stress barely affected the JA and MeJA concentrations in *res*, whereas root JA concentration and the leaf MeJA/JA ratio increased strongly in WT plants.

The high salt tolerance exhibited by *res* correlated with increased root-to-shoot transport of K^+ and maintenance of low Na^+/K^+ ratios, a key trait of salt tolerance, as discussed above. The hypothesis that root JA signalling affects salt tolerance by regulating Na^+/K^+ homeostasis was supported by recent data from transgenic rice lines with either suppression or overexpression of *OsJAZ9*, functioning as a transcriptional regulator in JA signalling (Wu *et al.*, 2015). *OsJAZ9*, and thus JA signalling, was involved in salt tolerance by regulating the expression of ion transporters for K^+ homeostasis. Additional reciprocal grafting experiments with the *res* mutant (with a constitutively high root JA concentration) and/or the JA-deficient mutant *JL5* (Howe *et al.*, 1996), also designated defenseless

1 (*def1*), may support the role of root signalling in salt tolerance in tomato by regulating the expression of ion transporters for K^+ homeostasis.

Salicylic acid

Salicylic acid (SA) is a small, ubiquitous, phenolic phytohormone known mainly for its function as an endogenous signal mediating local and systemic plant defence responses against pathogens (Janda and Ruelland, 2015). SA can also mediate plant responses to abiotic stresses, such as drought, temperature (cold and heat), heavy-metal toxicity, UV-B radiation, ozone, salt and osmotic stress (reviewed by Miura and Tada, 2014; Jayakannan *et al.*, 2015a; Khan *et al.*, 2015).

SA acts as a signalling molecule in plant defence against microbial pathogens. Following pathogen detection, SA biosynthesis is essential for local and systemic acquired resistance, as well as the accumulation of pathogenesis-related proteins. Systemic acquired resistance (SAR) is a long-distance signalling mechanism by which plants gain a long-lasting resistance against a broad spectrum of secondary pathogen infections at sites distant from that of the initial pathogen attack. The nature of the mobile signal that travels through the phloem from the site of infection to establish systemic immunity has been sought for decades (Vlot *et al.*, 2008). Initially, SA was postulated to be this mobile signal because it: (i) induces defence responses when applied to plants; (ii) is found in phloem exudates of infected leaves; and (iii) is required in systemic tissues for SAR (Malamy *et al.*, 1990; Métraux *et al.*, 1990).

The most compelling evidence that SA plays a role in the signal transduction pathway leading to SAR was demonstrated by the use of transformed tobacco and *Arabidopsis* plants expressing a bacterial salicylate dehydroxylase gene (*NahG*) that degrades SA into catechol (Gaffney *et al.*, 1993). These SA-deficient transgenic plants are unable to manifest an SAR response following pathogen infection. However, grafting experiments demonstrated that infected transgenic *NahG* tobacco rootstocks could trigger SAR in WT scions, despite their inability to accumulate SA after infection with tobacco mosaic virus (Vernooij *et al.*, 1994). These experiments excluded SA acting as a mobile signal for the systemic induction of SAR. It took another 13 years before Park *et al.* (2007) elucidated that the SA-derivative methyl salicylate (MeSA) accumulates in *NahG* transgenic tobacco and acts as a long-distance mobile signal for SAR. MeSA itself is not biologically active, but its hydrolysis to SA by the MeSA esterase activity of SA binding protein 2 (SABP2) in the systemic tissue triggers SAR. Grafting studies with tobacco plants silenced for SABP2 revealed that SABP2 activity in scions, but not in rootstocks, is required for normal SAR (Park *et al.*, 2007). MeSA can also serve as an airborne plant-to-plant defence signal that is emitted from infected plants. It can pass through the cuticular layer and, after conversion into SA, induces expression of defence genes in neighbouring plants (Shulaev *et al.*, 1997). Besides MeSA, many other compounds are now proposed to serve as mobile phloem SAR signals (reviewed by Gao *et al.*, 2015).

Numerous studies have indicated that appropriate concentrations of SA, through seed soaking, addition to hydroponic culture/irrigation or foliar spraying, may enhance the biotic and abiotic stress tolerance of vegetable crops (Horváth *et al.*, 2007; Miura and Tada, 2014; Khan *et al.*, 2015, and references therein).

Generally, low concentrations or transient application of SA promote, whereas high concentrations (>1 mM) or continual application decrease the biotic and abiotic stress tolerance of plants. SA inhibits peroxidases, catalase and carbonic anhydrase, which results in the production of reactive oxygen species (ROS). Low ROS levels act as secondary signal molecules to enhance the activities of cellular protective ROS-scavenging enzymes, whereas high concentrations of ROS cause oxidative stress leading to cell death and decreased abiotic stress tolerance (Miura and Tada, 2014). Besides activating antioxidant systems, the proposed mechanisms underlying SA-mediated abiotic stress tolerance are through modulating osmolyte synthesis, producing secondary metabolites, optimizing mineral nutrient status, signalling cross-talk with other phytohormones and stimulating endophyte colonization (reviewed by Khan *et al.*, 2015). Therefore, it has been suggested that SA has great agronomic potential to improve crop production under abiotic stress conditions.

It is therefore remarkable that hardly any grafting studies have examined the impact of rootstocks with altered SA concentration on scion performance under abiotic stress. Many *Arabidopsis* genotypes with altered SA levels are described (e.g. Rivas-San Vicente and Plasencia, 2011), yet to our knowledge they were not used in grafting studies aimed at elucidating the role of SA in root-to-shoot signalling under abiotic stress. In tomato, an SA-deficient transgenic line that constitutively expresses the bacterial *NahG* gene (Brading *et al.*, 2000) was used in reciprocal grafting experiments to analyse the contribution of SA in suboptimal temperature (17/14°C) tolerance (Ntatsi *et al.*, 2013). In Arabidopsis, *NahG* and other mutants with impaired SA signalling (*npr1*) or levels (*eds5* and *cpr1*) indicated that accumulation or perception of SA inhibits growth at chilling temperature (Scott *et al.*, 2004). The faster relative growth rates of the *NahG Arabidopsis* mutant at 4°C resulted primarily from enhanced cell expansion, rather than continuous cell division (Scott *et al.*, 2004). The enlarged cell size of *NahG* is associated with an extensive endoreduplication, which correlated well with stimulation of the expression of a D-type cyclin, CYCD3;1 (Xia *et al.*, 2009). The transcription level of CYCD3;1 is a key regulator for the G_1-to-S-phase transition and associated with endoreduplication and cell size in *Arabidopsis* (Menges *et al.*, 2006). In contrast to the *Arabidopsis NahG* mutant at 4°C, grafted SA-deficient *NahG* tomato was not more tolerant than WT to suboptimal temperature when self-grafted or used as either the scion or the rootstock (Ntatsi *et al.*, 2013). Regardless of the rootstock genotype, SA-deficient *NahG* tomato scions showed stomatal limitation of photosynthesis and, although only at suboptimal temperature, decreased chlorophyll contents. Thus, in tomato there is an optimal shoot SA level essential for stomatal opening and preventing chlorophyll degradation at a suboptimal temperature. The *NahG* tomato mutant is actually not very useful to examine the role of SA in abiotic stress signalling, as SA levels in the *NahG* scion cannot be complemented by grafting on to a WT rootstock due to degradation by the bacterial NahG enzyme. Therefore, producing tomato mutants with altered SA levels or signalling would be desirable, or grafting studies should be performed with already available SA mutants in *Arabidopsis*. The recent finding that the NPR1 (non-expressor of pathogenesis-related gene 1)-dependent SA signalling pathway is a crucial component of salt and oxidative stress tolerance in *Arabidopsis* encourages such grafting studies to further examine the role of SA in root–shoot signalling during salt and oxidative stress (Jayakannan *et al.*, 2015b).

Cytokinins

Many different environmental stresses reduce cytokinin biosynthesis in plants, suggesting that low endogenous foliar cytokinin levels may limit crop yields. While cytokinins can be synthesized throughout the plant, as indicated by the expression patterns of isopentenyltransferase (*ipt*) genes (Hirose *et al.*, 2008), much attention has focused on their role as root-sourced signals of soil conditions. For example, when plants were grafted on to rootstocks from a recombinant inbred line population derived from a *S. lycopersicum* × *Solanum cheesmaniae* cross, elevated xylem cytokinin concentrations were associated with maintenance of the photosynthetic apparatus under salt stress (Albacete *et al.*, 2009).

However, cytokinins might have negative effects on root development, as root-specific expression of cytokinin-degrading cytokinin oxidase/dehydrogenase genes enhanced root growth and plant resistance to drought and to a low nutrient supply (Werner *et al.*, 2010). As cytokinin concentrations correlate positively with shoot growth (Werner *et al.*, 2008) but negatively with root development (Werner *et al.*, 2010), genotypes with enhanced root cytokinin export to shoots should have maximal whole-plant growth rates. The recently discovered cytokinin transporter that loads cytokinins from the roots into the xylem vessels (Ko *et al.*, 2014), thereby altering cytokinin distribution between roots and shoots, thus becomes an attractive research (and perhaps crop improvement) target. However, under stress conditions, high root cytokinin biosynthesis may also improve plant growth. For example, grafting WT tomato scions on to a transgenic *ipt* rootstock that constitutively upregulated cytokinin production increased fruit cytokinin concentrations, and under salt stress (75 mM NaCl) increased tomato fruit yield by 30% compared with WT self-grafts (Ghanem *et al.*, 2011a). This crop improvement strategy of enhancing root cytokinin biosynthesis may only work when root systems are constrained (as often occurs in greenhouse horticulture), as cytokinins inhibit root development (Werner *et al.*, 2010), and increased root growth has often been associated with increased yield under water deficit in the field (Manschadi *et al.*, 2006). Alternatively, open-field crops may benefit from transformation with cytokinin oxidase/dehydrogenase genes under the control of root-specific promoters (Vojta *et al.*, 2016) or grafting on to transgenic rootstocks that downregulate cytokinin production (e.g. constitutive expression of cytokinin oxidase/dehydrogenase genes), thereby allowing greater root proliferation and thus better nutrient/water access.

Gibberellins

Although several gibberellic acids (GAs) have been identified in xylem sap, and GA-deficient mutants exist in easily grafted species such as pea and tomato, relatively few studies have evaluated the role of long-distance GA signalling in regulating physiological responses. Recently, micrografting experiments in *Arabidopsis* revealed that the GA precursor GA_{12} is xylem mobile and triggers GA signalling in the shoot (Regnault *et al.* 2015). The ability of a WT rootstock to restore stem elongation of a GA-deficient scion depended on the position of the lesion in the GA biosynthesis pathway (reviewed by Dodd, 2005). As decreased GA biosynthesis is often associated with environmental stresses (Colebrook *et al.*, 2014), it would be interesting to determine whether the GA status of the rootstock altered scion responses (e.g. leaf expansion and stomatal conductance) to these stresses.

Strigolactones

Strigolactones (SLs) are important in regulating shoot branching and serve as molecular signals for interaction with other organisms, such as parasitic plants and arbuscular mycorrhizae (AM) (Matusova *et al.*, 2005; Besserer *et al.*, 2006; Foo and Davies, 2011). SLs are synthesized in the roots and shoots and move from roots to shoots but not vice versa, as confirmed by grafting experiments with pea, *Arabidopsis* and petunia (Foo *et al.*, 2001; Morris *et al.*, 2001; Turnbull *et al.*, 2002; Booker *et al.*, 2005; Simons *et al.*, 2007). SLs synthesized in the roots can be actively extruded from the roots to the soil by the specific transporter PLEIOTROPIC DRUG RESISTANCE 1 (PDR1) protein (Kretzschmar *et al.*, 2012) to stimulate AM colonization (Akiyama *et al.*, 2005). AM colonization is stimulated under phosphate deficiency due to increasing SL biosynthesis in the roots (Yoneyama *et al.*, 2007). However, it was also found that SLs are not required for regulation of AM and rhizobial symbioses in response to nutrient deficiency (Foo *et al.*, 2013). High exogenous phosphate supply reduced the expression of SL biosynthetic genes in petunia. However, inhibition of AM development was not exclusively due to low SL content, as applying the synthetic SL GR24 did not alleviate the repression of AM by phosphate (Breuillin *et al.*, 2010). The fact that increased SL biosynthesis in the roots is positively correlated with inhibition of tiller-bud outgrowth under phosphate deficiency suggests that SLs coordinate both phosphate acquisition efficiency through symbiosis with AM and phosphate utilization efficiency by regulating shoot branching (Umehara *et al.*, 2010). High variability in SL composition among plants and the different susceptibility of parasitic weeds and AM fungi to these different SLs opens up plant-breeding possibilities of combining suppression of root infection by parasitic plants while building symbioses with AM, which might be efficient for phosphate acquisition (Yoneyama *et al.*, 2015).

Brassinosteroids

The steroid hormones brassinosteroids (BRs) perform critical roles during various plant growth processes, including embryogenesis, seed germination, cell division and proliferation, cell elongation, microspore germination, growth of pollen tubes, membrane polarization, photosynthesis, leaf senescence and death (Clouse and Sasse, 1998). BRs confer tolerance to a broad range of abiotic stresses; however, the mechanisms underlying this tolerance have not yet been completely understood (Fariduddin *et al.*, 2014). BRs are widely distributed throughout the plant with significantly higher levels of bioactive BRs (e.g. castasterone and brassinolide) in reproductive (pollen > seeds > fruits) than vegetative (shoots > roots) tissues (Symons *et al.*, 2008). Reciprocal grafting experiments with BR-deficient mutants and WT plants in pea (*lkb*), tomato (*dx*) and *Arabidopsis* (*brx*) revealed that WT scions and rootstocks failed to rescue the BR Dwarf mutant phenotype in either organ (Symons and Reid, 2004; Montoya *et al.*, 2005; Beuchat *et al.*, 2010). Exogenous application of BRs to roots or leaves resulted in direct metabolism of this hormone with minimal losses due to absorption or translocation (Symons *et al.*, 2008). These studies indicate that: (i) neither basipetal nor acropetal transport of BRs has a direct role in plant growth and development or root–shoot signalling; and (ii) BR action is predominantly local (at the vicinity of its site of synthesis). Instead, BRs may have an indirect role in long-distance signalling by altering auxin transport (Symons and Reid, 2004; Symons *et al.*, 2008) or synthesis of polyamines (Terakado *et al.*, 2006) and nitric oxide (Cui *et al.*, 2011). Applying the bioactive brassinosteroid 24-epibrassinolide (24-EBL) to cucumber leaves induced systemic tolerance to photo-oxidative stress in untreated upper and lower leaves, and enhanced root resistance to Fusarium wilt (Xia *et al.*, 2011). Local 24-EBL treatment activated the continuous production of H_2O_2, and the auto-propagative nature of the ROS signal, in turn, mediates systemic 24-EBL induction. Recently, it was proposed that phloem mRNAs may function as long-distance signalling components involved in the 24-EBL-regulated systemic resistance in cucumber (Li *et al.*, 2013). Although BRs are thus not considered to act as long-distance signals like many other plant hormones, their role in rootstock performance is evident, as BRs exert a concentration-dependent effect on root growth and architecture (Wei and Li, 2016).

4.2.3 Metabolite profile of the xylem sap: xylomics

Xylem sap composition, termed 'xylomics', not only provides the shoot with optimal nutrition but also regulates the necessary adaptive responses in the different plant organs (leaves, roots and vascular system) to optimize resource acquisition under unfavourable conditions (Pérez-Alfocea *et al.*, 2011). Quantifying the level of individual metabolites reflects the integration of gene expression, protein interactions and other regulatory processes, and is thus closer to the phenotype than measuring mRNA transcripts or proteins alone. Among all '-omics' technologies, metabolomics is the most universal and has been applied successfully to study plant molecular phenotypes in response to biotic and abiotic stresses to find particular metabolite patterns associated with stress tolerance.

Apart from ions, hormones, proteins and nucleic acids, other plant metabolites have been measured for decades because of their fundamental roles in regulating plant biochemical pathways or as signalling molecules during stress (Lisec *et al.*, 2006). Historically, the measurement of metabolites has been achieved mostly by spectrophotometric assays that detect single metabolites, or by simple chromatographic separation of mixtures of low complexity. However, in recent decades, methods that offer both accuracy and sensitivity for the measurement of highly complex mixtures of compounds have been established (Fiehn *et al.*, 2000; Fernie *et al.*, 2004; Weckwerth, 2011). The combination of metabolite profiling and multivariate data mining is a powerful approach to identify and quantify metabolites in plant tissues at a given developmental stage. This is a challenging task due to the wide array of molecules with different structures and chemical properties. For instance, it is estimated that a single accession of *Arabidopsis* contains more than 5000 metabolites, most of them as yet uncharacterized (Arbona *et al.*, 2013). Unlike transcriptomics, there is no single approach to detect all compounds, and an adequate combination of extraction and detection techniques is needed to increase metabolite coverage. Gas chromatography coupled to mass spectrometry (GC-MS) has classically been used in metabolite profiling studies but is mainly intended for volatiles and primary metabolites (e.g. sugars, amino acids and tricarboxylic acid cycle intermediates) that have been chemically derivatized to produce volatile compounds. However, high-performance liquid chromatography coupled to mass spectrometry (HPLC-MS) is more flexible and can be adapted to detect a vast array of compounds, including secondary metabolites, without prior derivatization (Table 4.1). Additionally, the chemical (and biochemical) identification of analytes represents an arduous task that can be approached by coupling (ultra)-HPLC to high-resolution mass spectrometry, to provide a rapid and powerful technology for structural identification of metabolites in complex plant matrices, and is the most broadly used current analytical technique for metabolite profiling. In most studies on xylem sap composition, only a small number of compounds have been characterized simultaneously (Alvarez *et al.*, 2008). New methods for small-molecule and protein characterization now allow a more comprehensive approach to studying how xylem sap composition changes in response to environmental perturbations. There has been relatively little work in analysing xylem sap metabolite profiles under different environmental stresses. By using high-throughput ultra-HPLC-MS technology, about 800 metabolites in a molecular mass to charge (m/z) range of 90–500 have been detected in the root xylem sap of grafted tomato plants (Albacete *et al.*, 2014, and references therein). However, the potential biological effect of many of these compounds in the scion has yet to be addressed.

Table 4.1. Biochemical groups of primary and secondary metabolism that can be measured in the m/z range 85–900 that can be transported in the xylem between the root and the shoot.

Metabolic pathway	Biochemical group
Primary metabolites	Amino acids
	Sugars
	Sugar phosphates
	Organic acids
	Fatty acids
	Polyols
Secondary metabolites	Alkaloids
	Flavonoids
	Glucosinolates
	Isoprenes
	Oxylipins
	Phenylpropanoids
	Pigments
	Saponins

Xylomics under abiotic stress conditions

Identifying root-derived chemical signals and understanding their specific effects in shoot tissues could open up new strategies to improve crop yield stability under abiotic stress conditions (Pérez-Alfocea *et al.*, 2011). Detailed analysis of xylem sap composition in relation to abiotic stress can reveal physiological mechanisms that are correlated with specific ionic (Albacete *et al.*, 2009) and biophysical (e.g. Dodd *et al.*, 2010) stresses in the rhizosphere, and reveal unexpected or previously uncharacterized biochemical interactions between different metabolites.

Allocation of specific amino compounds (especially glutamine) within the whole plant is associated with plant nitrogen balance, thereby influencing plant growth under adverse conditions. Coupling xylem amino-nitrogen concentration measurements with long-term sap flow measurements in *Eucalyptus regnans* demonstrated that total stand nitrogen transport was based largely on amino compounds of the glutamate group (>90%), and that these molecules are good markers of growth potential under different environmental conditions, in contrast to total nitrogen concentrations, which often show little variation (Pfautsch *et al.*, 2009). Xylem sap metabolome analysis of *B. oleracea* demonstrated the presence of many organic compounds such as sugars, organic acids and amino acids, which showed different responses following salt stress (Fernández-García *et al.*, 2011). Of these, amino acid concentrations (particularly glutamine, the most abundant amino acid) were substantially reduced by salt stress. It was suggested that some of these compounds could be used as biomarkers to study xylem differentiation and lignification under salt stress (Fernández-García *et al.*, 2011). In contrast, other studies have shown that total and individual (e.g. asparagine, glutamine and proline, and aspartic, glutamic and aminobutyric acids) amino acid and soluble sugar concentration increased by twofold in the xylem of tomato plants exposed to salt stress (Pérez-Alfocea *et al.*, 2000; Martínez-Andújar, 2006). This differential response highlights that the xylem sap metabolome changes under salt stress conditions depending on the plant species and/or the constitutive major nitrogen forms transported in the xylem (e.g. NO_3^- versus amino acids).

The xylem sap metabolome also changes in response to drought stress. Drought increased the concentrations of several metabolites (proline, valine, threonine, homoserine, *myo*-inositol, γ-aminobutyrate and nicotinic acid) that are important for signalling and adaptation to environmental stress in the xylem sap of pea (Charlton *et al.*, 2008). Progressively increasing drought stress altered the concentration of 31 compounds, especially several phenylpropanoid compounds, such as coumaric, caffeic and ferulic acids, in maize xylem sap (Alvarez *et al.*, 2008). Although the effects of different rootstocks (mostly commercial) on water uptake and/or drought tolerance in relation to plant performance are known, it is unknown whether changes in xylem metabolite composition are involved in the physiological responses. These studies suggest that changes in the concentration of a range of xylem sap compounds may alter development and signalling during adaptation to drought, but further experiments are needed to independently manipulate their concentration (and measure physiological responses) to establish causality.

Although plant adaptation to nutrient deficiency/toxicity may involve different metabolic changes, little is known about their effects on the metabolic profiles at the root, xylem, leaf and fruit levels. Nevertheless, iron deficiency affected the concentration of more than 40 metabolites (compared with iron-replete plants), with 26 being common in both tomato and pea (including several carboxylates, amino acids and ribose). The main chemical changes in the xylem sap consisted of decreases in amino acids, nitrogen-related metabolites and carbohydrates, and increases in tricarboxylic acid cycle metabolites (Rellán-Álvarez *et al.*, 2011). Although only a few reports have shown the impact of heavy-metal stress on xylem sap metabolite profile, exposing hydroponically grown cucumber plants to high arsenate concentrations (up to 1400 μg kg^{-1}) significantly changed the metabolite profile of xylem sap (Uroic *et al.*, 2012). Mass spectrometry detected four (unidentified) compounds that were significantly upregulated, while isoleucine concentrations were downregulated.

Xylomics under biotic stress conditions

Plants use inducible defence mechanisms to fend off harmful organisms. Resistance that is induced in response to local attack is often expressed systemically in organs that are not yet damaged. Long-distance signals can directly activate defence or can prime for subsequent stronger and faster induction of defence (Heil and Ton, 2008). While there is little information about applying non-targeted metabolomic approaches to study xylem sap composition under abiotic or elicitor-induced (using extrinsic molecules) stresses, even less is reported about plant–pathogen interaction studies. Indeed, this is 'an even greater challenge' (Allwood *et al.*, 2008) because plant–pathogen interactions generate extremely heterogeneous and biochemically complex samples, which further complicate the identification of whether metabolites are derived from the plant or from the interacting pathogen (e.g. Allwood *et al.*, 2006; Barsch *et al.*, 2006; Choi *et al.*, 2006; Keon *et al.*, 2007).

Only recently have changes in xylem sap metabolite composition under biotic stress started to be studied. In the sap of *Vitis vinifera* infected by the fungi *Phaeomoniella chlamydospora*, *Togninia minima* and *Fomitiporia mediterranea*, several exopolysaccharides were quantitatively affected, particularly glucogalactomannan and arabinogalactan (Bruno and Sparapano, 2007). Scytalone and isosclerone, usually produced *in vitro* by *P. chlamydospora* and *T. minima*, were also detected in the sap of vines infected by these diseases. Four phenolics belonging to three classes (benzoic acid derivatives, stilbenes and flavonol glycosides) were separated and identified by HPLC in the xylem sap of infected vines with three esca-associated fungi (Bruno and Sparapano, 2007). Other authors found that bacterial (*Xylella fastidiosa*) infection induced phenolic compounds in the xylem sap of *V. vinifera* 'Thompson Seedless' (Wallis and Chen, 2012). Two months post-inoculation with *X. fastidiosa*, catechin, digalloylquinic acid and astringin were found at greater levels in xylem sap; multiple catechins, procyanidins and stilbenoids were found at greater levels in xylem tissues; and precursors to lignin and condensed tannins were found at greater levels in xylem cell walls. Presumably, grapevines produce high levels of phenolic compounds following infection because these compounds act to inhibit, impair or otherwise kill pathogens (Wallis and Chen, 2012).

In field-grown soybean (*Glycine max*) plants, diurnal changes in the xylem sap metabolite profile revealed that asparagine and aspartate were the two predominant amino acids, while pinitol (D-3-O-methyl-chiro-inositol) was the most abundant carbohydrate. Their abundance may therefore represent a protective signal against disease attack and high temperatures (Krishnan *et al.*, 2011).

In conclusion, biotic and abiotic stresses are necessarily associated with strong metabolic modifications. Xylem sap profiling is likely to provide physiological markers to assist root-targeted breeding for resistance to individual or combined biotic and/or abiotic stresses. Therefore, the use of -omics techniques to underpin rootstock-mediated plant improvement provides new opportunities for yield stability under changing environmental conditions.

4.2.4 Physical signalling

In addition to chemical and molecular signalling via ions, phytohormones, peptides, small RNAs and assimilates, roots and shoots communicate via physical signals. Physical signalling includes hydraulic and electric signals, and propagating Ca^{2+} and ROS waves (Shabala *et al.*, 2016). These signalling mechanisms operate at different timescales: physical signalling operates over seconds to minutes, whereas chemical and molecular signalling occurs over hours to days. An example where physical signalling is proposed to play a significant role in root-to-shoot communication is the rapid stomatal closure in response to decreased soil water status (Christmann *et al.*, 2007; Grams *et al.*, 2007). This reaction is much faster than a chemical or molecule can be transmitted via the transpiration stream or within the phloem. Physical signalling is described only briefly in this chapter as: (i) no examples of grafting between specific combinations of scions and rootstocks are available to highlight the key role of specific physical signals; and (ii) its mechanisms have recently been reviewed elsewhere (Christmann *et al.*, 2013; Shabala *et al.*, 2016; Huber and Bauerle, 2016).

Hydraulic signalling comprises changes in the water potential (Ψ_w) induced by changes in xylem water tension, turgor or osmotic potential (Christmann *et al.*, 2013). Hydraulic signalling can be induced by local changes in water availability or wounding that affect the hydraulic properties of the vasculature, rapidly leading to a pressure signal (*see* Plate 7) (Choi *et al.*, 2016). Such a hydraulic signal moves in the rigid xylem vessels at the speed of sound (Christmann *et al.*, 2013). Root hydraulic signals can possibly be sensed and decoded in the shoot by: (i) osmo-sensing; (ii) turgor sensing; or (iii) mechanosensory channels (Shabala *et al.*, 2016). The perception of the hydraulic signal may lead to the conversion of the physical signal into a chemical one (e.g. ABA or JA), which mediates adaptive physiological responses (*see* Plate 7) (Christmann *et al.*, 2013).

Electrical signals in plants are defined as an ion imbalance across the plasma membrane leading to a voltage transient (Huber and Bauerle, 2016). They allow rapid responses to environmental stimuli such as wounding or re-irrigation, and are involved in root-to-shoot communication in plants under drought stress (Zimmermann and Mithöfer, 2013; Gallé *et al.*, 2015). Two main types of long-distance electrical signals are distinguished: rapid action potentials, with velocities of 0.5–30,000 cm s^{-1}, and slow-wave potentials or variation potentials, with velocities in the range of 0.1–1.0 cm s^{-1} (Gallé *et al.*, 2015; Huber and Bauerle, 2016). Action potentials are propagated in the phloem, whereas slow-wave potentials are transmitted through dead xylem tissue. The molecular basis and the cellular and physiological roles of electrical signals have been described in detail in recent reviews (Gallé *et al.*, 2015; Choi *et al.*, 2016; Shabala *et al.*, 2016).

Miller *et al.* (2009) described a new mechanism for rapid long-distance cell-to-cell communication in which a systemic autopropagating wave of ROS production was used to signal a range of abiotic stresses at a maximal rate of approximately 8 cm min^{-1}. ROS waves seem to function as a general priming signal, alerting remote tissues to the occurrence of a localized specific abiotic stress stimulus (Gilroy *et al.*, 2014). During ROS wave propagation, superoxide is produced in the apoplast by NADPH oxidases (also known as respiratory burst oxidase homologues or RBOHs), after which it is converted to H_2O_2 by superoxide dismutase (Baxter *et al.*, 2014; Choi *et al.*, 2016).

In addition to a direct role of extracellular Ca^{2+} as a mobile systemic signal in the xylem, as discussed previously, it was recently highlighted that propagating waves of increased cytosolic Ca^{2+} levels may operate in long-distance signalling in plants (Gilroy *et al.*, 2014). Ca^{2+} waves, which may operate at a rate of 400 μm s^{-1}, could be induced by salt stress and mechanical wounding, indicating that they are a general element in long-distance signalling. The apoplastic influx of Ca^{2+} across the plasma membrane might be a key event in both the machinery linking the perception of stress to the initial Ca^{2+} wave generation and the sustained transmission of the Ca^{2+} wave (Choi *et al.*, 2016).

4.2.5 Proteins

Several proteomic studies have shown changes in xylem and phloem sap protein composition in response to biotic and abiotic stresses. Acting alone or by interacting with RNAs, proteins are also considered long-distance signalling macromolecules that can contribute to the bidirectional rootstock–scion communication, with potential applications in improving vegetable production. Classically, a role in long-distance inter-organ signalling has been attributed to phloem-mobile proteins influencing plant development, defence and stress responses. Indeed, the best-characterized long-distance (source leaf to shoot apex) trafficking phloem-borne protein is FLOWERING LOCUS T (FT), which controls the onset of flowering in several species such as *Arabidopsis* (Corbesier *et al.*, 2007), rice (*Oryza sativa*; Tamaki *et al.*, 2007) and pumpkin (Lin *et al.*, 2007). Although little is still known about the role(s) of these macromolecules, new proteins with potential applications are being discovered. For instance, the cyclophilin SlCyp1, involved in auxin signalling, moves from the scion to the rootstock, integrating photosynthesis, root growth and root/shoot homeostasis in tomato (Spiegelman *et al.*, 2015). However, proteins are also present in the xylem sap, and their role in communication between the root and the scion has attracted increasing attention from basic and applied researchers (reviewed by Haroldsen *et al.*, 2012). Transgenic rootstocks overproducing anti-pathogenic compounds such as polygalacturonase-inhibiting proteins can reduce bacterial and fungal infection in grape and tomato fruits through graft-transmissible xylem transport (Dutt *et al.*, 2007; Gray *et al.*, 2007; Haroldsen *et al.*, 2012).

4.2.6 Small RNAs

Exchanges of both RNA and DNA molecules between rootstock and scion partners through graft-transmissible long-distance mRNA translocation in the phloem and xylem (Lucas *et al.*, 2001; Brosnan *et al.*, 2007; Molnar *et al.*, 2010; Brosnan and Voinnet, 2011) or through horizontal bidirectional gene transfer via either large DNA pieces or an entire plastid (Stegemann and Bock, 2009) or nuclear (Fuentes *et al.*, 2014) genomes help to explain the molecular basis for grafting-induced genotypic and phenotypic variation. Long-distance shoot-to-root and root-to-shoot transport of small interfering RNAs (siRNAs) and mRNAs, and their potential role in organ communication, has become a major research domain in recent years (Palauqui *et al.*, 1997; Lough and Lucas, 2006; Omid *et al.*, 2007; Harada, 2010; Spiegelman *et al.*, 2013; reviewed by Goldschmidt, 2014). Mobile RNAs have been identified in diverse species with known functions in hormone response, transcriptional control and organ development (reviewed by Hannapel, 2010). The potential of this communication mechanism to overcome the genetic barriers between different taxa (as in the first RNA mobilome (mobile genetic elements) that moved over long distances between donor *Arabidopsis* (rosids) rootstock and recipient *Nicotiana benthamiana* (asterids) scion; Notaguchi *et al.*, 2015) could expand the genetic diversity that could be exploited through grafting. For example, RNA interference is likely to be applied to produce virus-resistant transgenic plants (Lemgo *et al.*, 2013), if silenced rootstocks can efficiently transmit the silencing signal to non-transformed scions (Kasai *et al.*, 2011; Zhao and Song, 2014). Similarly, the upregulation of genes associated with DNA methylation and histone modification is the basis of epigenetic mechanisms and can be affected by siRNAs (Cookson and Ollat, 2013), thus providing a powerful strategy for plants to adapt to environmental change (Meyer, 2015). Therefore, plants implement the means by which RNA is transported over long distances through the vascular system to specific target organs. The major challenges that lie ahead for plant biologists are to identify the numerous RNA–protein complexes that mediate this long-distance transport network and to understand how they function in this unique delivery system. Special efforts to understand the role of those mobile signals from the rootstock, and their association with phenotypic changes in the scion, may allow agronomic exploitation of these rootstock × scion interactions in crop improvement.

4.3 Conclusions

Recent improvements in metabolomic analytical techniques offer exciting opportunities to understand the nature of root-to-shoot and shoot-to-root signalling *in planta*. However, care needs to be exercised in interpreting results based on changes in xylem and phloem sap composition to ensure minimal contamination of these transport fluids by damaged cell debris (Dodd, 2005). Consequently, phenotyping the physiological responses of grafted plants (where the rootstock or scion show genetic differences in the regulation of specific biochemical pathways) will continue to provide a powerful demonstration of the importance of root–shoot communication. Moreover, applying such fundamental knowledge within horticultural cropping systems (many of which use grafted plants) will provide opportunities to increase crop productivity and resource use efficiency.

References

Akiyama, K., Matsuzaki, K. and Hayashi, H. (2005) Plant sesquiterpenes induce hyphal branching in arbuscular mycorrhizal fungi. *Nature* 435, 824–827.

Albacete, A., Ghanem, M.E., Martínez-Andújar, C., Acosta, M., Sánchez-Bravo, J., Martínez, V., Lutts, S., Dodd, I.C. and Pérez-Alfocea, F. (2008) Hormonal changes in relation to biomass partitioning and shoot growth impairment in salinized tomato (*Solanum lycopersicum* L.) plants. *Journal of Experimental Botany* 59, 4119–4131.

Albacete, A., Martínez-Andújar, C., Ghanem, M.E., Acosta, M., Sánchez-Bravo, J., Asins, M.J., Cuartero, J., Lutts, S., Dodd, I.C. and Pérez-Alfocea, F. (2009) Rootstock-mediated changes in xylem ionic and hormonal status are correlated with delayed leaf senescence, and increased leaf area and crop productivity in salinized tomato. *Plant, Cell and Environment* 32, 928–938.

Albacete, A., Martínez-Andújar, C. and Pérez-Alfocea, F. (2014) Hormonal and metabolic regulation of source–sink relations under salinity and drought: from plant survival to crop yield stability. *Biotechnology Advances* 32, 12–30.

Albacete, A., Martínez-Andújar, C., Martínez-Pérez, A., Thompson, A.J., Dodd, I.C. and Pérez-Alfocea, F. (2015) Unravelling rootstock × scion interactions to improve food security. *Journal of Experimental Botany* 66, 2211–2226.

Allario, T., Brumos, J., Colmenero-Flores, J.M., Iglesias, D.J., Pina, J.A., Navarro, L., Talon, M., Ollitrault, P. and Morillon, R. (2013) Tetraploid Rangpur lime rootstock increases drought tolerance via enhanced constitutive root abscisic acid production. *Plant, Cell and Environment* 36, 856–868.

Allwood, J.W., Ellis, D.I., Heald, J.K., Goodacre, R. and Mur, L.A.J. (2006) Metabolomic approaches reveal that phosphatidic and phosphatidyl glycerol phospholipids are major discriminatory non-polar metabolites in responses by *Brachypodium distachyon* to challenge by *Magnaporthe grisea*. *The Plant Journal* 46, 351–368.

Allwood, J.W., Ellis, D.I. and Goodacre, R. (2008) Metabolomic technologies and their application to the study of plants and plant–host interactions. *Physiologia Plantarum* 132, 117–135.

Alvarez, S., Marsh, E.L., Schroeder, S.G. and Schachtman, D.P. (2008) Metabolomic and proteomic changes in the xylem sap of maize under drought. *Plant, Cell and Environment* 31, 325–340.

Arbona, V., Manzi, M., de Ollas, C. and Gómez-Cadenas, A. (2013) Metabolomics as a tool to investigate abiotic stress tolerance in plants. *International Journal of Molecular Sciences* 14, 4885–4911.

Atkinson, C.J., Mansfield, T.A. and Davies, W.J. (1990) Does calcium in xylem sap regulate stomatal conductance. *New Phytologist* 116, 19–27.

Ballaré, C.L. (2011) Jasmonate-induced defences: a tale of intelligence, collaborators and rascals. *Trends in Plant Science* 16, 249–257.

Barsch, A., Tellström, V., Patschkowski, T., Küster, H. and Niehaus, K. (2006) Metabolite profiles of nodulated alfalfa plants indicate that distinct stages of nodule organogenesis are accompanied by global physiological adaptations. *Molecular Plant–Microbe Interactions* 19, 998–1013.

Bauer, H., Ache, P., Lautner, S., Fromm, J., Hartung, W., Al-Rasheid, K.A., Sonnewald, S., Sonnewald, U., Kneitz, S., Lachmann, N., Mendel, R.R., Bittner, F., and Hetherington, A.M. (2013) The stomatal response to reduced relative humidity requires guard cell-autonomous ABA synthesis. *Current Biology* 23, 53–57.

Baxter, A., Mittler, R. and Suzuki, N. (2014) ROS as key players in plant stress signalling. *Journal of Experimental Botany* 65, 1229–1240.

Belimov, A.A, Dodd, I.C., Hontzeas, N., Theobald, J.C., Safronova, V.I., Davies, W.J. (2009) Rhizosphere bacteria containing ACC deaminase increase yield of plants grown in drying soil via both local and systemic hormone signalling. *New Phytologist* 181, 413–423.

Besserer, A., Puech-Pages, V., Kiefer, P., Gomez-Roldan, V., Jauneau, A., Roy, S., Portais, J.C., Roux, C., Becard, G. and Sejalon-Delmas, N. (2006) Strigolactones stimulate arbuscular mycorrhizal fungi by activating mitochondria. *PLoS Biology* 4, 1239–1247.

Beuchat, J., Scacchi, E., Tarkowska, D., Ragni, L., Strnad, M. and Hardtke, C.S. (2010) *BRX* promotes Arabidopsis shoot growth. *New Phytologist* 188, 23–29.

Booker, J., Sieberer, T., Wright, W., Williamson, L., Willett, B., Stirnberg, P., Turnbull, C., Srinivasan, M., Goddard, P. and Leyser, O. (2005) *MAX1* encodes a cytochrome P450 family member that acts downstream of *MAX3/4* to produce a carotenoid-derived branch-inhibiting hormone. *Developmental Cell* 8, 443–449.

Bosch, M., Wright, L.P., Gershenzon, J., Wasternack, C., Hause, B., Schaller, A. and Stintzi, A. (2014) Jasmonic acid and its precursor 12-oxyphytodienoic acid control different aspects of constitutive and increased herbivore defences in tomato. *Plant Physiology* 166, 396–410.

Bradford, K.J. and Yang, S.F. (1980) Xylem transport of 1-amino-cyclopropane-1-carboxylic acid, an ethylene precursor, in waterlogged tomato plants. *Plant Physiology* 65, 322–326.

Brading, P.A., Hammond-Kosack, K.E., Parr, A. and Jones, J.D.G. (2000) Salicylic acid is not required for Cf-2 and Cf-9-dependent resistant of tomato to *Cladosporium fulvum*. *The Plant Journal* 23, 305–318.

Breuillin, F., Schramm, J., Hajirezaei, M., Ahkami, A., Favre, P., Druege, U., Hause, B., Bucher, M., Kretzschmar, T., Bossolini, E., Kuhlemeier, C., Martinoia, E., Franken, P., Scholz, U. and Reinhardt, D. (2010) Phosphate systemically inhibits development of arbuscular mycorrhiza in *Petunia hybrida* and represses genes involved in mycorrhizal functioning. *The Plant Journal* 64, 1002–1017.

Brosnan, C.A. and Voinnet, O. (2011) Cell-to-cell and long-distance siRNA movement in plants: mechanisms and biological implications. *Current Opinion in Plant Biology* 14, 580–587.

Brosnan, C.A., Mitter, N., Christie, M., Smith, N.A., Waterhouse, P.M. and Carroll, B.J. (2007) Nuclear gene silencing directs reception of long-distance mRNA silencing in *Arabidopsis*. *Proceedings of the National Academy of Sciences USA* 104, 14741–14746.

Bruno, G. and Sparapano, L. (2007) Effects of three esca-associated fungi on *Vitis vinifera* L.: V. Changes in the chemical and biological profile of xylem sap from diseased cv. Sangiovese vines. *Physiological and Molecular Plant Pathology* 71, 210–229.

Campos, M.L., Kang, J.-H. and Howe, G.A. (2014) Jasmonate-triggered plant immunity. *Journal of Chemical Ecology* 40, 657–675.

Cassaniti, C., Giuffrida, F., Scuderi, D. and Leonardi, C. (2011) The effect of rootstock and nutrient solution concentration on eggplant grown in soilless system. *Journal of Food, Agriculture and Environment* 9, 252–256.

Charlton, A., Donarski, J., Harrison, M., Jones, S., Godward, J., Oehlschlager, S., Arques, J., Ambrose, M., Chinoy, C., Mullineaux, P. and Domoney, C. (2008) Responses of the pea (*Pisum sativum* L.) leaf metabolome to drought stress assessed by nuclear magnetic resonance spectroscopy. *Metabolomics* 4, 312–327.

Chauvin, A., Caldelari, D., Wolfender, J.-L. and Farmer, E.E. (2013) Four 13-lipoxygenases contribute to rapid jasmonate synthesis in wounded *Arabidopsis thaliana* leaves: a role for lipoxygenase 6 in responses to long-distance wound signals. *New Phytologist* 197, 566–575.

Choi, Y.H., Kim, H.K., Linthorst, H.J.M., Hollander, J.G., Lefeber, A.W.M., Erkelens, C., Nuzillard, J.-M. and Verpoorte, R. (2006) NMR Metabolomics to revisit the tobacco mosaic virus infection in *Nicotiana tabacum* leaves. *Journal of Natural Products* 69, 742–748.

Choi, W.-G., Hilleary, R., Swanson, S.J., Kim, S.-H. and Gilroy, S. (2016) Long-distance electrical and calcium signaling in plants. *Annual Review of Plant Biology* 67, 287–307.

Christmann, A., Weiler, E.W., Steudle, E. and Grill, E. (2007) A hydraulic signal in root-to-shoot signalling of water shortage. *Plant Journal* 52, 167–174.

Christmann, A., Grill, E. and Huang, J. (2013) Hydraulic signals in long-distance signalling. *Current Opinion in Plant Biology* 16, 293–300.

Clouse, S.D. and Sasse, J.M. (1998) Brassinosteroids: essential regulators of plant growth and development. *Annual Review Plant Physiology and Plant Molecular Biology* 49, 427–451.

Clarkson, D.T. (1984) Calcium transport between tissues and its distribution in the plant. *Plant, Cell and Environment* 7, 449–456.

Colebrook, E.H., Thomas, S.G., Phillips, A.L. and Hedden, P. (2014) The role of gibberellin signalling in plant responses to abiotic stress. *Journal of Experimental Biology* 217, 67–75.

Colla, G., Rouphael, Y., Rea, E. and Cardarelli, M. (2012) Grafting cucumber plants enhance tolerance to sodium chloride and sulfate salinization. *Scientia Horticulturae* 135, 177–185.

Cookson, S.J. and Ollat, N. (2013) Grafting with rootstocks induces extensive transcriptional re-programming in the shoot apical meristem of grapevine. *BMC Plant Biology* 13, 147.

Corbesier, L., Vincent, C., Jang, S.H., Fornara, F., Fan, Q., Searle, I., Giakountis, A., Farrona, S., Gissot, L., Turnbull, C., and Coupland, G. (2007) FT protein movement contributes to long-distance signalling in floral induction of *Arabidopsis*. *Science* 16, 1030–1033.

Crawford, N.M. and Glass, A.D.M. (1998) Molecular and physiological aspects of nitrate uptake in plants. *Trends in Plant Science* 3, 389–395.

Creelman, R.A. and Mullet, J.E. (1995) Jasmonic acid distribution and action in plants: regulation during development and response to biotic and abiotic stress. *Proceedings of the National Academic of Sciences USA* 92, 4114–4119.

Cui, J.-X., Zhou, Y.-H., Ding, J.G., Xia, X.J., Shi, K., Chen, S.C., Asami, T., Chen, Z. and Yu, J.Q. (2011) Role of nitric oxide in hydrogen peroxide-dependent induction of abiotic stress tolerance by brassinosteroids in cucumber. *Plant, Cell and Environment* 34, 347–358.

Dar, T.A., Uddin, M., Khan, M.A., Hakeem, K.R. and Jaleel, H. (2015) Jasmonates counter plant stress: a review. *Environmental and Experimental Botany* 115, 49–57.

de Ollas, C. And Dodd, I.C. (2016) Physiological impacts of ABA–JA interactions under water-limitation. *Plant Molecular Biology* 92, 641–650.

de Ollas, C., Hernando, B., Arbona, V. and Gómez-Cadenas, A. (2013) Jasmonic acid transient accumulation is needed for abscisic acid increase in citrus roots under drought stress conditions. *Physiologia Plantarum* 147, 296–306.

de Ollas, C., Arbona, V. and Gómez-Cadenas, A. (2015) Jasmonoyl isoleucine accumulation is needed for abscisic acid build-up in roots of Arabidopsis under water stress conditions: JA–ABA interaction in roots. *Plant, Cell and Environment* 38, 2157–2170.

Dodd, I.C. (2005) Root-to-shoot signalling: assessing the roles of 'up' in the up and down world of long-distance signalling *in planta*. *Plant and Soil* 274, 257–275.

Dodd, I.C. (2013) Abscisic acid and stomatal closure: a hydraulic conductance conundrum? *New Phytologist* 197, 6–8.

Dodd, I.C., Egea, G., Watts, C.W. and Whalley, W.R. (2010) Root water potential integrates discrete soil physical properties to influence ABA signalling during partial rootzone drying. *Journal of Experimental Botany* 61, 3543–3551.

Donas-Ucles, F., Jimenez-Luna, M.D.M., Gongora-Corral, J.A., Perez-Madrid, D., Verde-Fernandez, D. and Camacho-Ferre, F. (2014) Influence of three rootstocks on yield and commercial quality of "Italian sweet" pepper. *Ciencia e Agrotecnologia* 6, 538–545.

Du, H., Liu, H. and Xioung, L. (2013) Endogenous auxin and jasmonic acid levels are differently modulated by abiotic stresses in rice. *Frontiers in Plant Science* 4, 397.

Dutt, M., Li, Z.T., Kelley, K.T., Dhekney, S.A., van Aman, M., Tattersall, J. and Gray D.J. (2007) Transgenic rootstock protein transmission in grapevines. *Acta Horticulturae* 738, 749–753.

Else, M.A. and Jackson, M.B. (1998) Transport of 1-aminocyclopropane-1-carboxylic acid (ACC) in the transpiration stream of tomato (*Lycopersicon esculentum*) in relation to foliar ethylene production and petiole epinasty. *Australian Journal of Plant Physiology* 25, 453–458.

English, P.J., Lycett, G.W., Roberts, J.A. and Jackson, M.B. (1996) Increased 1-aminocyclopropane-1-carboxylic acid oxidase activity in shoots of flooded tomato plants raises ethylene production to physiologically active levels. *Plant Physiology* 109, 1435–1440.

Erb, M., Lemk, C., Degenhardt, J. and Turlings, T.C.J. (2009) The underestimated role of roots in defense against leaf attackers. *Trends in Plant Science* 14, 653–659.

Ernst, L., Goodger, J.Q.D., Alvarez, S., Marsh, E.L., Berla, B., Lockhart, E., Jung, J., Li, P., Bohnert, H.J. and Schachtman, D.P. (2010) Sulphate as a xylem-borne chemical signal precedes the expression of ABA biosynthetic genes in maize roots. *Journal of Experimental Botany* 61, 3395–3405.

Fariduddin, Q., Yusuf, M., Ahmad, I. and Ahmad, A. (2014) Brassinosteroids and their role in response of plants to abiotic stress. *Biologia Plantarum* 58, 9–17.

Farmer, E.E., Gasperini, D. and Acosta, I.F. (2014) The squeeze cell hypothesis for the activation of jasmonate synthesis in response to wounding. *New Phytologist* 204, 282–288.

Felle, H.H., Herrmann, A., Huckelhoven, R. and Kogel, K.H. (2005) Root-to-shoot signalling: apoplastic alkalinization, a general stress response and defence factor in barley (*Hordeum vulgare*). *Protoplasma* 227, 17–24.

Fernández-García, N., Martínez, V., Cerdá, A. and Carvajal, M. (2002) Water and nutrient uptake of grafted plants grown under saline conditions. *Journal of Plant Physiology* 159, 899–905.

Fernández-García, N., Hernández, M., Casado-Vela, J., Bru, R., Elortza, F., Hedden, P. and Olmos, E. (2011) Changes to the proteome and targeted metabolites of xylem sap in *Brassica oleracea* in response to salt stress. *Plant, Cell and Environment* 34, 821–836.

Fernie, A.R., Trethewey, R.N., Krotzky, A.J. and Willmitzer, L. (2004) Metabolite profiling: from diagnostics to systems biology. *Nature Reviews Molecular Cell Biology* 5, 763–769.

Fiebig, A. and Dodd, I.C. (2016) Inhibition of tomato shoot growth by over-irrigation is linked to nitrogen deficiency and ethylene. *Physiologia Plantarum* 156, 70–83.

Fiehn, O., Kopka, J., Dörmann, P., Altmann, T., Trethewey, R.N. and Willmitzer, L. (2000) Metabolite profiling for plant functional genomics. *Nature Biotechnology* 18, 1157–1161.

Foo, E. and Davies, N.W. (2011) Strigolactones promote nodulation in pea. *Planta* 234, 1073–1081.

Foo, E., Turnbull, C.G.N. and Beveridge, C.A. (2001) Long-distance signaling and the control of branching in the *rms1* mutant of peal. *Plant Physiology* 126, 203–209.

Foo, E., Yoneyama, K., Hugill, C.J., Quittenden, L.J. and Reid, J.B. (2013) Strigolactones and the regulation of pea symbioses in response to nitrate and phosphate deficiency. *Molecular Plant* 6, 76–87.

Forde, B.G. (2002a) The role of long-distance signalling in plant responses to nitrate and other nutrients. *Journal of Experimental Botany* 53, 39–43.

Forde, B.G. (2002b) Local and long-range signaling pathways regulating plant responses to nitrate. *Annual Review of Plant Biology* 53, 203–224.

Fragoso, V., Rothe, E., Baldwin, I.T. and Kim, S.-G. (2014) Root jasmonic acid synthesis and perception regulate folivore-induced shoot metabolites and increase *Nicotiana attenuata* resistance. *New Phytologist* 202, 1335–1345.

Fuentes, I., Stegeman, S., Golczyk, H., Karcher, D. and Bock, R. (2014) Horizontal genome transfer as an asexual path to the formation of new species. *Nature* 511, 232–235.

Gaffney, T., Friedrich, L., Vernooij, B., Negrotto, D., Nye, G., Uknes, S., Ward, E., Kessmann, H. and Ryals, J. (1993) Requirement of salicylic acid for the induction of systemic acquired resistance. *Science* 261, 754–756.

Gallé, A., Lautner, S., Flexas, J. and Fromm, J. (2015) Environmental stimuli and physiological responses: the current view on electrical signalling. *Environmental and Experimental Botany* 114, 15–21.

Gao, Q.-M., Zhu, S., Kachroo, P. and Kachroo, A. (2015) Signal regulators of systemic acquired resistance. *Frontiers in Plant Science* 6, 228.

Garcia-Abellan, J.O., Fernandez-Garcia, N., Lopez-Berenguer, C., Egea, I., Flores, F.B., Angosto, T., Capel, J., Lozano, R., Pineda, B., Moreno, V., Olmos, E. and Bolarin, M.C. (2015) The tomato *res* mutant which accumulates JA in roots in non-stressed conditions restores cell structure alterations under salinity. *Physiologia Plantarum* 155, 296–312.

Gasperini, D., Chauvin, A., Acosta, I.F. Kurenda, A. and Stolz, S. (2015) Axial and radial oxylipin transport. *Plant Physiology* 169, 2244–2254.

Ghanem, M.E., Albacete, A., Smigocki, A.C., Frebort, I., Pospíšilová, H., Martínez-Andújar, C., Acosta, M., Sanchez-Bravo, J., Lutts, S., Dodd, I.C. and Pérez-Alfocea, F. (2011a) Root-synthesised cytokinins improve shoot growth and fruit yield in salinized tomato (*Solanum lycopersicum* L.). *Journal of Experimental Botany* 62, 125–140.

Ghanem, M.E., Hichri, I., Smigocki, A.C., Albacete, A., Fauconnier, M.L., Diatloff, E., Martinez-Andujar, C., Lutts, S., Dodd, I.C. and Pérez-Alfocea, F. (2011b) Root-targeted biotechnology to mediate hormonal signalling and improve crop stress tolerance. *Plant Cell Reports* 30, 807–823.

Gilroy, S., Suzuki, N., Miller, G., Choi, W.-G. and Toyota, M. (2014) A tidal wave of signals: calcium and ROS at the forefront of rapid systemic signaling. *Trends in Plant Science* 19, 623–630.

Giuffrida, F., Cassaniti, C. and Leonardi, C. (2013) The influence of rootstock on growth and ion concentrations in pepper (*Capsicum annuum* L.) under saline conditions. *Journal of Horticultural Science and Biotechnology* 88, 110–116.

Giuffrida, F., Cassaniti, C. and Leonardi, C. (2014) Tomato and eggplant scions influence the effect of rootstock under Na_2SO_4 salinity. *Acta Agriculturae Scandinavica B: Soil and Plant Science* 64, 700–709.

Giuffrida, F., Cassaniti, C., Agnello, M. and Leonardi, C. (2015) The influence of rootstock on growth and ion concentration in eggplant under different saline conditions. *Acta Horticulturae* 1086, 161–166.

Glass, A.D.M. and Siddiqi, M.Y. (1995) Nitrogen absorption by plant roots. In: Srivastava, H.S. and Singh, R.P. (eds) *Nitrogen Nutrition in Higher Plants*. Associated Publishing Co., New Delhi, India, pp. 21–56.

Gloser, V., Korovetska, H., Martin, A.I., Hájičková, M., Prokop, Z., Wilkinson, S. and Davies, W.J. (2016) The dynamics of xylem sap pH under drought: a universal response in herbs? *Plant and Soil* (in press), doi:10.1007/s11104-016-2962-6.

Goldschmidt, E.G. (2014) Plant grafting: new mechanisms, evolutionary implications. *Frontiers in Plant Science* 17, 1–9.

Gómez, S., Ferreiri, R.A., Schueller, M. and Orians, C.M. (2010) Methyl jasmonate elicits rapid changes in carbon and nitrogen dynamics in tomato. *New Phytologist* 188, 835–844.

Goodger, J.Q.D., Sharp, R.E., Marsh, E.L. and Schachtman, D.P. (2005) Relationships between xylem sap constituents and leaf conductance of well-watered and water-stressed maize across three xylem sap sampling techniques. *Journal of Experimental Botany* 56, 2389–2400.

Grams, T.E.E., Kozioloek, C., Lautner, S., Matyssek, R. and Fromm, J. (2007) Distinct roles of electric and hydraulic signals on the reaction of leaf gas exchange upon irrigation in *Zea mays*. *Plant, Cell and Environment* 30, 79–84.

Gray D., Li Z., Dhekney S., Dutt M., Hopkins D. and Zimmerman T. (2007) Field testing of transgenic grapevine for bacterial and fungal disease resistance. *HortScience* 42, 858–858.

Grebner, W., Stingl, N.E., Oenel, A., Mueller, M.J. and Berger, S. (2013) Lipoxygenase6-dependent oxylipin synthesis in roots is required for abiotic and biotic stress resistance of Arabidopsis. *Plant Physiology* 161, 2159–2170.

Gregory, P.J. (2006) *Plant Roots: Growth,Activity and Interaction with Soil.* Blackwell Publishing, Oxford, UK.

Guichard, S., Bertin, N., Leonardi, C. and Gary, C. (2001) Tomato fruit quality in relation to water and carbon fluxes. *Agronomie* 21, 385–392.

Hannapel, D.J. (2010) A model system of development regulated by the long-distance transport of mRNA. *Journal of Integrative Plant Biology* 52, 40–52.

Harada, T. (2010) Grafting and RNA transport via phloem tissue in horticultural plants. *Scientia Horticulturae* 125, 545–550.

Haroldsen, V.M., Szczerba, M.W., Aktas, H., Lopez-Baltazar, J., Odias, M.J., Chi-Ham, C.L., Labavitch, J.M., Bennett, A.B. and Powell, A.L. (2012) Mobility of transgenic nucleic acids and proteins within grafted rootstocks for agricultural improvement. *Frontiers in Plant Science* 3, 39.

Hasegawa, S., Sogabe, Y., Asano, T., Nakagawa, T., Nakamura, H., Kodama, H., Ohta, H., Yamaguchi, K., Mueller, M.J. and Nishiuchi, T. (2011) Gene expression analysis of wounding-induced root-to-shoot communication in *Arabidopsis thaliana*. *Plant, Cell and Environment* 34, 705–716.

Heil, M. and Ton, J. (2008) Long-distance signalling in plant defence. *Trends in Plant Science* 13, 264–272.

Herde, O., Atzorn, R., Fisahn, J., Wasternack, C., Willmitzer, L. and Pena-Cortes, H. (1996) Localized wounding by heat initiates the accumulation of proteinase inhibitor II in abscisic acid-deficient plants by triggering jasmonic acid biosynthesis. *Plant Physiology* 112, 853–860.

Hirose, N., Takei, K., Kuroha, T., Kamada-Nobusada, T., Haywashi, H. and Sakakibara, H. (2008) Regulation of cytokinin biosynthesis, compartmentalization and translocation. *Journal of Experimental Botany* 59, 75–83.

Ho, L.C., Belda, R., Brown, M., Andrews, J. and Adams, P. (1993) Uptake and transport of calcium and the possible causes of blossom end rot in tomato. *Journal of Experimental Botany* 44, 509–518.

Holbrook, N.M., Shashidhar, V.R., James, R.A. and Munns, R. (2002) Stomatal control in tomato with ABA-deficient roots: response of grafted plants to soil drying. *Journal of Experimental Botany* 53, 1503–1514.

Horváth, E., Szalai, G. and Janda, T. (2007) Induction of abiotic stress tolerance by salicylic acid signalling. *Journal Plant Growth Regulation* 26, 290–300.

Howe, G.A., Lightner, J., Browse, J. and Ryan, C.A. (1996) An octadecanoid pathway mutant (*JL5*) of tomato is compromised in signalling for defense against insect attack. *Plant Cell* 8, 2067–2077.

Hu, Y., Jiang, L., Wang, F. and Yu, D. (2013) Jasmonate regulates the INDUCER OF CBF EXPRESSION-C-REPEAT BINDING FACTOR/DRE BINDING FACTOR 1 cascade and freezing tolerance in *Arabidopsis*. *Plant Cell* 25, 2907–2924.

Huang, Y., Li, J., Hua, B., Liu, Z., Fan, M. and Bie, Z. (2013) Grafting onto different rootstocks as a means to improve watermelon tolerance to low potassium stress. *Scientia Horticulturae* 149, 80–85.

Huber, L.A. and Bauerle, T.L. (2016) Long-distance plant signalling pathways in response to multiple stressors: the gap in knowledge. *Journal of Experimental Botany* 67, 2063–2079.

Ivashikina, N.V. and Feyziev, Y.M. (1998) Regulation of nitrate uptake in maize seedlings by accompanying cations. *Plant Science* 131, 25–34.

Jackson, M.B. (1993) Are plant hormones involved in root to shoot communication? *Advances in Botanical Research* 19, 103–187.

Jackson, M.B., Saker, L.R., Crisp, C.M., Else, M.A. and Janowiak, F. (2003) Ionic and pH signalling from roots to shoots of flooded tomato plants in relation to stomatal closure. *Plant and Soil* 253, 103–113.

Janda, M. and Ruelland, E. (2015) Magical mystery tour: salicylic acid signalling. *Environmental and Experimental Botany* 114, 117–128.

Jayakannan, M., Bose, J., Babourina, O., Rengel, Z. and Shabala, S. (2015a) Salicylic acid in plant salinity stress signaling and tolerance. *Plant Growth Regulation* 76, 25–40.

Jayakannan, M., Bose, J., Babourina, O., Shabala, S., Massart, A., Poschenrieder, C. and Rengel, Z. (2015b) The NPR1-dependent salicylic acid signalling pathway is pivotal for enhanced salt and oxidative stress tolerance in *Arabidopsis*. *Journal of Experimental Botany* 66, 1865–1875.

Jeschke, W.D. and Pate, J.S. (1991) Modelling of the uptake, flow and utilization of C, N and H_2O within whole plants of *Ricinus communis* L. based on empirical data. *Journal of Plant Physiology* 137, 488–498.

Jia, W. and Davies, W.J. (2007) Modification of leaf apoplastic pH in relation to stomatal sensitivity to root-sourced abscisic acid signals. *Plant Physiology* 143, 68–77.

Kasai, A., Bai, S., Li, T. and Harada, T. (2011) Graft-transmitted siRNA signal from the root induces visual manifestation of endogenous post-transcriptional gene silencing in the scion. *PLoS ONE* 6, e16895.

Kazan, K. (2015) Diverse roles of jasmonates and ethylene in abiotic stress tolerance. *Trends in Plant Science* 20, 219–229.

Keon, J., Antoniw, J., Carzaniga, R., Deller, S., Ward, J.L., Baker, J.M., Beal, M.H., Hammond-Kosack, K. and Rudd, J.J. (2007) Transcriptional adaptation of *Mycosphaerella graminicola* to programmed cell death (PCD) of its susceptible wheat host. *Molecular Plant–Microbe Interactions* 20, 178–193.

Khan, M.I.R., Fatma, M., Per, T.S., Anjum, N.A. and Khan, N.A. (2015) Salicylic acid-induced abiotic stress tolerance and underlying mechanisms in plants. *Frontiers in Plant Science* 6, 462.

Ko, D., Kang, J., Kiba, T., Park, J., Kojima, M., Do, J., Kim, K.Y., Kwon, M., Endler, A., Song, W.Y., Martinoia, E., Sakakibara, H. and Lee, Y. (2014) *Arabidopsis* ABCG14 is essential for the root-to-shoot translocation of cytokinin. *Proceedings of the National Academy of Sciences USA* 111, 7150–7155.

Koo, A.J., Gao, X., Jones, A.D. and Howe, G.A. (2009) A rapid wound signal activates the systemic synthesis of bioactive jasmonates in Arabidopsis. *The Plant Journal* 59, 974–986.

Korovetska, H., Novák, O., Jůza, O. and Gloser, V. (2014) Signalling mechanisms involved in the response of two varieties of *Humulus lupulus* L. to soil drying: I. changes in xylem sap pH and the concentrations of abscisic acid and anions. *Plant and Soil* 380, 375–387.

Krapp, A., David, L.C., Chardin, C., Girin, T., Marmagne, A., Leprince, A., Chaillou, S., Ferrario-Méry, S., Meyer, C. and Daniel-Vedele, F. (2014) Nitrate transport and signalling in *Arabidopsis*. *Journal of Experimental Botany* 65, 789–798.

Kretzschmar, T., Kohlen, W., Sasse, J., Borghi, L., Schlegel, M., Bachelier, J.B., Reinhardt, D., Bours, R., Bouwmeester, H.J. and Martinoia, E. (2012) A petunia ABC protein controls strigolactone-dependent symbiotic signalling and branching. *Nature* 483, 341–344.

Krishnan, H., Natarajan, S., Bennett, J. and Sicher, R. (2011) Protein and metabolite composition of xylem sap from field-grown soybeans (*Glycine max*). *Planta* 233, 921–931.

Larrieu, A., Champion, A., Legrand, J., Lavenus, J., Mast, D., Brunoud, G., Oh, J., Guyomarc'h, S., Pizot, M., Farmer, E.E., Turnbull, C., Vernoux, T., Bennet, M.J. and Laplaze, L. (2015) A fluorescent hormone biosensor reveals the dynamics of jasmonate signalling in plants. *Nature Communications* 6, 6043.

Lee, T.-M., Lur, H.-S., Lin, Y.-H. and Chu, C. (1996) Physiological and biochemical changes related to methyl jasmonate-induced chilling tolerance of rice (*Oryza sativa* L.) *Plant, Cell and Environment* 19, 65–74.

Lemgo, G., Sabbadini, S., Pandolfini, T. and Mezzetti, B. (2013) Biosafety considerations of RNAi-mediated virus resistance in fruit-tree cultivars and in rootstock. *Transgenic Research* 22, 1073–1088.

Leonardi, C. and Giuffrida, F. (2006) Variation of plant growth and macronutrient uptake in grafted tomatoes and eggplant on three different rootstocks. *European Journal of Horticultural Science* 71, 97–101.

Lexa, M. and Cheeseman, J.M. (1997) Growth and nitrogen relations in reciprocal grafts of wild-type and nitrate reductase-deficient mutants of pea (*Pisum sativum* L. var. Juneau). *Journal of Experimental Botany* 48, 1241–1250.

Li, L., Li, C., Lee, G.I. and Howe, G.A. (2002) Distinct role for jasmonate synthesis and action in the systemic wound response of tomato. *Proceedings of the National Academy of Sciences USA* 99, 6416–6421.

Li, D.-M., Guo, Y.-K., Zhang, J., Wang, X.J. and Bai, J.-G. (2012) The pretreatment of cucumber with methyl jasmonate regulates antioxidant enzyme activities and protects chloroplast and mitochondrial ultrastructure in chilling-stressed leaves. *Scientia Hortuculturae* 143, 135–143.

Li, P., Zhou, Y., Xia, X., Shi, K., Chen, Z. and Yu, J. (2013) Brassinosteroids-induced systemic stress tolerance was associated with increased transcripts of several defence-related genes in the phloem in *Cucumis sativus*. *PLoS ONE* 8, e66582.

Lin, M.K., Belanger, H., Lee, Y.J., Varkonyi-Gasic, E., Taoka, K., Miura, E., Xoconostle-Cázares, B., Gendler, K., Jorgensen, R.A., Phinney, B., Lough, T.J. and Lucas, W.J. (2007) FLOWERING LOCUS T protein may act as the long-distance florigenic signal in the cucurbits. *The Plant Cell* 19, 488–1506.

Lisec, J., Schauer, N., Kopka, J., Willmitzer, L. and Fernie, A.R. (2006) Gas chromatography mass spectrometry-based metabolite profiling in plants. *Nature Protocols* 1, 387–396.

Llanes, A., Masciarelli, O., Ordóñez, R., Isla, M.I. and Luna, V. (2014) Differential growth responses to sodium salts involve different abscisic acid metabolism and transport in *Prosopis strombulifera*. *Biologia Plantarum* 58, 80–88.

Lough, T.J. and Lucas, W.J. (2006) Integrative plant biology: role of phloem long-distance trafficking. *Annual Review of Plant Biology* 57, 203–232.

Lucas, W.J., Yoo, B.-C. and Kragler, F. (2001) RNA as a long-distance information macromolecule in plants. *Nature Reviews Molecular Cell Biology* 2, 849–857.

Mabood, F. and Smith, D.L. (2005) Pre-incubation of *Bradyrhizobium japonicum* with jasmonates accelerates nodulation and nitrogen fixation in soybean (*Glycine max*) at optimal and suboptimal root-zone temperatures. *Physiologia Plantarum* 125, 311–323.

Malagoli, P., Britto, D.T., Schulze, L.M. and Kronzucker, H.J. (2008) Futile Na^+ cycling at the root plasma membrane in rice (*Oryza sativa* L.): kinetics, energetics, and relationship to salinity tolerance. *Journal of Experimental Botany* 59, 4109–4117.

Malamy, J., Carr, J.P., Klessig, D.F. and Raskin, I. (1990) Salicylic acid: a likely endogenous signal in the resistance response of tobacco to viral infection. *Science* 250, 1002–1004.

Manschadi, A.M., Christopher, J., Devoil, P. and Hammer, G.L. (2006) The role of root architectural traits in adaptation of wheat to water-limited environments. *Functional Plant Biology* 33, 823–837.

Martínez-Andújar, C. (2006) Análisis funcional de una asparragina sintetasa dependiente de amonio en tomate (*Solanum lycopersicum* L.) cultivado en condiciones de salinidad y diferente nutrición nitrogenada. PhD thesis, Universidad de Murcia, Murcia, Spain.

Martínez-Ballesta, M.C., Alcaraz-López, C., Muries, B., Mota-Cadenas, C. and Carvajal, M. (2010) Physiological aspects of rootstock–scion interactions. *Scientia Horticulturae* 127, 112–118.

Martínez-Rodríguez, M.M., Estañ, M.T., Moyano, E., García-Abellan, J.O., Flores, F.B., Campos, J.F., Al-Azzawi, M.J., Flowers, T.J. and Bolarín, M.C. (2008) The effectiveness of grafting to improve salt tolerance in tomato when an 'excluder' genotype is used as scion. *Environmental and Experimental Botany* 63, 392–401.

Matsuura, H., Takeishi, S., Kiatoka, N., Sato, C., Sueda, K., Masuta, C. and Nabeta, K. (2012) Transportation of *de novo* synthesized jasmonyl isoleucine in tomato. *Phytochemistry* 83, 25–33.

Matusova, R., Rani, K., Verstappen, F.W.A., Franssen, M.C.R., Beale, M.H. and Bouwmeester, H.J. (2005) The strigolactone germination stimulants of the plant-parasitic *Striga* and *Orobanche* spp. are derived from the carotenoid pathway. *Plant Physiology* 139, 920–934.

McAdam, S.A.M., Sussmilch, F.C. and Brodribb, T.J. (2016) Stomatal responses to vapour pressure deficit are regulated by high speed gene expression in angiosperms. *Plant, Cell and Environment* 39, 485–491.

McDonnell, L., Plett, J.M., Andersson-Gunneras, S., Kozela, C., Dugardeyn, J., Van Der Straeten, D., Glick, B.R., Sundberg, B. and Regan, S. (2009) Ethylene levels are regulated by a plant encoded 1-aminocyclopropane-1-carboxylic acid deaminase. *Physiologia Plantarum* 136, 94–109.

McLaughlin, S.B. and Wimmer, R. (1999) Calcium physiology and terrestrial ecosystem processes. *New Phytologist* 142, 373–417.

Menges, M., Samland, A.K., Planchais, S. and Murray, A.H. (2006) The D-type cyclin CYCD3;1 is limiting for the G1-to-S-phase transition in Arabidopsis. *Plant Cell* 18, 893–906.

Métraux, J.-P., Signer, H., Ryals, J.A., Ward, E., Wyss-Benz, M., Gaudin, J., Raschdorf, K., Schmidt, E., Blum, W. and Inverardi, B. (1990) Increase in salicylic acid at the onset of systemic acquired resistance in cucumber. *Science* 250, 1004–1006.

Meyer, P. (2015) Epigenetic variation and environmental change. *Journal of Experimental Botany* 66, 3541–3548.

Miller, A.J., Fan, X.R., Orsel, M., Smith, S.J. and Wells, D.M. (2007) Nitrate transport and signalling. *Journal of Experimental Botany* 58, 2297–2306.

Miller, G., Schlauch, K., Tam, R., Cortes, D., Torres, M.A., Shulaev, V., Dangl, J.L. and Mittler, R. (2009) The plant NADPH oxidase RBOHD mediates rapid systemic signaling in response to diverse stimuli. *Science Signaling* 2, ra45.

Mistrik, I. and Ullrich, C.I. (1996) Mechanism of anion uptake in plant roots: quantitative evaluation of H^+/NO_3^- and $H^+/H_2PO_4^-$ stoichiometries. *Plant Physiology and Biochemistry* 34, 629–636.

Miura, K. and Tada, Y. (2014) Regulation of water, salinity, and cold stress responses by salicylic acid. *Frontiers in Plant Science* 5, 4.

Molnar, A., Melnyk, C.W., Bassett, A., Hardcastle, T.J., Dunn, R. and Baulcombe, D.C. (2010) Small silencing RNAs in plants are mobile and direct epigenetic modification in recipient cells. *Science* 328, 872–875.

Montoya, T., Nomura, T., Yokota, T., Farrar, K., Harrison, K., Jones, J.D., Kaneta, T., Kamiya, Y., Szekeres, M. and Bishop, G.J. (2005) Patterns of *Dwarf* expression and brassinosteroid accumulation in tomato reveal the importance of brassinosteroid synthesis during fruit development. *The Plant Journal* 42, 262–269.

Morris, S.E., Turnbull, C.G.N., Murfet, I.C. and Beveridge, C.A. (2001) Mutational analysis of branching in pea. Evidence that *Rms1* and *Rms5* regulate the same novel signal. *Plant Physiology* 126, 1205–1213.

Notaguchi, M., Higashiyama, T. and Suzuki, T. (2015) Identification of mRNAs that move over long distances using an RNA-Seq analysis of Arabidopsis/*Nicotiana benthamiana* heterografts. *Plant and Cell Physiology* 56, 311–321.

Ntatsi, G., Savvas, D., Druege, U. and Schwarz, D. (2013) Contribution of phytohormones in alleviating the impact of sub-optimal temperature stress on grafted tomato. *Scientia Horticulturae* 149, 28–38.

Oeller, P.W., Min-Wong, L., Taylor, L.P., Pike, D.A. Theologis, A. (1991) Reversible inhibition of tomato fruit senescence by antisense RNA. *Science* 254, 437–439.

Omid, A., Keilin, T., Glass, A., Leshkowitz, D. and Wolf, S. (2007) Characterization of phloem-sap transcription profile in melon plants. *Journal of Experimental Botany* 58, 3645–3656.

Orsini, F., Sanoubar, R., Oztekin, G.B., Kappel, N., Tepecik, M., Quacquarelli, C., Tuzel, Y., Bona, S. and Gianquinto, G (2013) Improved stomatal regulation and ion partitioning boosts salt tolerance in grafted melon. *Functional Plant Biology* 40, 628–636.

Palauqui, J.C., Elmayan, T., Pollien, J.M. and Vaucheret, H. (1997) Systemic acquired silencing: transgene-specific post-transcriptional silencing is transmitted by grafting from silenced stocks to non-silenced scions. *EMBO Journal* 16, 4738–4745.

Park, S.-W., Kaimoyo, E., Kumar, D., Mosher, S. and Klessing, D.F. (2007) Methyl salicylate is a critical mobile signal for plant systemic acquired resistance. *Science* 318, 113–116.

Pedranzani, H., Racagni, G., Alemano, S., Miersch, O., Ramírez, I., Penas-Cortés, H., Taleisnik, E., Machado-Domenech, E. and Abdala, G. (2003) Salt tolerant tomato plants show increased levels of jasmonic acid. *Plant Growth Regulation* 41, 149–158.

Pérez-Alfocea, F., Estañ, M.T., Caro, M. and Bolarín, M.C. (1993) Response of tomato cultivars to salinity. *Plant and Soil* 150, 203–211.

Pérez-Alfocea, F., Balibrea, M.E., Santa-Cruz, A. and Estañ, M.T. (1996) Agronomical and physiological characterization of salinity tolerance in a commercial tomato hybrid. *Plant and Soil* 180, 251–257.

Pérez-Alfocea, F., Balibrea, M.E., Alarcón, J.J. and Bolarín, M.C. (2000) Composition of xylem and phloem exudates in relation to the salt-tolerance of domestic and wild tomato species. *Journal of Plant Physiology* 156, 367–374.

Pérez-Alfocea, F., Albacete, A., Ghanem, M.E. and Dodd, I.C. (2010) Hormonal regulation of source–sink relations to maintain crop productivity under salinity: a case study of root-to-shoot signalling in tomato. *Functional Plant Biology* 37, 592–603.

Pérez-Alfocea, F., Ghanem, M.E., Gómez-Cadenas, A. and Dodd, I.C. (2011) Omics of root-to-shoot signaling under salt stress and water deficit. *OMICS* 15, 893–901.

Peuke, A.D. (2010) Correlations in concentrations, xylem and phloem flows, and partitioning of elements and ions in intact plants. A summary and statistical re-evaluation of modelling experiments in *Ricinus communis*. *Journal of Experimental Botany* 61, 635–655.

Pfautsch, S., Gessler, A., Adams, M.A. and Rennenberg, H. (2009) Using amino-nitrogen pools and fluxes to identify contributions of understory *Acacia* spp. to overstory *Eucalyptus regnans* and stand nitrogen uptake in temperate Australia. *New Phytologist* 183, 1097–1113.

Pierce, M. and Raschke, K. (1981) Synthesis and metabolism of abscisic acid in detached leaves of *Phaseolus vulgaris* L. after loss and recovery of turgor. *Planta* 153, 156–165.

Rahayu, Y.S., Walch-Liu, P., Neumann, G., Römheld, V., von Wirén, N. and Bangerth, F. (2005) Root-derived cytokinins as long-distance signals for NO_3^--induced stimulation of leaf growth. *Journal of Experimental Botany* 56, 1143–1152.

Rajendran, K.A., Tester, M. and Roy, S.J. (2009) Quantifying the three main components of salinity tolerance in cereals. *Plant, Cell and Environment* 32, 237–249.

Regnault, T., Daviere, J.M., Wild, M., Sakvarelidze-Achard, L., Heintz, D., Bergua, E.C., Lopez-Diaz, I., Gong, F., Hedden, P. and Achard, P. (2015) The gibberellin precursor GA_{12} acts as a long-distance growth signal in *Arabidopsis*. *Nature Plants* 6, 15073.

Reid, D.E., Hayashi, S., Lorenc, M., Stiller, J., Edwards, D., Gresshoff, P.M. and Ferguson, B.J. (2012) Identification of systemic responses in soybean nodulation by xylem sap feeding and complete transcriptome sequencing reveal a novel component of the autoregulation pathway. *Plant Biotechnology Journal* 10, 680–689.

Reinoso, H., Sosa, L., Reginato, M. and Luna, V. (2005) Histological alterations induced by sodium sulphate in the vegetative anatomy of *Prosopis strombulifera* (Lam.) Benth. *World Journal of Agricultural Sciences* 1, 109–119.

Rellán-Álvarez, R., El-Jendoubi, H., Wohlgemuth, G., Abadía, A., Fiehn, O., Abadía, J. and Álvarez-Fernández, A. (2011) Metabolite profile changes in xylem sap and leaf extracts of Strategy I plants in response to iron deficiency and resupply. *Frontiers in Plant Science* 2, 66.

Rivas-San Vicente, M. and Plasencia, J. (2011) Salicylic acid beyond defence: its role in plant growth and development. *Journal of Experimental Botany* 62, 3321–3338.

Rodriguespousada, R.A., Derycke, R., Dedonder, A., Vancaeneghem, W., Engler, G., van Montagu, M. and van der Straeten, D. (1993) The Arabidopsis 1-aminocyclopropane-1-carboxylate synthase gene-1 is expressed during early development. *Plant Cell* 5, 897–911.

Rothwell, S.A., Elphinstone, E.D. and Dodd, I.C. (2015) Liming can decrease legume crop yield and leaf gas exchange by enhancing root-to-shoot ABA signalling. *Journal of Experimental Botany* 66, 2335–2345.

Ruiz, L.P., Atkinson, C.J. and Mansfield, T.A. (1993) Calcium in the xylem and its influence on the behaviour of stomata. *Philosophical Transactions of the Royal Society B: Biological Sciences* 341, 67–74.

Ruiz, J.M., Belakbir, A., López-Cantarero, I. and Romero, L. (1997) Leaf-macronutrient content and yield in grafted melon plants. A model to evaluate the influence of rootstock genotype. *Scientia Horticulturae* 71, 227–234.

Sacher, R.F., Staples, R.C. and Robinson, R.W. (1982) Saline tolerance in hybrids of *Lycopersicon esculentum* × *Solanum pennellii* and selected breeding lines. In: San Pietro, A. (ed.) *Biosaline Research: A Look to the Future*. Plenum Press, New York, pp. 325–336.

Sachs, T. (2005) Auxin's role as an example of the mechanisms of shoot/root relations. *Plant and Soil* 268, 13–19.

Sakakibara, H., Takei, K. and Hirose, N. (2006) Interactions between nitrogen and cytokinin in the regulation of metabolism and development. *Trends in Plant Science* 11, 440–448.

Salvador-Recatalà, V., Tjallingii, W.F. and Farmer, E.E. (2014) Real-time, *in vivo* intracellular recordings of caterpillar-induced depolarization waves in sieve elements using aphid electrodes. *New Phytologist* 203, 674–684.

Sánchez-Romera, B., Ruiz-Lozano, J.M., Li, G., Luu, D.-T., Martínez-Ballesta, Mdel.C., Carvajal, M., Zamarreno, A.M., Garciá-Mina, J.M., Maurel, C. and Aroca, R. (2014) Enhancement of root hydraulic conductivity by methyl jasmonate and the role of calcium and abscisic acid in this process. *Plant, Cell and Environment* 37, 995–1008.

Savchenko, T. and Dehesh, K. (2014) Drought stress modulates oxylipin signature by eliciting 12-OPDA as a potent regulator of stomatal aperture. *Plant Signalling and Behaviour* 9, e28304.

Savchenko, T., Kolla, V.A., Wang, C.-Q., Nasafi, Z., Hicks, D.R., Phadungchob, B., Chehab, W.E., Brandizzi, F., Froehlich, J. and Dehesh, K. (2014) Functional convergence of oxylipin and abscisic acid pathways controls stomatal closure in response to drought. *Plant Physiology* 164, 1151–1160.

Schachtman, D.P. and Goodger, J.Q. (2008) Chemical root to shoot signaling under drought. Trends in Plant Science 13, 281–287.

Schjoerring, J.K., Husted, S., Mäck, G. and Mattsson, M. (2002) The regulation of ammonium translocation in plants. *Journal of Experimental Botany* 53, 883–890.
Sharp, R.G. and Davies, W.J. (2009) Variability among species in the apoplastic pH signalling response to drying soils. *Journal of Experimental Botany* 60, 4363–4370.
Scott, I.M., Clarke, S.M., Wood, J.E. and Mur, L.A.J. (2004) Salicylate accumulation inhibits growth at chilling temperature in Arabidopsis. *Plant Physiology* 135, 1040–1049.
Shabala, S., White, R.G., Djordjevic, M.A., Ruan, Y.-L. and Mathesius, U. (2016) Root-to-shoot signalling: integration of diverse molecules, pathways and functions. *Functional Plant Biology* 43, 87–104.
Shi, Y., An, L., Li, X., Huang, C. and Chen, G. (2011) The octadecanoic signalling pathway participates in the chilling-induced transcription of ω-3 fatty acid desaturases in *Arabidopsis*. *Plant Physiology and Biochemistry* 49, 208–215.
Shulaev, V., Silverman, P. and Raskin, I. (1997) Airborne signaling by methyl salicylate in plant pathogen resistance. *Nature* 385, 718–721.
Siebrecht, S., Herdel, K., Schurr, U. and Tischner, R. (2003) Nutrient translocation in the xylem of poplar – diurnal variations and spatial distribution along the shoot axis. *Planta* 217, 783–793.
Simons, J.L., Napoli, C.A., Janssen, B.J., Plummer, K.M. and Snowden, K.C. (2007) Analysis of the *DECREASED APICAL DOMINANCE* genes of petunia in the control of axillary branching. *Plant Physiology* 143, 697–706.
Spiegelman, Z., Golan, G. and Wolf, S. (2013) Don't kill the messenger: long-distance trafficking of mRNA molecules. *Plant Science* 213, 1–8.
Spiegelman, Z., Ham, B.K., Zhang, Z., Toal, T.W., Brady, S.M., Zheng, Y., Fei, Z., Lucas, W.J. and Wolf, S. (2015) A tomato phloem-mobile protein regulates the shoot-to-root ratio by mediating the auxin response in distant organs. *The Plant Journal* 83, 853–863.
Stahlberg, R. and Cosgrove, D.J. (1995) Comparison of electric and growth responses to excision in cucumber and pea seedlings. II. Long-distance effects are caused by the release of xylem pressure. *Plant, Cell and Environment* 18, 33–41.
Stegemann, S. and Bock, R. (2009) Exchange of genetic material between cells in plant tissue grafts. *Science* 324, 649–651.
Stratmann, J.W. (2003) Long distance run in the wound response – jasmonic acid is pulling ahead. *Trends in Plant Science* 8, 247–250.
Symons, G.M. and Reid, J.B. (2004) Brassinosteroids do not undergo long-distance transport in pea. Implications for the regulation of endogenous brassinosteroid levels. *Plant Physiology* 135, 2196–2206.
Symons, G.M., Ross, J.J., Jager, C.E. and Reid, J.B. (2008) Brassinosteroid transport. *Journal of Experimental Botany* 59, 17–24.
Tamaki, S., Matsuo, S., Wong, H.L., Yokoi, S. and Shimamoto, K. (2007) Hd3a protein is a mobile flowering signal in rice. *Science* 316, 1033–1036.
Tamogami, S., Noge, K., Abe, M., Agrawal, G.K. and Rakwal, R. (2012) Methyl jasmonate is transported to distal leaves via vascular process metabolizing itself into JA-Ile and triggering VOCs emission as defensive metabolites. *Plant Signaling and Behavior* 7, 1378–1381.
Terakado, J., Yoneyama, T. and Fujihara, S. (2006) Shoot-applied polyamines suppress nodule formation in soybean (*Glycine max*). *Journal of Plant Physiology* 163, 497–505.
Thompson, A.J., Andrews, J., Mulholland, B.J., McKee, J.M., Hilton, H.W., Horridge, J.S., Farquhar, G.D., Smeeton, R.C., Smillie, I.R., Black, C.R. and Taylor, I.B. (2007) Overproduction of abscisic acid in tomato increases transpiration efficiency and root hydraulic conductivity and influences leaf expansion. *Plant Physiology* 143, 1905–1017.
Tudela, D. and Primo-Millo, E. (1992) 1-Aminocyclopropane-1-carboxylic acid transported from roots to shoots promotes leaf abscission in Cleopatra mandarin (*Citrus reshni* Hort. ex Tan.) seedlings rehydrated after water-stress. *Plant Physiology* 100, 131–137.

Turnbull, C.G.N., Booker, J.P. and Leyser, H.M.O. (2002) Micrografting techniques for testing long-distance signalling in *Arabidopsis*. *The Plant Journal* 32, 255–262.

Turner, J.G., Ellis, C. and Devoto, A. (2002) The jasmonate signal pathway. *Plant Cell* 13, S153–S164.

Tytgat, T.O.G., Verhoeven, K.J.F., Jansen, J.J., Raaijmakers, C.E., Bakx-Schotman, T., McIntyre, L.M., van der Putten, W.H., Biere, A. and van Dam, N.M. (2013) Plants know where it hurts: root and shoot jasmonic acid induction elicit differential responses in *Brassica oleracea*. *PLoS ONE* 8, e65502.

Umehara, M., Hanada, A., Magome, H., Takeda-Kamiya, N. and Yamaguchi, S. (2010) Contribution of strigolactones to the inhibition of tiller bud outgrowth under phosphate deficiency in rice. *Plant and Cell Physiology* 51, 1118–1126.

Uroic, K., Salaun, P., Raab, A. and Feldmann, J. (2012) Arsenate impact on the metabolite profile, production and arsenic loading of xylem sap in cucumbers (*Cucumis sativus* L.). *Frontiers in Physiology* 3, 55.

van de Poel, B. and van der Straeten, D. (2014) 1-aminocyclopropane-1-carboxylic acid (ACC) in plants: more than just the precursor of ethylene! *Frontiers in Plant Science* 5, 640.

Vernooij, B., Friedrich, L., Morse, A., Reist, R., Kolditz-Jawhar, R., Ward, E., Uknes, S., Kessmann, H., and Ryals, J. (1994) Salicylic acid is not the translocated signal responsible for inducing systemic acquired resistance but is required in signal transduction. *Plant Cell* 6, 959–965.

Villalta, I., Reina-Sánchez, A., Bolarín, M.C., Cuartero, J., Belver, A., Venema, K., Carbonell, E.A. and Asins, M.J. (2008) Genetic analysis of Na^+ and K^+ concentrations in leaf and stem as physiological components of salt tolerance in tomato. *Theoretical and Applied Genetics* 116, 869–880.

Vlot, A.C., Klessing, D.F. and Park, S.-W. (2008) Systemic acquired resistance: the elusive signal(s). *Current Opinion in Plant Biology* 11, 436–442.

Vojta, P., Kokáš, F., Husičková, A., Grúz, J., Bergougnoux, V., Marchetti, C.F., Jiskrová, E., Ježilová, E., Mik, V., Čudejková, M. and Galuszka, P. (2016) Whole transcriptome analysis of transgenic barley with altered cytokinin homeostasis and increased tolerance to drought stress. *New Biotechnology* 33, 676–679.

Wallis, C.M. and Chen, J. (2012) Grapevine phenolic compounds in xylem sap and tissues are significantly altered during infection by *Xylella fastidiosa*. *Phytopathology* 102, 816–826.

Wasternack, C. (2007) Jasmonates: an update on biosynthesis, signal transduction and action in plant stress response, growth and development. *Annals of Botany* 100, 681–697.

Wasternack, C. (2014) Action of jasmonates in plant stress responses and development – applied aspects. *Biotechnology Advances* 32, 31–39.

Wasternack, C. and Hause, B. (2013) Jasmonates: biosynthesis, perception, signal transduction and action in plant stress response, growth and development. An update to the 2007 review in *Annals of Botany*. *Annals of Botany* 111, 1021–1058.

Wasternack, C., Goetz, S., Hellwege, A., Forner, S., Strnad, M. and Hause, B. (2012) Another JA/COI1-independent role of OPDA detected in tomato embryo development. *Plant Sigalling and Behaviour* 7, 10.

Weckwerth, W. (2011) Unpredictability of metabolism – the key role of metabolomics science in combination with next-generation genome sequencing. *Analytical and Bioanalytical Chemistry* 400, 1967–1978.

Wei, Z. and Li, J. (2016) Brassinosteroids regulate root growth, development, and symbiosis. *Molecular Plant* 9, 86–100.

Werner, T., Holst, K., Poers, Y., Guivarc'h, A., Mustroph, A., Chriqui, D., Grimm, B. and Schmuelling, T. (2008) Cytokinin deficiency causes distinct changes of sink and source parameters in tobacco shoots and roots. *Journal of Experimental Botany* 59, 2659–2672.

Werner, T., Nehnevajova, E., Kollmer, I., Novak, O., Strnad, M., Kramer, U. and Schmulling, T. (2010) Root-specific reduction of cytokinin causes enhanced root growth, drought tolerance, and leaf mineral enrichment in *Arabidopsis* and tobacco. *Plant Cell* 22, 3905–3920.

White, P.J. (2001) The pathways of calcium movement to the xylem. *Journal of Experimental Botany* 52, 891–899.

Wilkinson, S. (1999) pH as a stress signal. *Plant Growth Regulation* 29, 87–99.

Wilkinson, S. and Davies, W.J. (2002) ABA-based chemical signalling: the co-ordination of responses to stress in plants. *Plant, Cell and Environment* 25, 195–210.

Wu, H., Ye, H., Yao, R., Zhang, T. and Xioung, L. (2015) OsJAZ9 acts as a transcriptional regulator in jasmonate signalling and modulates salt stress tolerance in rice. *Plant Science* 232, 1–12.

Xia, J., Zhao, H., Liu, W., Li, L. and He, Y. (2009) Role of cytokinin and salicylic acid in plant growth at low temperatures. *Plant Growth Regulation* 57, 211–221.

Xia, X.-J., Zhou, Y.-H., Ding, J., Shi, K., Asami, T., Chen, Z. and Yu, J.-Q. (2011) Induction of systemic stress tolerance by brassinosteroid in *Cucumis sativus*. *New Phytologist* 191, 706–720.

Yoneyama, K., Xie, X., Kusumoto, D., Sekimoto, H., Sugimoto, Y., Takeuchi, Y. and Yoneyama, K. (2007) Nitrogen deficiency as well as phosphorus deficiency in sorghum promotes the production and exudation of 5-deoxystrigol, the host recognition signal for arbuscular mycorrhizal fungi and root parasites. *Planta* 227, 125–132.

Yoneyama, K., Arakawa, R., Ishimoto, K., Kim, H.I., Kisugi, T., Xie, X., Nomura, T., Kanampiu, F., Yokota, T., Ezawa, T. and Yoneyama, K. (2015) Difference in *Striga*-susceptibility is reflected in strigolactone secretion profile, but not in compatibility and host preference in arbuscular mycorrhizal symbiosis in two maize cultivars. *New Phytologist* 206, 983–989.

Zhang, J.H. and Davies, W.J. (1991) Antitranspirant activity in xylem sap of maize plants. *Journal of Experimental Botany* 42, 317–321.

Zhang, C. and Huang, Z. (2013) Effects of endogenous abscisic acid, jasmonic acid, polyamines, and polyamine oxidase activity in tomato seedlings under drought stress. *Scientia Horticulturae* 159, 172–177.

Zhang, Z.P. and Baldwin, I.T. (1997) Transport of [2-^{14}C] jasmonic acid from leaves to roots mimics wound-induced changes in endogenous jasmonic acid pools in *Nicotiana sylvestris*. *Planta* 203, 436–441.

Zhao, D. and Song, G.Q. (2014) Rootstock-to-scion transfer of transgene derived small interfering RNAs and their effect on virus resistance in non-transgenic sweet cherry. *Plant Biotechnology Journal* 12, 1319–1328.

Zhao, M.-L., Wang, J.-N., Shan, W., Fan, J.-G., Kuang, J.F., Wu, K.Q., Li, X.-P., Chen, W.-C., He, F.-Y., Chen, J.Y. and Lu, W.-J. (2013) Induction of jasmonate signalling regulators MaMYC2s and their physical interactions with MaICE1 in methyl jasmonate-induced chilling tolerance in banana fruit. *Plant, Cell and Environment* 36, 30–51.

Zimmermann, M.R. and Mithöfer, A. (2013) Electrical long-distance signalling in plants. In: Baluška, F. (ed.) *Long-Distance Systemic Signalling and Communication in Plants*. Springer-Verlag, Berlin/Heidelberg, Germany, pp. 291–308.

Zimmermann, P., Hirsch-Hofmann, M., Henning, L. and Gruissem, W. (2004) GENEVESTIGATOR. Arabidopsis microarray database and analysis toolbox. *Plant Physiology* 136, 2631–2632.

5 Physiological and Molecular Mechanisms Underlying Graft Compatibility

Ana Pina,[1*] Sarah Jane Cookson,[2] Angeles Calatayud,[3] Alessandra Trinchera[4] and Pilar Errea[1]

[1]Instituto Agroalimentario de Aragón (CITA Universidad de Zaragossa), Zaragoza, Spain; [2]Institut des Sciences de la Vigne et du Vin, INRA/ Université de Bordeaux, Bordeaux, France; [3]Instituto Valenciano de Investigaciones Agrarias, Valencia, Spain; [4]Centro di ricerca Agricoltura e Ambiente, CREA, Rome, Italy

5.1 Introduction

Grafting has been used for millennia to increase uniformity, vigour and resistance to biotic and abiotic stresses (e.g. low soil temperature, drought, salinity and flooding) of vegetatively propagated plants (Hartmann *et al.*, 2002; Lee and Oda, 2003; Mudge *et al.*, 2009). The grafting technique plays an important role in the production of horticultural crops, including fruit trees, ornamental plants and vegetable crops (Lee and Oda, 2003); it can be considered an important and alternative innovative practice of integrated pest management and a promising alternative for soil fumigants in vegetable production (Guan *et al.*, 2012). Currently, although the use of grafted plants is increasing rapidly, the practice of grafting remains limited, mainly due to inconsistent grafting success (Hartmann *et al.*, 2002). Considering the variation of graft compatibility between even closely related species, it is necessary to evaluate graft compatibility before considering the use of a rootstock with a specific scion genotype (Lee *et al.*, 2010; Guan *et al.*, 2012). Although a number of rootstocks are available, little information is provided related to scion–rootstock compatibility. This issue should be considered, as the performance of a grafted plant depends on the properties of the scion and rootstock genotypes, the compatibility of the rootstock with the scion, environmental conditions and cultivation methods (Andrews and Serrano-Marquez, 1993; Lee, 1994). Given the relative importance of graft compatibility throughout the world,

* apina@aragon.es

there has been surprisingly little research dedicated to the study of this phenomenon. The large number of genotypes that can be combined by grafting produces a wide range of different physiological, biochemical and anatomical interactions to study, slowing down the progress in this research area.

The aim of this chapter is to review the biology of grafting and the factors influencing graft compatibility, focusing on the molecular and physiological aspects affecting graft development and success in herbaceous and woody plants.

5.2 Anatomical and Physiological Steps During Graft Union Development

5.2.1 Graft establishment between compatible and incompatible combinations

The general cellular events that occur after grafting are quite well known and are common to woody and herbaceous plants. These events include adhesion of the two graft partners, callus cell proliferation at the graft interface and cross-bridge formation of the vascular bundle to establish a functional vascular connection. All of these steps can determine the future of a grafted plant.

When the cut tissues of a scion and a rootstock come into intimate contact, the formation of parenchymatous cells capable of interlocking them starts, favouring the development of scion–rootstock connections. Studies on artichoke–cardoon grafts showed that, when scion–rootstock compatibility is good, filamentous connections between the rootstock and scion are formed 3 days after grafting (DAG), with the appearance of horizontal bridges over time (Trinchera *et al.*, 2013). The formation of these horizontal bridges corresponds to the beginning of scion–rootstock connectivity and represents the first skeleton of the undifferentiated cell callus development in response to wounding (Fig. 5.1).

These early cell interconnections, with a tubular feature similar to cell-wall projections, suggest an active role of cell recognition in promoting cambium tissue contact (Miller and Barnett, 1993). In artichoke–cardoon grafts, large-diameter structures (3–8 μM) described as 'macrotubules' were observed at 3 DAG, indicating that these structures are formed during early contact between both graft partners (Fig. 5.2). These structures probably facilitate water and nutrient flow, and promote the exchange of signalling molecules across the grafting interface (Trinchera *et al.*, 2013).

Another aspect related to the parenchymatous callus formation at the graft junction is the adhesion of the two partners, mediated in higher plants by the activation of the dictyosomes of Golgi bodies, which secrete a mixture of pectins, carbohydrates and proteins at the grafting interface (Moore and Walker, 1981). The accumulation of pectic carbohydrate material on the graft interface tissues during the first days after grafting has been shown in herbaceous (Fig. 5.2; Trinchera *et al.*, 2013) and woody (Miller and Barnett, 1993; Pina *et al.*, 2012) plants.

From the newly formed callus, new cambial cells differentiate, forming a continuous cambial connection between the scion and rootstock. Furthermore, prior to the binding of vascular cambium across the callus bridge, the initial xylem

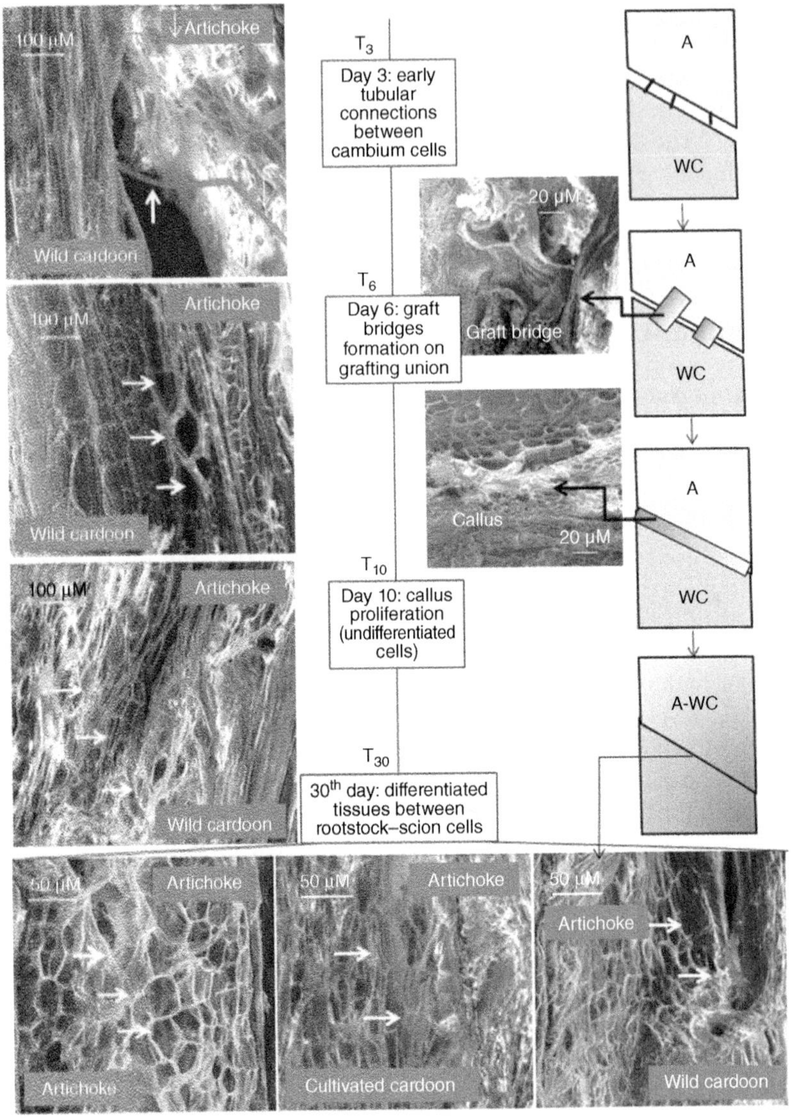

Fig. 5.1. Schematic representation of graft union formation between globe artichoke (A) and wild cardoon (WC), and related images by scanning electron microscopy showing the variable pressure of graft union formation in artichoke–artichoke, artichoke–cultivated cardoon and artichoke–wild cardoon combinations. (From Trinchera *et al.*, 2013, with permission.)

and phloem can be differentiated, permitting the vascular connection between the scion and the rootstock (Hartmann *et al.*, 2002). In melon–pumpkin grafts, Aloni *et al.* (2008) suggested two stages for graft union formation: an initial stage, in which vascular connections from the two partners are reunited (14 DAG) and a second phase in which graft incompatibility or compatibility is expressed

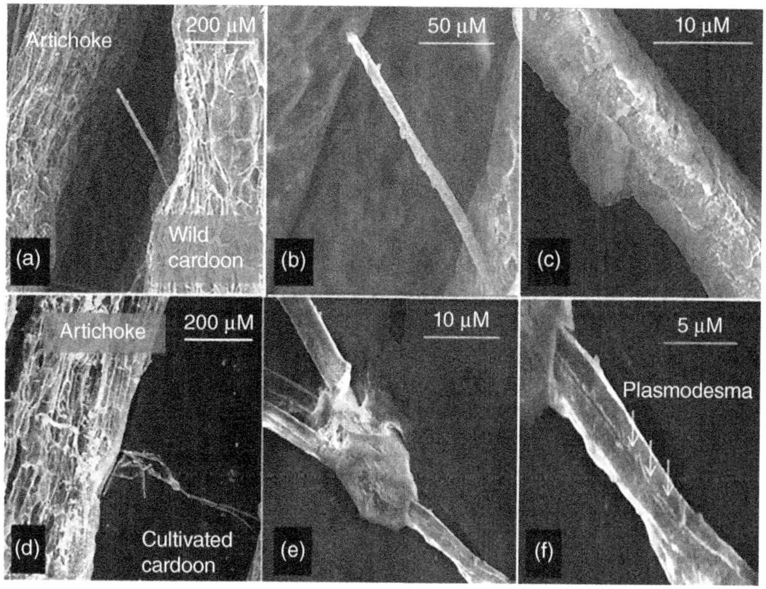

Fig. 5.2. Scanning electron microscopy variable pressure images of interconnections between artichoke–wild cardoon (a–c) and artichoke–cultivated cardoon (d–f) graft combinations, at 3 DAG. (From Trinchera *et al.*, 2013, with permission.)

(24 DAG). Incompatible grafts showed lower water uptake and root sugar concentration, increased H_2O_2 and superoxide levels, lower peroxidase and superoxide dismutase (SOD) activities (Aloni *et al.*, 2008) and a decrease in the photosynthetic activity in the grafting zone (Calatayud *et al.*, 2013). Practical information related to vegetable graft compatibility is given in Table 5.1. Growth inhibition and high mortality were observed in severely incompatible grafted plants such as pepper–tomato and pepper–aubergine grafts at 21 DAG; this was associated with higher concentration of sugars and starch above compared with below the graft union (Kawaguchi *et al.*, 2008). These observations were correlated with narrow and irregular xylem connections between the scion and rootstock (Kawaguchi *et al.*, 2008). Similarly, in other vegetable crops (tomato, cabbage and radish) a necrotic, shrunken cell layer is formed in some areas at the graft interface at 3 DAG. The necrotic layer is formed from the collapsed cells and cytoplasmic remnants at the graft interface. Subsequently, some scion parenchyma cells adjacent to the necrotic layer start to divide along the inner surface of the wounded tissues, generating the callus. This cell proliferation process is strongly influenced by the grafting compatibility. In fact, in homografting, the necrotic layer disappears at the moment of cell division in correspondence with the callus formation, while, in incompatible grafting (as in *Arabidopsis*–tomato grafts), the necrotic layer seems to directly or indirectly inhibit the vascular tissue differentiation, thus blocking full vascular graft union formation between the two graft partners (Flaishman *et al.*, 2008). In addition, the relationship between necrotic layer dissolution and grafting compatibility was evaluated by measuring the electrical resistance at the graft junction in compatible–incompatible combinations. In compatible

Table 5.1. Graft compatibility from different vegetable rootstock–scion combinations

Scion	Rootstock	Rootstock advantages	Compatibility	Reference
Melon 'Arava'	Pumpkin hybrid genotype RS59	*Fusarium* resistance	Compatible	Edelstein *et al.* (2004)
Melon 'Arava'	Pumpkin hybrid genotype RS62	*Fusarium* resistance	Incompatible	Edelstein *et al.* (2004)
Pepper 'Scotch Bonnet'	Aubergine 'Black Beauty'	Salt tolerant	High compatibility	Ives *et al.* (2012)
Pepper 'Scotch Bonnet'	Tomato 'Akash'	Salt tolerant	Moderate compatibility	Ives *et al.* (2012)
Pepper 'Scotch Bonnet'	*Solanum nigrum* (bitter gumma)	Salt tolerant	Low compatibility	Ives *et al.* (2012)
Tomato 'Roma'	Pepper 'Long Red Cayenne'	Disease resistance	Severely incompatible	Kawaguchi *et al.* (2008)
Tomato 'Roma'	Aubergine 'Market Supreme'	Disease resistance	Moderately incompatible	Kawaguchi *et al.* (2008)
Pepper 'Dous des Landes'	Tomato 'Saint Pierre'	Disease resistant radicular system	Compatible	Deloire and Hébant (1982)
Pepper 'Yollowonder' and 'Floridae'	Tomato 'Saint Pierre'	Disease resistant radicular system	Incompatible	Deloire and Hébant (1982)
Artichoke 'Romolo'	Wild cardoon 'Sardo'	Resistance to Verticillium	High compatibility	Trinchera *et al.* (2013)
Artichoke 'Romolo'	Cultivated cardoon 'Belgio'	Resistance to Verticillium	Moderate compatibility	Trinchera *et al.* (2013)
Artichoke 'Istar'	Cultivated cardoon 'Bianco avorio'	Resistance to Verticillium	High compatibility	Temperini *et al.* (2013)
Artichoke 'Istar'	Wild cardoon 'Sardo'	Resistance to Verticillium	Low compatibility	Temperini *et al.* (2013)
Pepper 'Adige'	Pepper (*Capsicum annuum*) 'Serrano'	Salt and drought resistance	Low compatibility	Penella *et al.* (2013)
Pepper 'Adige'	Pepper (*Capsicum baccatum*) 'BOL-558'	Salt and drought resistance	High-moderately compatibility	Penella *et al.* (2015)

scion–rootstock grafts, the electrical resistance increases rapidly with the necrotic layer development in the first 3 DAG and then steadily decreases from 3 to 8 DAG due to callus proliferation. Callus proliferation leads to the breakdown of the necrotic layer, formation of secondary plasmodesmata between adjacent scion–rootstock cells and the differentiation of vascular elements in the callus. Nevertheless, in incompatible grafting combinations, such as amaranth–tomato grafts, the electrical resistance at the graft junction remained high, because of the persistence of the necrotic layer (Yang *et al.*, 1992). Furthermore, the incompatibility is evident through the lethal cellular senescence at the grafting surface, characterized by cell-wall suberization, vesiculation of the cytoplasm, degeneration of cellular organelles, loss of membrane integrity and, finally, death and collapse of the cell.

In comparison with vegetable grafts, in which there is a high percentage of failure of successful grafts or where premature death of the graft union take place, in some fruit tree grafts, the incompatibility is characterized by anatomical alterations in the graft union area, which do not prevent the initial growth of the tree during the first few years. After callus cell proliferation at the graft interface, the differentiation of new cambial cells does not take place to the same extent in compatible and incompatible grafts, where some of the callus remains undifferentiated and becomes parenchymatous tissue. These parenchymatous regions intermingle with vascular tissue, and the union that occurs is mechanically weak, which causes the subsequent failure of the graft union due to a defective structural development (Errea *et al.*, 1994a). In pear–quince (Ermel *et al.*, 1997, 1999), apricot–peach and apricot–plum combinations (Errea *et al.*, 1994a), incompatibility is caused by limited and/or not fully functional vascular reconnection between the scion and rootstock at the graft interface. In other combinations, some biochemical influence moves across the graft interface, causing phloem degeneration, and consequently carbohydrate movement from the scion to the rootstock is restricted at the graft union. In some peach–plum combinations, incompatibility appears generally during the beginning of the first summer after grafting (Zarrouk *et al.*, 2006), and is expressed by tree growth cessation and premature tree defoliation (Moing *et al.*, 1987). Alternatively, delayed incompatibility has been described in *Prunus avium–Prunus cerasus* grafts that failed after a period of years of apparently successful growth (Andrews and Serrano-Marquez, 1983; Hartmann and Kester, 1983; Treutter and Feucht, 1988). This delayed incompatibility is often foreshadowed by one or more typical symptoms developing prior to tree mortality (Webster and Schmidt, 1996). Sweet cherry scions showing incompatibility with *P. cerasus* rootstocks formed less phloem, the sieve plate had narrower pores and a high proportion of these pores were blocked with callose when compared with fully compatible grafts (Webster and Schmidt, 1996). Such responses, however, are based on translocation and thus could still be classified as translocated incompatibility.

5.2.2 Translocation between grafted partners

Grafted plants may show increased uptake of water and minerals compared with self-rooted plants as a consequence of the vigorous root system selected as a rootstock. The vascular connections at the scion–rootstock interface may determine water and nutrient translocation (Martínez-Ballesta *et al.*, 2010). The investigation of water relations by hydraulic conductance (K) measurements can indicate the water status of grafted plants. In tomato grafts, the gradual increase of K from 4 to 8 DAG indicated the establishment of new vascular connections (Turquois and Malone, 1996; Fernández-García *et al.*, 2004). K measurements below and above the graft union gave similar values in various grafted vegetables, suggesting that the graft union is not necessarily a barrier to water passage for a compatible scion–rootstock combination (Martínez-Ballesta *et al.*, 2010). Hence, the graft union may act like a continuous unit with respect to water movement. Regarding nutrient uptake, different authors have related the physical characteristics of the root system (lateral and vertical development) to attributes for enhancing the mineral content of the shoot (Jang, 1992; Penella *et al.*, 2015). In graft union development, the formation of sieve-tube connections and symplastic continuity of sieve tubes are prerequisites for effective assimilate translocation from the scion to the rootstock (Schöning and Kollmann, 1997). Studies on phloem translocation of [^{14}C]sucrose showed effective translocation across the graft interface in compatible graft unions (tomato–potato) correlating with the increasing numbers of sieve-tube connections, but this was not observed in incompatible grafts (faba bean–sunflower), suggesting apoplastic transport (Schöning and Kollmann, 1997).

Different compatibility/incompatibility mechanisms have been proposed, with special reference to the possibility that a phenomenon of cellular recognition must occur as part of adhesion and the events that follow in successful graft union formation (Moore and Walker, 1981; Jeffree and Yeoman, 1983). In grafts of cucumber–squash, it has been suggested that changes in protein banding may be due to polypeptides migrating symplastically across the graft union via the connecting phloem in such a way that translocation of signalling molecules, such as polypeptides in the phloem, could be significant in cell recognition and compatibility between the graft partners (Tiedemann and Carsens-Behrens, 1994). Indeed, a long-distance protein, mRNA and small RNA graft-transmissible signals are currently emerging as novel mechanisms to regulate nutritional and developmental root–shoot relationships and may play a pivotal role in grafting physiology (Goldschmidt, 2014). However, as yet there is no evidence for a specific biochemical or immunological recognition/rejection mechanism between the graft components.

5.3 Role of Secondary Metabolites at the Interface in Graft Incompatibility

When pear cultivars are grafted on to quince rootstocks, a cyanogenic glycoside, prunasin, which is normally found in quince but not in pear, is translocated into the pear phloem. Pear tissues breakdown the prunasin with a glycosidase, which liberates toxic hydrocyanic acid at the graft interface. The presence of cyanide causes the death of cells and destroys xylem and phloem tissues at the graft interface (Gur *et al.*, 1968). However, the presence of cyanogenic glycosides in woody plants is restricted to relatively few genera, and this reaction cannot be considered as a universal cause of graft failure.

The effect of grafting on the content of phenolic compounds in graft unions marked by differing degrees of compatibility has been investigated in many species, and relationships between scion–rootstock compatibility and phenolic acid accumulation have been demonstrated (Errea, 1998; Usenik *et al.*, 2006; Hudina *et al.*, 2014; Canas *et al.*, 2015). Phenolic compounds constitute a distinctive and unique group of plant metabolites that play important roles in higher plants. They participate in the structure of the plants and are involved in a great number of metabolic pathways (Vermerris and Nicholson, 2006; Lattanzio *et al.*, 2008). These compounds are synthesized by plants in response to physical injury or under the influence of several biotic and abiotic stress situations (Treutter, 2006). Orchard management can also modify the phenol content of plant tissues

(Mikulic-Petkovsek *et al.*, 2010), in addition to differences linked to their developmental stage (Usenik *et al.*, 2006).

Diverse phenolic compounds have been implicated in processes of division, development and differentiation into new tissues. These processes are involved in the usual sequence of events in the healing of a graft union (Errea, 1998; Hartmann *et al.*, 2002). Therefore, phenolic compounds could be implicated during the steps of incompatible graft establishment causing insufficient callus proliferation, cell necrosis and accumulation of certain metabolites (Errea, 1998). Feucht *et al.* (1988) noticed that prunin application to *Prunus* callus cultures affected growth and cellular differentiation and initiated proanthocyanidin accumulation. Some studies have also shown quantitative and qualitative differences in the accumulation of phenols at the graft interface of different scion–rootstock combinations. It has been suggested that these metabolites are associated with different degrees of compatibility (Errea *et al.*, 2001; Usenik *et al.*, 2006; Mng'omba *et al.*, 2008; Hudina *et al.*, 2014). When graft unions are established between two different genotypes, physiological disorders may appear at the graft interface (Feucht and Treutter, 1991). In *Prunus* combinations, the biochemical basis for this assumption is that the phenolic compounds produced by heterospecific *Prunus* graft partners are quantitatively and qualitatively different (Geibel and Feucht, 1991; Errea *et al.*, 1994b), and that a steep gradient is therefore established at the line of the union. Among the numerous polyphenols found in nature, proanthocyanidins or condensed tannins (oligomers and polymers composed of elementary flavan-3-ol units) are considered the second most abundant group of natural phenolics after lignins (He *et al.*, 2008). These polyphenols are not only involved in normal processes of stress and wounding but can also play a prominent role in graft union formation by influencing lignification processes and inhibiting auxin transport, and by their protein-precipitating features (Lockard and Schneider, 1981; Hagerman, 2012), consequently impairing a good grafting union.

The effect of grafting on the content of phenolic compounds in graft unions marked by differing degrees of compatibility has been investigated by several lines of research, which found that the relationships between the scion and rootstock are affected by phenol accumulation. Elevated production of phenylpropanoid metabolites is a well documented fact in graft-incompatible combinations from different species, such as apricot (Errea *et al.*, 1992; Usenik *et al.*, 2006; Pina and Errea, 2008a), plum (Rodrigues *et al.*, 2002; Zarrouk *et al.*, 2010), cherry (Gebhardt and Feucht, 1982), pear (Musacchi *et al.*, 2000; Hudina *et al.*, 2014), loquat (Mng'omba *et al.*, 2008), eucalyptus (de Cooman *et al.*, 1996) and grapevine (Canas *et al.*, 2015) grafted on to different rootstocks. The large number of incompatible scion–rootstock combinations with quantitative alterations in the phenolic compound profiles at the graft interface suggests that this is a conserved marker of graft incompatibility. According to Musacchi *et al.* (2000), compounds potentially more suited as indicators of graft incompatibility are in general those found in both scion and rootstock and also accumulate at the graft interface. In apricot, accumulation of monomeric and oligomeric flavan-3-ol compounds were related to problems in the differentiation of the callus (Errea *et al.*, 1994b) and disorganization at the subcellular level (Errea *et al.*, 1994a). The presence of phenolic compounds is generally associated with small cells in incompatible combinations, which do not differentiate and form successful unions (Errea *et al.*, 2001) (*see* Plate 9). Similarly, the flavanol monomers (–)-epicatechin and (+)-catechin and the dimer procyanidin B2 showed the highest graft interface concentration in incompatible quince (Musacchi *et al.*, 2000) and apricot–plum combinations (Errea *et al.*, 2000). More recent studies also performed on pear–quince combinations indicated that not only catechin and procyanidin B1 but also arbutin and several flavonols could be associated with graft incompatibility (Hudina *et al.*, 2014). By analogy with incompatible fruit tree grafts, an abnormal phenolic accumulation (gallic, ellagic, gentisic, *p*-coumaric acid and (+)-catechin) in the graft interface zone could represent the first wound-induced response at the graft unions of incompatible micrografted *Eucalyptus gunnii* (de Cooman *et al.*, 1996). Catechins have been reported to reduce or block stress-induced lipid peroxidation of the membranes (Gadkari and Balaramana, 2015; Martínez-Ballesta *et al.*, 2010). Obviously, flavanols can be considered to form part of a complex fail-safe system that is active under highly oxidative and deleterious conditions occurring in stressed tissues (Elstner *et al.*, 1994). Furthermore, regarding the mechanical stability of the union observed in *Prunus* and *Pyrus* combinations, a role has been proposed for a number of phenolics such as cinnamic acid derivatives or proanthocyanidins as cross-linking and cell-wall-stabilizing compounds (Pizzi and Cameron, 1986). Likewise, in response to grafting, the scion–rootstock interface of incompatible *Arabidopsis*–tomato grafts was composed of a dark red enriched material, similar to polyphenols. It appeared that this material may have restricted, or at least been involved in, preventing vascular bridge formation between the two species, a typical incompatibility response (Flaishman *et al.*, 2008). Taken as a whole, most authors have reported that a significant accumulation of (+)-catechin above the graft union could be used as a biochemical marker of graft incompatibility (Errea, 1998; de Cooman *et al.*, 1996; Musacchi *et al.*, 2000; Usenik *et al.*, 2006; Hudina *et al.*, 2014; Canas *et al.*, 2015).

Phenolic compounds interact with endogenous oxidative enzymes and are subject to rapid oxidation and further polymerization. In situations of graft incompatibility, polyphenols may escape from the vacuolar compartment into the cytoplasm due to loss of vacuolar membrane integrity by alteration of lipid peroxidation. Phenolics released from the vacuole into the cytosol are oxidized by peroxidases and polyphenol oxidases (Errea, 1998). The quinones and the polymers originated by these reactions can then interact with proteins, and as a consequence, grafting may be totally unsuccessful. In this respect, it is interesting to mention the importance of peroxidase and polyphenol oxidases in grafting incompatibility in trees. The peroxidase enzyme activity is associated with differentiating xylem (Sterjiades *et al.*, 1993), as well as with lignification processes (Harkin and Obst, 1973), and it was demonstrated that higher peroxidase activity at the graft interface was related to graft incompatibility in *Prunus* sp. (Rodrigues *et al.*, 2002) and peach–plum combinations (Zarrouk *et al.*, 2010), as well as in incompatible *E. gunnii* micrografts (de Cooman *et al.*, 1996). It was also suggested that differences in peroxidase isozyme banding patterns between the rootstock and scion may be the cause of bad vascular connections (Santamour, 1992). Some findings have also revealed that so-called monophenols stimulate indole-3-acetic acid (IAA) oxidases such as naringenin-7-glucoside (prunin), which accumulates above the graft union of

1

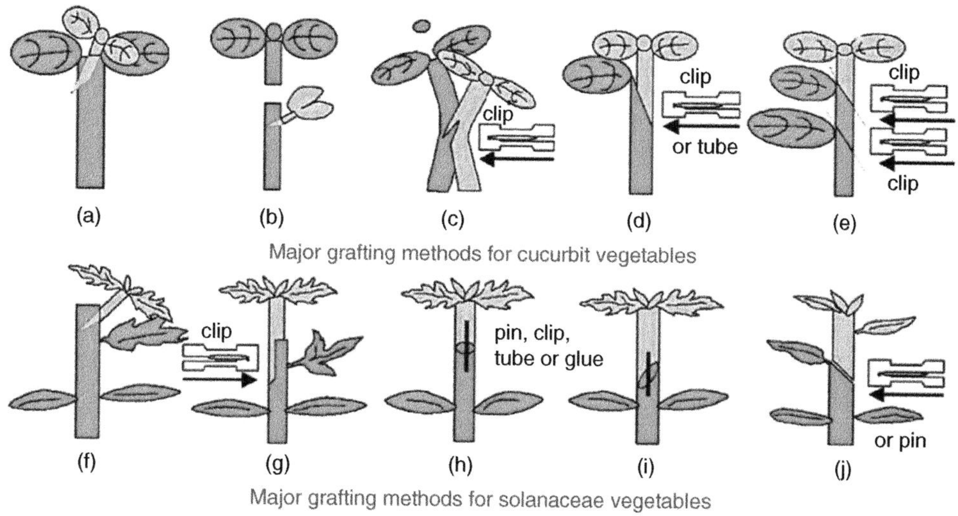

2

3

4

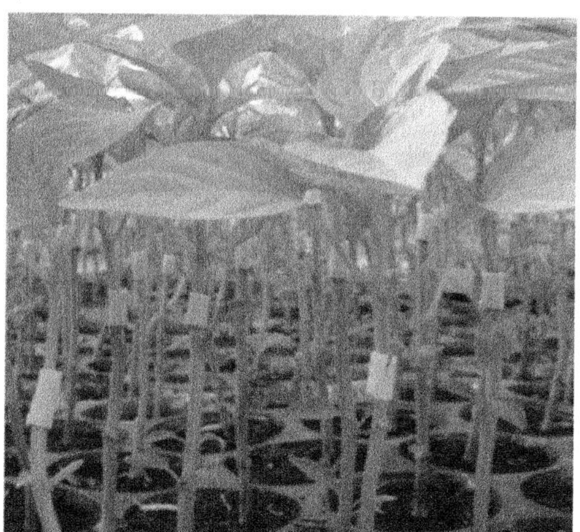

5

Plate 1. Main methods for grafting vegetables. (a, b) Hole insertion grafting; (c) approach grafting; (d, e, j) splice grafting; (f, g) cleft grafting; and (h, i) pin grafting. (From Lee *et al.*, 2010, with permission.)

Plate 2. Grafted watermelon transplants produced by the hole insertion method.

Plate 3. Grafted melon transplants produced by the one-cotyledon splice grafting method.

Plate 4. Grafted pepper transplants produced by the splice grafting method.

Plate 5. Grafted aubergine transplants produced by the cleft grafting method.

6

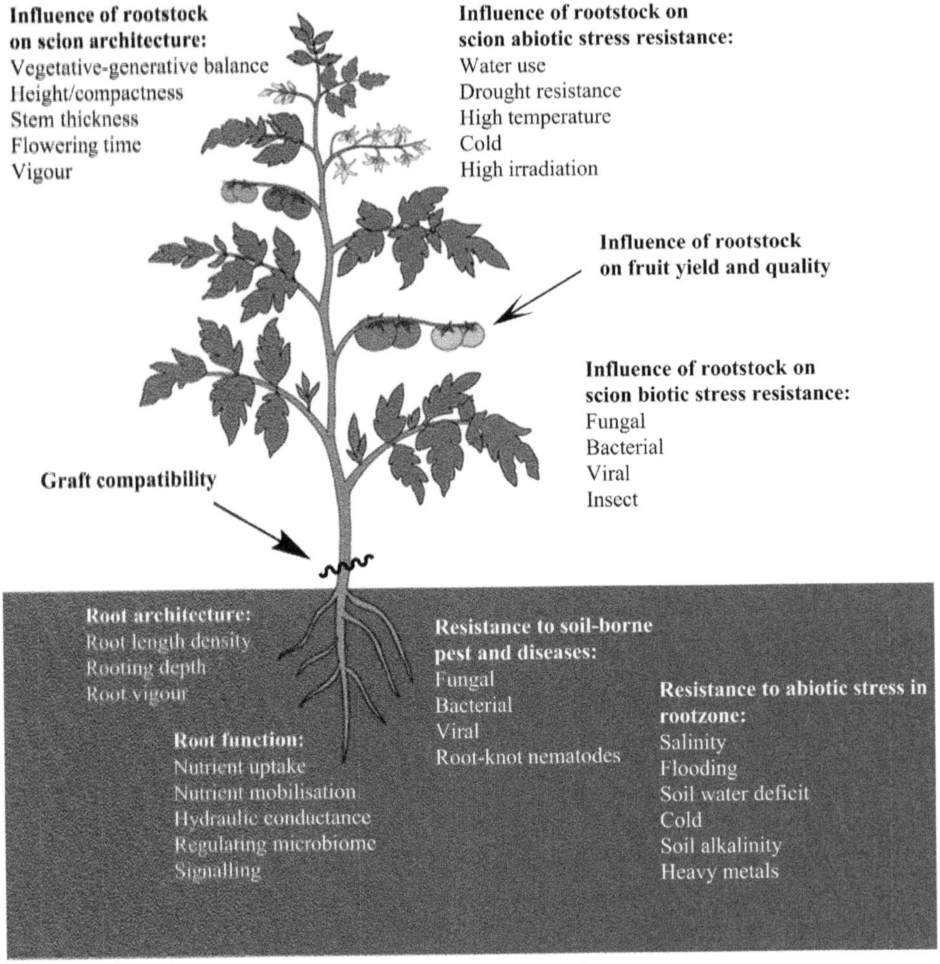

Plate 6. Diverse target traits for rootstock breeding. (Image courtesy of TomExpress, http://gbf.toulouse.inra.fr/tomexpress/www/welcomeTomExpress.php)

7

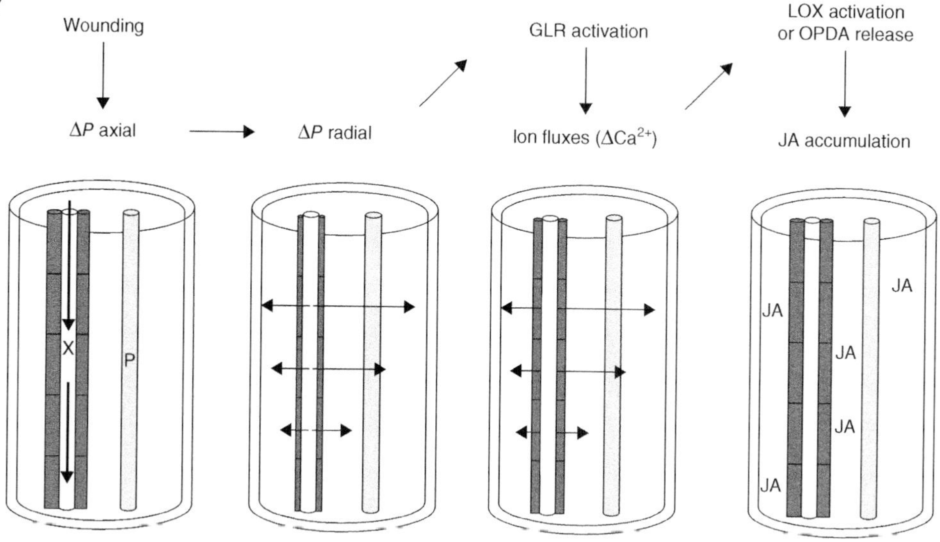

8

Plate 7. Model of the squeeze cell hypothesis within the bundle sheath. This hypothesis states that wounding-induced axial and radial pressure changes (ΔP) in the xylem directly or indirectly modulate glutamine receptor-like (GLR) protein activity, leading to Ca^{2+} fluxes and accumulation of JA in and beyond contact cells distal to the wounding site. This proposed mechanism, by which GLRs and/or Ca^{2+} promote JA synthesis, includes lipoxygenase (LOX) activation or *cis*-(+)-12-oxo-phytodienoic acid (OPDA) release. (From Farmer *et al.*, 2014, with permission.)

Plate 8. Multiple puncture probe guided by an electronic texture analyser performing firmness test on cross-sectioned watermelon.

9

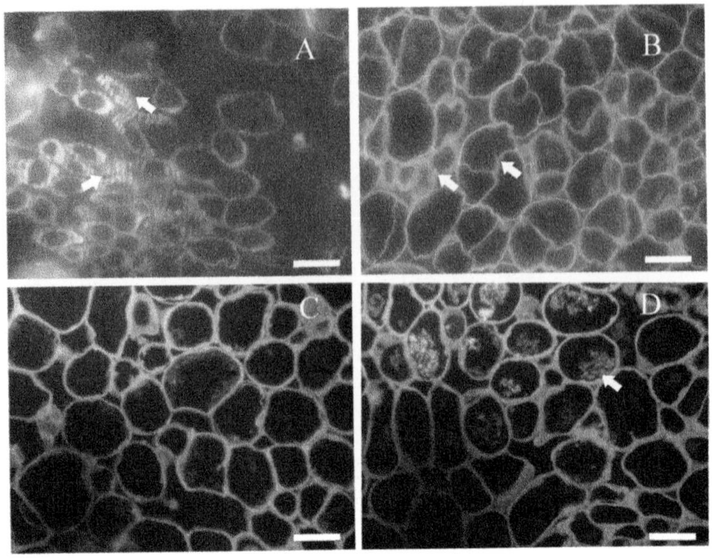

10

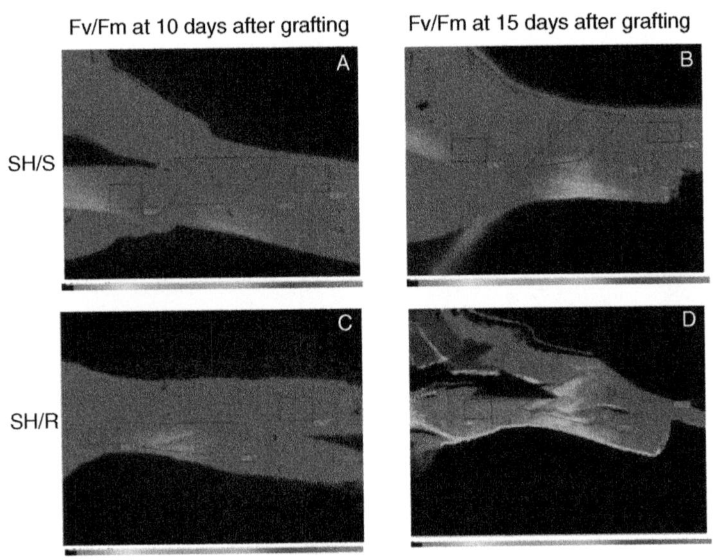

Plate 9. Histological and histochemical responses of the graft interface of compatible (a, c) and incompatible (b, d) *Prunus* scion–rootstock combinations. (a) A compatible union showing the central area of the callus bridge containing isolated tracheids at 10 DAG (arrows). (b) An incompatible union at 10 DAG. Cell-wall thickening is irregular and the cell walls are bent (arrows), resulting in a collapsed cell shape. Cells in (a) and (b) were stained with calcofluor white. (c, d) Compatible (c) and incompatible (d) combinations stained with Naturstoff reagent at 15 DAG. The response to the stain indicated the occurrence of phenolic compounds (white arrow) in vacuoles of callus cells from incompatible unions. The intense yellow fluorescence of the incompatible combination clearly indicated significant accumulation of phenols at the graft interface. Bars, 50 μm. (From Pina *et al.*, 2012, with permission.)

Plate 10. Chlorophyll fluorescence imaging of Fv/Fm after dark adaptedion for SH–S and SH–R plants at 10 and 15 DAG. The false colour code depicted at the bottom of each image ranges from 0.000 (black) to 1.000 (pink). Images were taken from a single plant. The three areas in each image are accompanied by a red box displaying the average Fv/Fm values within these areas. The areas were defined by PAM software.

11

12

13
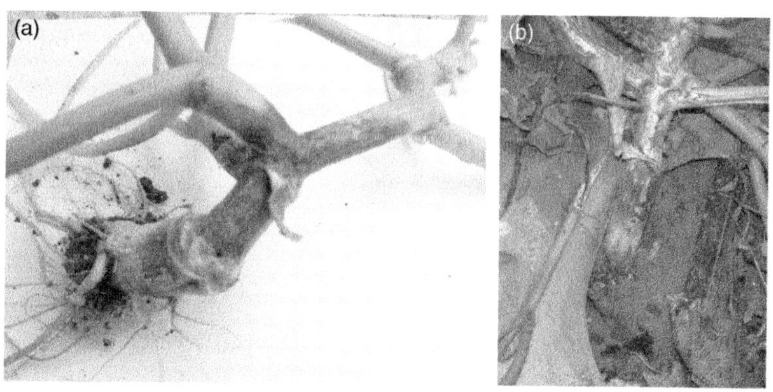

Plate 11. Effect of soil temperature on the development of grafted melons. (a) Damaged grafted melons grown under transparent plastic, allowing overheating of the soil. (b) Yellow plastic prevents overheating of the soil, allowing proper development of the grafted plants. (Photographs courtesy of A. Koen, Hishtil Nurseries, Israel.)

Plate 12. Nematode damage in a small non-grafted melon plant (left) and a melon grafted on to *Cucurbita* rootstock (right). Although the melon and the *Cucurbita* roots were damaged, the size of the *Cucurbita* roots compensated for the damage.

Plate 13. Non-damaged *Cucurbita* rootstock and lesion development on the scion. (a) Grafted melon infected with *Macrophomina phaseolina*. (b) Melon infected with *Monosporascus cannonballus* and *Rhizoctonia solani*. The pathogens can penetrate the rootstock, move toward the scion and can damage the scion only.

14

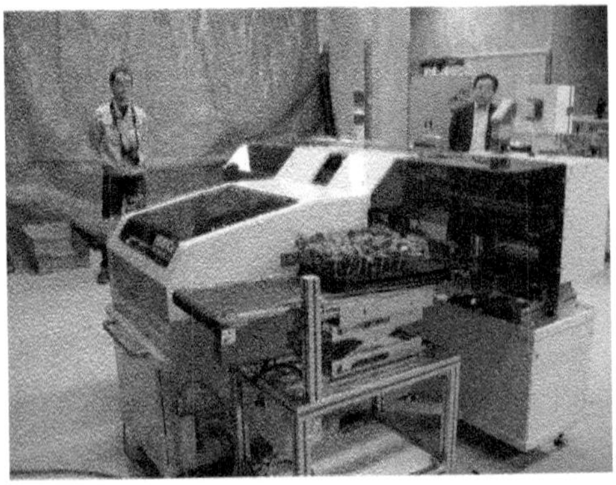

15

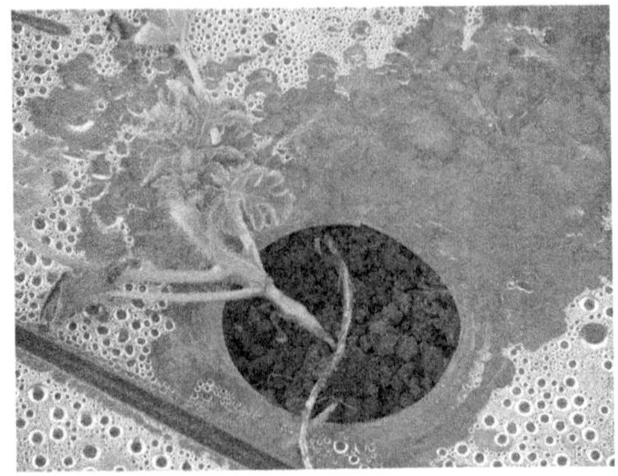

16

18

17

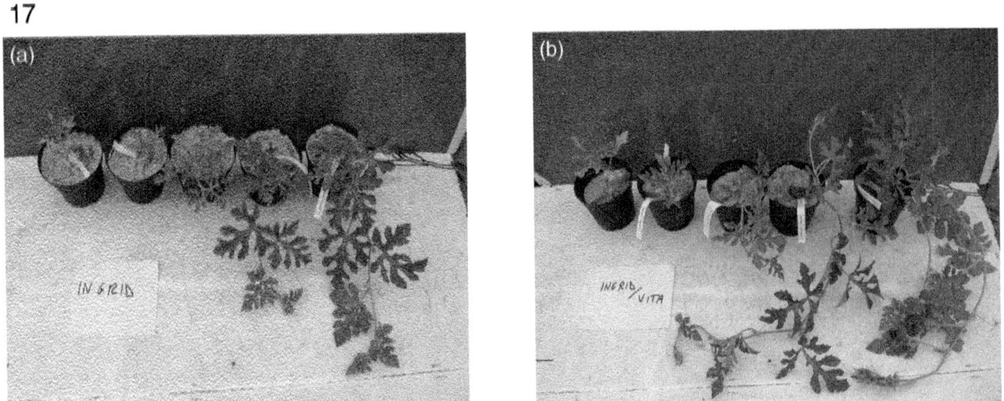

Plate 14. Grafting robot. (Photograph courtesy of Chieri Kubota, University of Arizona, USA.)

Plate 15. Effect of heat stress under transparent polyethylene on development of *Cucurbita* rootstock 'TZ-148'. The stem has constricted and cannot support the scion (watermelon). Avigdor village, Israel, June 2008.

Plate 16. Cucumber grafted (back) and non-grafted (front) with salt injury symptoms at Tuscia University, Italy.

Plate 17. Effects of increasing root nitrate availability (from left to right) in non-grafted watermelon (a) and in watermelon grafted on to a *Lagenaria* rootstock (b) at Tuscia University, Italy.

Plate 18. Effect of flooding on non-grafted aubergine (left) and aubergine grafted on to the tomato rootstock 'Minuet' (right) in Shikmim, Israel, 2008.

19

20

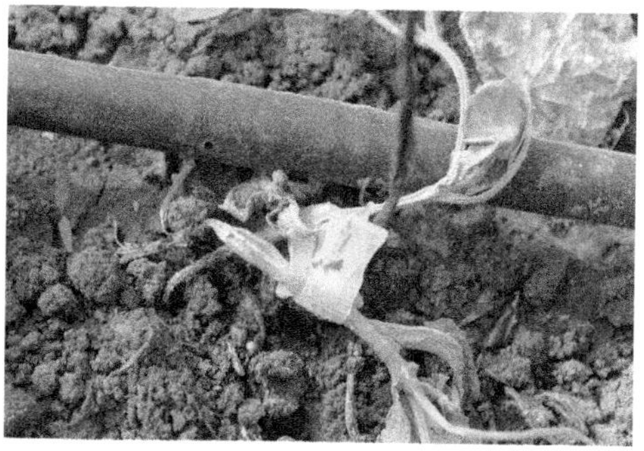

21

22

23

24

Plate 19. Using old rootstock seedlings for grafting hampers grafted seedling establishment. Consequently, the rootstock degenerates, followed by rooting from the scion.

Plate 20. Internal rooting in grafted Cucurbitaceae. The sensitive scion has rooted through the resistant rootstock and been infected by pathogenic *Fusarium* spp.

Plate 21. Effect of transplanting process on grafted melon establishment. Correct transplanting depth and soil tightening (left), compared with the wrong transplanting depth and the presence of air pockets (right).

Plate 22. Grafted watermelon under different roof coverings. Heat stress generated by high radiation and clear polyethylene mainly under transparent soil mulch retards the plantlets (front).

Plate 23. Effect of soil mulching on the establishment of grafted melon in Ein-Tamar, Israel, 2008. High radiation with transparent polyethylene (right) results in heat stress, causing rootstock retardation. Grafted melons under opaque-yellow mulch (left) have improved growth.

Plate 24. A grafted watermelon burned from reflected radiation after transplanting into small holes in a silver mulch.

25

26

27

28

(a)

28

(b)

(c)

Plate 25. A grafted tomato plant decapitated above the cotyledons with the axillary shoots of the cotyledons grown into two main shoots.

Plate 26. A grafted tomato plant decapitated above the second leaf with the axillary shoots of the first and second leaf grown into two main shoots.

Plate 27. Cleft grafting in globe artichoke grafted on to cardoon rootstock at Tuscia University, Italy.

Plate 28. *Phaseolus coccineus* cvs 'Aintree' (P1) (a), 'White Emergo' (P2) (b) and Portuguese landrace 'Seven-year bean' (P3) (c) as rootstocks of green bean *Phaseolus vulgaris* cultivars. (Splice grafted plants obtained from the Aromas e Flores-Hortofloricultura Lda. nursery, Portugal.)

incompatible *P. avium–P. cerasus* graftings (Treutter *et al.*, 1985, 1986) and may destroy auxin (Feucht *et al.*, 1983). Moreover, other studies working with tomato reported that peroxidase activity is implicated in the grafting process and that the enzyme was located mainly in the graft region (Fernández-García *et al.*, 2004).

5.4 Cell-to-cell Communication Between Graft Partners

5.4.1 Plant growth regulator and graft union formation

In recent decades, there has been substantial progress in understanding hormonal signalling processes in graft communication during wound healing and vascular regeneration (reviewed by Chen *et al.*, 2014). Plant hormones are important endogenous factors that play an important role in scion–rootstock communication (reviewed by Aloni *et al.*, 2010). Several lines of evidence suggest that indole acetic acid (IAA) (the predominant auxin in higher plants) undergoes both an active, polar transport from cell to cell through whole plants and plant organs and long-distance translocation in the xylem and phloem systems (Benjamins and Scheres, 2008). In normal development, auxin is synthesized in shoot apical tissues and is actively transported towards the base of the plant, where it affects root development, morphology and functioning (Aloni, 1995; Lomax *et al.*, 1995; Yin *et al.*, 2012; Wang *et al.*, 2014; Spiegelman *et al.*, 2015). For example, the failure of auxin to cross the graft interface in the case of dwarfing rootstocks in apple leads to reduced rootstock xylem formation and hence a poor supply of water and minerals to the scion, producing the dwarfing effect (Lockard and Schneider, 1981; Soumelidou *et al.*, 1994).

Several studies have further highlighted the role of auxin in controlling graft union development and graft incompatibility processes in different species. Auxin translocation from the scion to the rootstock was found to accelerate the formation of a successful graft in cactus (Shimomura and Fujihara, 1977). Working with interspecific *Cucurbita* spp. rootstocks with differential compatibilities when grafted with melon, Aloni *et al.* (2008) proposed that the main cause of graft incompatibility is the occurrence of hormonal imbalance, primarily of auxin and ethylene, in the root system following the establishment of the grafting connections. This study supported the notion that auxin produced in the scion is translocated downwards to the root after the grafting connection is established, and that when reaching a threshold concentration, auxin triggers degradative processes, causing root decay in the incompatible pumpkin rootstock. Furthermore, several reports have suggested that root inhibition by high auxin concentrations may be caused by auxin-induced ethylene production (Mulkey *et al.*, 1982; Rahman *et al.*, 2001). As shown by Aloni *et al.* (2010), incompatibility may result from basipetal auxin transport to the rootstock where it induces ethylene production and oxidative stress. In compatible grafting, antioxidative mechanism(s) may be activated, reducing oxidative stress in the root and enabling its growth. Exogenous application of IAA transport blockers, abscisic acid, anti-ethylene agents or compounds with antioxidant properties may reduce grafting incompatibility.

Genomics studies have identified genes with defined functions in either auxin influx or efflux differentially expressed during graft union development. Several authors have highlighted the role of auxin at early stages of graft union formation in *Arabidopsis* (Yin *et al.*, 2012; Wang *et al.*, 2014), and Cookson *et al.* (2013) identified the induction of one gene encoding an auxin influx carrier at 3 DAG in the graft interface zone of grapevine. Recently, transcriptomic analysis using hypocotyl micrografting in *Arabidopsis* described the stages of graft union development and suggested a role for cell-to-cell communication processes during the early stages (Yin *et al.*, 2012). In this study, activation of the ethylene and jasmonic acid biosynthesis pathways and the stimulus–response categories at 1 DAG confirmed that plants utilized the wound-induced programme to accomplish successful graft union formation. Nevertheless, a complete understanding of the hormonal control of grafting success is far from complete (Aloni *et al.*, 2010).

5.4.2 Cell-to-cell communication at the graft interface

It is known that, before the scion can become dependent on the rootstock in grafted plants, cellular contact must be established to enable the formation of a symplasmic and apoplasmic transport system between the graft partners. One factor that contributes to a successful vegetable or fruit tree graft is the establishment of symplastic cell connections, which facilitate the transfer of compounds between the scion and rootstock (Hartmann *et al.*, 2002). Increasing evidence indicates that cell recognition and direct cellular communication between the callus of the scion and the rootstock is crucial for grafting success (Jeffree and Yeoman, 1983; Kollmann *et al.*, 1985; Pina *et al.*, 2009, 2012; Yin *et al.*, 2012; Trinchera *et al.*, 2013). During grafting, plasmodesmata, one of the main communication pathways, are formed *de novo* across existing cell walls between opposing surfaces of the scion and rootstock (Kollmann and Glockmann, 1991; Yang *et al.*, 1992). Research on the mechanisms of plasmodesmata formation have shown differences in the development of interspecific plasmodesmata between graft partners, suggesting that cell recognition and functional coordination between the two different genotypes at the graft interface is involved in graft formation (Kollmann and Glockmann, 1985; Kollmann *et al.*, 1985). One hypothesis on the development of rootstock–scion incompatibility is that a late rejection is predetermined at the initial step of callus contact. It has been shown that insufficient coordination between adjacent cells at the graft interface may lead to the formation of mismatching, half plasmodesmata in incompatible faba bean–sunflower heterografts (Kollmann *et al.*, 1985). However, in compatible cucumber–figleaf gourd heterografts and tomato homografts, the wound response is followed by dissolution of the necrotic layer, a prerequisite to the formation of continuous secondary plasmodesmata between cells of both graft partners (Tiedemann, 1989; Yang *et al.*, 1992). Late graft rejection in fruit trees has been associated with limited plasmodesmal coupling at the time of grafting within one of the partners (Pina *et al.*, 2009). The problem of incompatibility in grafted *Prunus* spp. trees may be related to low intercellular transport capacity – a low size-exclusion limit or a low number of functional plasmodesmata – in either rootstock or scion callus tissues or between the two (Pina *et al.*, 2009, 2012). In addition, Trinchera *et al.* (2013)

confirmed *de novo* formation of many plasmodesmata between scions and rootstocks with high compatibility, particularly in the globe artichoke–wild cardoon combination. Overall, it has been proposed that a novel control factor of connectivity could reach the graft partner and changes its innate rate of communication. Furthermore, gene expression differences suggest the involvement of plasmodesmata in graft union success. The expression levels of plasmodesmata-located protein PDLP1A (AT5G43980) and AT2G41870 were elevated at early stages of development in *Arabidopsis* homografts (Yin *et al.*, 2012). It is known that PDLP1A targets the plasmodesmata and participates in plasmodesmal trafficking (Thomas *et al.*, 2008). These results also imply that plasmodesmata may contribute to cell-to-cell communication in graft union development.

5.5 Understanding the Molecular Mechanisms Involved in Graft Union Formation and Compatibility

5.5.1 Genes differentially expressed during graft union formation

The cellular events at the graft interface have been well characterized by histological studies in woody (apple tree and *Prunus* spp., Soumelidou *et al.*, 1994; Errea *et al.*, 2001; Pina *et al.*, 2009) and non-woody (*Arabidopsis*, Flaishman *et al.*, 2008; artichoke, Trinchera *et al.*, 2013) plants. However, the molecular mechanisms for graft union development are still largely unknown. In recent years, more attention has been paid to global changes in gene expression during the process of graft formation. Gene expression studies have been used to shed light on the mechanisms behind graft union development in various species. In homografts, the genes differentially expressed at the graft interface are presumably the same as those induced during wound-healing processes. cDNA-amplified fragment length polymorphism analysis was used to examine the gene expression in homografts of hickory (*Carya cathayensis*) at four time points (0, 3, 7 and 14 DAG) during graft union development (Zheng *et al.*, 2010). In this study, the authors identified 49 genes differentially expressed during graft union formation in hickory and revealed that 19 genes of known function were involved in IAA transport, the cell cycle, signal transduction, water metabolism, nuclear metabolism, amino acid metabolism, protein metabolism, carbon metabolism and secretion of substances. Among these, 12 genes, potentially related to cambium formation and cell growth, were induced by grafting at the transcriptional level, suggesting that *C. cathayensis* undergoes a complex change in metabolism in the first 14 DAG. Similarly, graft union development in *Arabidopsis* hypocotyl micrografts has been studied recently at the transcriptional level at 1 DAG, which showed the involvement of wound and hormone signalling (Yin *et al.*, 2012). A set of stimulus–response categories was significantly overrepresented, suggesting that a wound-induced programme was initiated in the graft union developmental processes. In a recent study, homografting of overwintering stems of grapevine resulted in the upregulation of many genes involved in cell-wall synthesis, wound responses, secondary metabolism, hormone signalling, and phloem and xylem development from 3 to 28 DAG (Cookson *et al.*, 2013). Overlap of genes differentially expressed at the scion–rootstock interface and regulation of gene expression over time suggested that some similar processes may take place between the reactivation of stem growth in the spring and the graft union formation process (Cookson *et al.*, 2013). However, the majority of genes differentially regulated in the graft interface were specific to graft union formation. In this sense, numerous graft interface-specific genes were identified as being associated with defence and wound responses, such as genes encoding MYB-like 102, germin-like proteins, peroxidases and chitinases (Cookson *et al.*, 2013). In addition, graft union formation triggered the differential expression of transcription factors involved in the callus maintenance in the graft interface compared with the rootstock tissue, suggesting that callus formation requires considerable regulation at the transcriptional level. In particular, it was found that the *LATERAL BOUNDARY DOMAIN 4* (*LBD4*) gene was induced at 3 DAG and could be related to the formation and maintenance of non-differentiated callus cells (Fan *et al.*, 2012; Ikeuchi *et al.*, 2013). In addition, a late embryogenesis abundant protein with structural similarity to animal fibronectin domain proteins was also induced (Cookson *et al.*, 2013). Because fibronectins from animal cells have been shown to be involved in cell adhesion, motility, wound healing and maintenance of cell shape, these results potentially suggested a role in graft union formation (Singh *et al.*, 2005).

In a later study, Cookson *et al.* (2014) identified many genes induced by grafting in two different grapevine genotypes (heterografts) in comparison with the wound-like gene expression changes in homografts during the first month after grafting. In general, heterografting triggered the differential expression of a large number of genes involved in plant defence and stress responses. Compared with homografting, heterografting resulted in the rapid induction of oxidative stress responses and pathogenesis-related proteins, and was followed by the induction of many other genes involved in plant stress responses (Fig. 5.3).

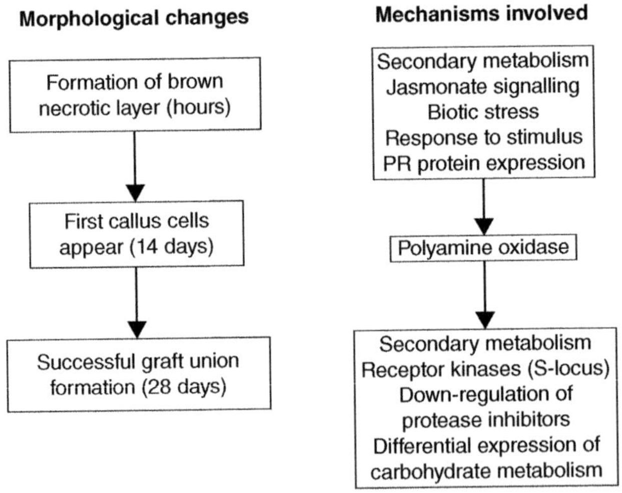

Fig. 5.3. Relationship between morphological changes and molecular mechanisms involved in grafting together two different genotypes. (From Cookson *et al.*, 2014, with permission.)

At this early stage of development, genes encoding enzymes involved in oxidative stress were also identified (e.g. glutathione *S*-transferases, ascorbate oxidase, polyphenol oxidases and peroxidases) and could be associated with the induction of an oxidative burst at the graft interface. By 7 and 14 DAG, the functional category polyamine oxidase was enriched, which has been shown to be one of the key elements for oxidative burst to induce programme cell death (Yoda *et al.*, 2006). The sustained induction of many genes involved in secondary metabolism enzymes (cell-wall precursors, lignin and flavonol biosynthesis) and repression of protease kinases towards the end of graft union formation (14–28 DAG) were also observed (Cookson *et al.*, 2014). This could suggest that the cells at the graft interface are able to sense the presence of an intraspecific or interspecific heterograft, which could trigger an immune-type response.

5.5.2 Genes differentially expressed between compatible and incompatible graft combinations

It is widely known that the more closely the plants are related botanically, the better the chances are for the graft union to be successful (Hartmann *et al.*, 2002). Although graft incompatibility has been widely reported (reviewed by Hartmann *et al.*, 2002; Pina and Errea, 2005; Goldschmidt, 2014), the molecular causes underlying graft incompatibility are not well known. Recent studies have indicated that as well as physiological and anatomical changes during graft formation, molecular changes may also be involved in generating different behaviours between compatible–incompatible combinations (Aloni *et al.*, 2008; Pina and Errea, 2008a,b; Nocito *et al.*, 2010; dos Santos Pereira *et al.*, 2014; Irisarri *et al.*, 2015). Most molecular studies have been conducted with fruit trees. In certain tree scion–rootstock combinations, incompatibility may appear several years after grafting (Errea *et al.*, 2001; Zarrouk *et al.*, 2010), whereas in herbaceous plants (e.g. *Cucurbita* spp.), apparently successful grafts may become incompatible at 21 DAG (Edelstein *et al.*, 2004; Aloni *et al.*, 2008).

Incompatibility between different genotypes has been associated with the induction of phenylalanine ammonia lyase (PAL) in *in vitro* callus unions. PAL expression was further increased by incompatible heterografts (apricot–plum) in comparison with both homograft controls (Pina and Errea, 2008a). Two isoforms of PAL have been identified in the genus *Prunus* (PAL1 and PAL2); the PAL1 isoform has a greater influence in the scion–rootstock incompatibility process (dos Santos Pereira *et al.*, 2014). UDP-glucose pyrophosphorylase (UGPase) was also identified in *Prunus* spp. as a promising protein that could be related to the graft compatibility/incompatibility responses (Pina and Errea, 2008b). This protein is involved in sucrose metabolism and the biosynthesis of cell-wall polysaccharides, and it was found that UGPase mRNA transcript and enzyme activity levels were diminished in the rootstock from incompatible heterografts (apricot–plum) at early stages of graft development compared with the corresponding autografts (Pina and Errea, 2008b).

Oxidative stress has been also implicated in the grafting incompatibility response of herbaceous and woody plants (Aloni *et al.*, 2008, Nocito *et al.*, 2010;

Irisarri *et al.*, 2015). As the production of reactive oxygen species (ROS) is one of the primary and general events following wounding in higher plants (Slesak *et al.*, 2008), coping with oxidative stress could be an important factor in achieving successful grafts. Although exposing plants to stress situations, such as grafting, would trigger the antioxidant defence systems, there are indications that, within incompatible scion–rootstock interfaces, either the level of ROS is increased or there is a less efficient detoxification system (Aloni *et al.*, 2008; Irisarri *et al.*, 2015). For instance, non-successful grafts show signs of cellular stress, such as high accumulation of H_2O_2, superoxide radicals (O_2^-) and increases in the levels of 2-thiobarbituric acid reactive metabolites (Aloni *et al.*, 2008; Nocito *et al.*, 2010). Incompatible melon–aubergine combinations showed the lowest level of antioxidant enzymes (SOD and peroxidase) and higher levels of ROS at the graft interface than compatible grafts at 14 and 24 DAG, and could be responsible for degradation of the grafting zone (Aloni *et al.*, 2008). At earlier stages of development, Nocito *et al.* (2010) reported that the activity of five antioxidant enzymes (ascorbate peroxidase (APX), dehydroascorbate reductase (DHR), glutathione reductase (GR), SOD and catalase (CAT)) was increased in incompatible pear–quince heterografts in comparison with the compatible pear–pear autografts after a 7-day co-culture period. In addition, specific isoforms of genes involved in oxidative stress were identified in pear–pear homografts and pear–quince heterografts. Genes differentially expressed between compatible and incompatible *in vitro* callus heterografts were identified. The transcript levels of six antioxidant genes (*SOD1*, *SOD3*, *APX3*, *APX6*, *CAT1* and *CAT3*) were repressed at 10 DAG in the incompatible heterograft in comparison with the compatible graft (Irisarri *et al.*, 2015). These results suggested decreased oxidative protection during the graft union development in the incompatible heterograft. It has been proposed that the oxidative stress symptoms may represent a belated response to the auxin imbalance (Aloni *et al.*, 2010; Goldschmidt, 2014).

In summary, important metabolic pathways have been identified as responsible for physiological failure in graft-incompatible scion–rootstock combinations in different plant species, such as the phenylpropanoid and cell-wall biosynthesis pathways. A large number of genes have been associated with this agronomic trait, suggesting that the genetic control of graft (in)compatibility is complex. The establishment of plant models for various levels of graft incompatibility with *Arabidopsis* will facilitate the study of the fundamental cellular genetic and molecular aspects involved in this important agricultural trait to a greater extent than any other model system (Flaishman *et al.*, 2008). Although not a crop, *Arabidopsis* may provide valuable insights into the genetic control of graft formation by allowing the identification of many genes and cellular processes involved in graft union formation and incompatibility (Flaishman *et al.*, 2008; Yin *et al.*, 2012).

5.6 Methods for Examining Graft Union Development and Compatibility

5.6.1 *In vitro* techniques

The use of tissue culture techniques may be an interesting approach to study compatibility/incompatibility phenomena, as the physiological and cellular responses of *in vitro* graft systems mimic those that occur *in vivo*, and could reduce drastically the amount of time and effort necessary for determining the compatibility response. In this sense, previous studies have shown that *in vitro* graft systems closely match *in vivo* graft systems, in both herbaceous grafts for tomato–potato (Wang and Kollmann, 1996) and faba bean–sunflower (Schöning and Kollmann, 1997) and fruit species (Richardson *et al.*, 1996; Errea *et al.*, 2001; Espen *et al.*, 2005). These tissue culture techniques include *in vitro* shoot tip grafting, grafting of an explanted stem segment *in vitro* and callus grafting. *In vitro* shoot tip grafting, also called micrografting, consists of aseptically grafting a small shoot tip on to an *in vitro* rootstock seedling and is often used to produce virus-free plants. *In vitro* grafting methods using explanted stem segments consist of cutting a part of a plant stem into segments, grafting these together and culturing them in sterile medium. To establish callus grafts, or *in vitro* callus fusion, callus tissue is obtained from stem fragments cultured in *vitro*, and the grafts are established by making two clean-cut callus pieces and placing one on top of the other under sterile conditions in the same culture medium and conditions as used for callus initiation. Advances in micrografting methods for studying graft union formation have been particularly been done in *Arabidopsis*, as reported by Yin *et al.* (2012). This method facilitated sample analysis and yielded higher success rates than previous methods, and could easily be applied to other dicotyledonous plants, such as tomato, lucerne and tobacco.

As pointed out by Moore (1991), an incompatible response does not need to be associated with any particular stage of tissue development, as the incompatibility factor would be produced by 'ground tissues' such as callus or suspension cultures, and successful grafting of callus masses *in vitro* indicates that graft compatibility does not require the presence of more highly differentiated cells (Moore, 1983).

5.6.2 Histological studies

The cellular events at the graft interface have been well characterized by histological techniques in various herbaceous and woody plants from the study of plasmodesmata formation at the nanometer level to the macroscopic tissue organization of the callus (Soumelidou *et al.*, 1994; Pina *et al.*, 2009; Errea *et al.*, 2001; Trinchera *et al.*, 2013; Irisarri *et al.*, 2015). Light and electron microscopy have given detailed information related to graft union formation in tomato autografts (Jeffree and Yeoman, 1983), pear–quince grafts (Ermel *et al.*, 1997) and in artichoke (Trinchera *et al.*, 2013) and apricot (Errea *et al.*, 1994a, 2001) grafts, as well as in the model plant *Arabidopsis* (Flaishman *et al.*, 2008). Scanning electron microscopy can be used effectively in anatomical studies on graft union development, without prior sample preparation, saving time and avoiding possible artefacts, which could be introduced with specimen preparation. Likewise, the development of confocal bio-imaging tools constitutes a valuable non-invasive method to study plasmodesmal permeability in living tissues with high temporal and spatial resolution (Martens *et al.*, 2004). Photoactivation and photobleaching techniques have been applied to study cell-to-cell communication via plasmodesmata in *in vitro* grown callus. It was demonstrated that callus cells in compatible and incompatible grafts differed in their ability to establish symplasmic connections (Pina *et al.*, 2009, 2012). In addition, three-dimensional visualization of the graft interface using X-ray tomography has provided new insights into the spatial tissue organization of the graft interface in grapevine (Milien *et al.*, 2012).

5.6.3 Chlorophyll fluorescence imaging as a diagnostic technique

Grafting causes stress in plants: mechanical wounding in the scion and rootstock results in localized cell death, loss of water and solutes, and disruption of the vascular system. The activation of repair mechanisms place a high metabolic demand on the grafting area to supply carbon skeletons, synthesize new molecules or increase antioxidative enzyme activity (Fernández-García *et al.*, 2004; Pina and Errea, 2005; Aloni *et al.*, 2008). Many of these processes can be supported by photosynthetic activity (Calatayud *et al.*, 2013). Information on changes in the efficiency of photochemistry can be obtained by measuring and visualizing the yield of chlorophyll *a* fluorescence (Maxwell and Johnson, 2000). Variations in fluorescence parameters can be visualized through images that reveal spatial–temporal changes in grafted plants using imaging fluorometers, which provide a quick and non-invasive technique. This method has been validated in grafted melon plants (Calatayud *et al.*, 2013), where compatibility/incompatibility studies were performed using melon cultivars ('Ricura' (R) and 'Sancho' (S), *Cucumis melo* var. *saccharinus*) grafted on to *Cucurbita maxima* × *Cucurbita moschata* ('Shintoza' (SH)). Yield production showed than SH–S plants had higher compatibility than SH–R plants. The Fv/Fm ratio, which is proportional to the maximal quantum yield of photosystem II photochemistry, was a good indicator of compatibility degree. The values of the Fv/Fm ratio images were higher in SH–S than in SH–R plants at 10 and 15 DAG (Calatayud *et al.*, 2013) (*see* Plate 10). These studies showed that chlorophyll fluorescence imaging allows diagnosis of compatibility or incompatibility at an earlier phase after grafting.

5.7 Conclusions

Plant grafting is a widely used means of plant propagation, especially in fruit trees but also in vegetables, where the purpose is to add value to the elite varieties by improving different traits. It is of considerable importance in the deployment and regional adaptation of elite cultivars, but its range of application is restricted by anatomical, physiological and biochemical aspects that produce incompatible grafts. Graft incompatibility is a general phenomenon in many vegetable and woody species, and is a factor that makes rootstock and cultivar selection difficult, as the introduction of some new varieties requires knowledge of the extent and nature of incompatibility reactions. In this sense, early detection of graft compatibility substantially aids cultivar and rootstock breeding selection.

Some authors have reported that many anatomical and physiological changes are produced during the early graft development of compatible and incompatible unions. However, there is limited information on the molecular and cellular recognition mechanisms of incompatibility. Recently, genes known to be involved in wounding/healing, vascular differentiation, programmed cell death, ROS production/elimination, phenylpropanoid pathways and cell-wall biosynthesis, have been identified during graft union development and graft incompatibility reactions. More research is needed to investigate how to overcome graft incompatibility. The use of auxins, ethylene blockers, abscisic acid or its analogues, gibberellins and compounds with antioxidation properties, as well as inhibitors of enzymes involved in the phenylpropanoid pathway, are all candidates of interest.

References

Aloni, R. (1995) The induction of vascular tissues by auxin and cytokinin. In: Davies, P.J. (ed.) *Plant Hormones: Physiology, Biochemistry and Molecular Biology*, 2nd edn. Kluwer Academic Publishers, Dordrecht, The Netherlands, pp. 531–546.

Aloni, B., Karni, L., Deventurero, G., Levin, Z., Cohen, R., Katzir, N., Lotan-Pompan, M., Edelstein, M., Aktas, H., Turhan, E., Joel, D.M., Horev, C. and Kapulnik, Y. (2008) Physiological and biochemical changes at the rootstock–scion interface in graft combinations between *Curcurbita* rootstocks and melon scion. *Journal of Horticultural Science and Biotechnology* 83, 777–783.

Aloni, B., Cohen, R., Karni, L., Aktas, H. and Edelstein, M. (2010) Hormonal signaling in rootstock–scion interactions. *Scientia Horticulturae* 127, 119–126.

Andrews, P.K. and Serrano-Marquez, C. (1993) Graft incompatibility. *Horticultural Reviews* 15, 183–231.

Benjamins, R. and Scheres, B. (2008) Auxin: the looping star in plant development. *Annual Review of Plant Biology* 59, 443–465.

Calatayud, A., San Bautista, A., Pascual, B., Maroto, J.V. and López-Galarza, S. (2013) Use of chlorophyll fluorescence imaging as diagnostic technique to predict compatibility in melon graft. *Scientia Horticulturae* 149, 13–18.

Canas, S., Assunçao, M., Brazão, J., Zanol, G. and Eiras-Dias, J.E. (2015) Phenolic compounds involved in grafting incompatibility of *Vitis* spp: development and validation of an analytical method for their quantification. *Phytochemical Analysis* 26, 1–7.

Chen, J.J., Zhang, J. and He, X.Q. (2014) Tissue regeneration after bark girdling: an ideal research tool to investigate plant vascular development and regeneration. *Physiologia Plantarum* 151, 147–155.

Cookson, S.J., Clemente-Moreno, M.J., Hevin, C., Nyamba-Mendome, L.Z., Delrot, S., Trossat-Magnin, C. and Ollat, N. (2013) Graft union formation in grapevine induces transcriptional changes related to cell wall modification, wounding, hormone signalling, and secondary metabolism. *Journal of Experimental Botany* 64, 2997–3008.

Cookson, S.J., Clemente-Moreno, M.J., Hevin, C., Nyamba-Mendome, L.Z., Delrot, S., Magnin, N., Trossat-Magnin, C. and Ollat, N. (2014) Heterografting with nonself rootstocks induces genes involved in stress responses at the graft interface when compared with autografted controls. *Journal of Experimental Botany* 65, 2473–2481.

Deloire, A. and Hébant, C. (1982) Peroxidase activity and lignifications at the interface between stock and scion of compatible and incompatible grafts of *Capsicum* on *Lycopersicon*. *Annals of Botany* 49, 887–891.

de Cooman, L., Everaert, E., Curir, P. and Dolci, M. (1996) The possible role of phenolics in incompatibility expression in *Eucalyptus gunnii* micrografts. *Phytochemical Analysis* 7, 92–96.

dos Santos Pereira, I., da Silva Messias, R., Diniz Campos, A., Errea, P. and Pina, A. (2014) Growth characteristics and phenylalanine ammonia-lyase activity in peach grafted on different *Prunus* spp. *Biologia Plantarum* 58, 114–120.

Edelstein, M., Burger, Y., Horev, C., Porat, A., Meir, A. and Cohen, R. (2004) Assessing the effect of genetic and anatomic variation of *Cucurbita* rootstocks on vigour, survival and yield of grafted melons. *Journal of Horticultural Science and Biotechnology* 79, 370–374.

Elstner, E.F., Osswald, W., Volpert, R. and Schempp, H. (1994) Phenolic antioxidants. *Acta Horticulturae* 381, 304–335.

Ermel, F.F., Poëssel, J.L., Faurobert, M. and Catessons, A.M. (1997) Early scion/stock junction in compatible and incompatible pear/pear and pear/quince grafts: a histocytological study. *Annals of Botany* 79, 505–515.

Ermel, F.F., Kervella, J., Catesson, A.M. and Poëssel, J.L. (1999) Localized graft incompatibility in pear/quince (*Pyrus communis*/*Cydonia oblonga*) combinations: multivariate analysis of histological data from 5-month-old grafts. *Tree Physiology* 19, 645–654.

Errea, P. (1998) Implications of phenolic compounds in graft incompatibility in fruit tree species. *Scientia Horticulturae* 74, 195–205.

Errea, P., Treutter, D. and Feucht, W. (1992) Specificity of individual flavan-3-ols interfering with the grafting stress of apricots. *Angewandte Botanik* 66, 21–24.

Errea, P., Felipe, A. and Herrero, M. (1994a) Graft establishment between compatible and incompatible *Prunus* spp. *Journal of Experimental Botany* 45, 393–401.

Errea, P., Treutter, D. and Feucht, W. (1994b) Characterization of flavanol type-polyphenols in apricot cultivar and rootstocks. *Advances in Horticultural Science* 8, 165–169.

Errea, P. Gutmann, M. and Feucht, W. (2000) Physiological implications of flavan-3-ols in apricot-rootstock combinations. *Advances in Horticultural Science* 14, 126–134.

Errea, P., Garay, L. and Marin, J.A. (2001) Early detection of graft incompatibility in apricot (*Prunus armeniaca*) using in vitro techniques. *Physiologia Plantarum* 112, 135–141.

Espen, L., Cocucci, M. and Sacchi, G.A. (2005) Differentiation and functional connection of vascularelements in compatible and incompatible pear/quince internode micrografts. *Tree Physiology* 25, 1419–1425.

Fan, M.Z., Xu, C.Y., Xu, K. and Hu, Y. (2012) LATERAL ORGAN BOUNDARIES DOMAIN transcription factors direct callus formation in *Arabidopsis* regeneration. *Cell Research* 22, 1169–1180.

Fernández-García, N., Carvajal, M. and Olmos, E. (2004) Graft union formation in tomato plants: peroxidase and catalase involvement. *Annals of Botany* 93, 53–60.

Feucht, W. and Treutter, D. (1991) Phenol gradients in opposing cells of *Prunus* heterografts. *Advances in Horticultural Science* 5, 107–111.

Feucht, W., Schmid, P.P.S. and Christ, E. (1983) Compatibility in *Prunus avium*/*Prunus cerasus* grafts during initial phase. II. Reduction of cell number and peroxidases in the rootstock cambium. *Scientia Horticulturae* 21, 225–231.

Feucht, W., Treutter, D. and Schmid, P.P.S. (1988) Inhibition of growth and xylogenesis and promotion of vacuolation in *Prunus* callus by the flavanone prunin. *Plant Cell Report* 7, 189–192.

Flaishman, M.A., Loginovski, K., Golobowich, S. and Lev-Yadun, S. (2008) *Arabidopsis thaliana* as a model system for graft union development in homografts and heterografts. *Journal of Plant Growth Regulation* 27, 231–239.

Gadkari, P.V. and Balaramana, M. (2015) Catechins: sources, extraction and encapsulation: a review. *Food and Bioproducts Processing* 93, 122–138.

Gebhardt, K. and Feucht, W. (1982) Polyphenol changes at the union of *Prunus avium*/*Prunus cerasus* grafts. *Journal of Horticultural Science* 57, 253–258.

Geibel, M. and Feucht, W. (1991) Flavonoid 5-glucosides from *Prunus cerasus* bark and their characteristic weak glycosidic bonding. *Phytochemistry* 30, 1519–1521.

Goldschmidt, E.E. (2014) Plant grafting, new mechanisms, evolutionary implications. *Frontiers in Plant Science* 5, 727.

Guan, W., Zhao, X., Hassel, R. and Thies, J. (2012) Defense mechanisms involved in disease resistance of grafted vegetables. *HortScience* 47, 164–170.

Gur, A., Samish, R.M. and Lifshitz, E. (1968) The role of cyanogenic glycoside of the quince in the incompatibility between pear cultivars and quince rootstocks. *Horticultural Research* 8, 113–134.

Hagerman, A.E. (2012) Fifty years of polyphenol-protein complexes. In: Cheynier, V., Sarni-Manchado, P. and Quideau, S. (eds) *Recent Advances in Polyphenol Research*, Vol. 3. Wiley, Oxford, pp. 71–98.

Harkin, J.M. and Obst, T.R. (1973) Lignification in trees: indication of exclusive peroxidase participation. *Science* 180, 296–297.

Hartmann, H.T. and Kester, D.E. (1983) *Plant Propagation. Principles and Practices*, 4th edn. Prentice-Hall, London, pp. 726.

Hartmann, H.T., Kester, D.E., Davies, F.T. and Geneve, R.L. (2002) *Plant Propagation. Principles and Practices*, 7th edn. Prentice-Hall, Upper Saddle River, New Jersey.

He, F., Pan, Q.H., Shi, Y. and Duan, C.Q. (2008) Biosynthesis and genetic regulation of proanthocyanidins in plants. *Molecules* 13, 2674–2703.

Hudina, A., Orazen, M.,P., Jakopic, J. and Stampar, F. (2014) The phenolic content and their involvement in the graft incompatibility process of various pear rootstocks (*Pyrus communis* L.). *Journal of Plant Physiology* 171, 76–84.

Ikeuchi, M, Sugimoto, K. and Iwase1, A. (2013) Plant callus: mechanisms of induction and repression. *Plant Cell* 25, 3159–3173.

Irisarri, P., Binczycki, P., Errea, P., Martens, H.J. and Pina, A. (2015) Oxidative stress associated with rootstock–scion interactions in pear/quince combinations during early stages of graft development. *Journal of Plant Physiology* 176, 25–35.

Ives, L., Brathwaite, R., Barclay, G., Isaac, W.A. and Connor, C.B. (2012) Graft compatibility of scotch bonnet. *Journal of Agricultural Science and Technology* 2, 81–92.

Jang, K.U. (1992) *Utilization of Sap and Fruit Juice of Luffa cylindrica* L. Korean Ginseng and Tobacco Institute, Taejon, South Korea.

Jeffree, C.E. and Yeoman, M.M. (1983) Development of intercellular connections between opposing cells in a graft union. *New Phytologist* 93, 491–509.

Kawaguchi, M., Taji, A., Backhouse, D. and Oda, M. (2008) Anatomy and physiology of graft incompatibility in *Solanaceous* plants, *Journal of Horticultural Science and Biotechnology* 83, 581–588.

Kollmann, R. and Glockmann, C. (1985) Studies on graft unions. I. Plasmodesmata between cells of plants belonging to different unrelated taxa. *Protoplasma* 124, 224–235.

Kollmann, R. and Glockmann, C. (1991) Studies on graft unions. III. On the mechanism of secondary formation of plasmodesmata at the graft interface. *Protoplasma* 165, 71–85.

Kollmann, R., Yang, S. and Glockmann, C. (1985) Studies on graft unions. II. Continuous and half plasmodesmata in different regions of the graft interface. *Protoplasma* 126, 19–29.

Lattanzio, V., Kroon, P.A., Quideau, S. and Treutter, D. (2008) Plant phenolics – secondary metabolites with diverse functions. In: Daayf, F. and Lattanzio, V. (eds) *Recent Advances in Polyphenol Research*, Vol. 1. Wiley, Oxford, pp. 1–24.

Lee, J.M. (1994) Cultivation of grafted vegetables. I. Current status, grafting methods, and benefits. *HortScience* 29, 235–239.

Lee, J.M. and Oda, M. (2003) Grafting of herbaceous vegetable and ornamental crops. *Horticultural Reviews* 28, 61–124.

Lee, J.M., Kubota, C., Tsao, S.J., Bie, Z., Hoyos-Echevarria, P., Morra, L. and Oda, M. (2010) Current status of vegetable grafting: diffusion, grafting techniques, automation. *Scientia Horticulturae* 127, 93–105.

Lockard, R.G. and Schneider, G.W. (1981) Stock and scion relationships and the dwarfing mechanism in apple. *Horticultural Reviews* 3, 315–375.

Lomax, T.L., Muday, G.K. and Rubery, P.H. (1995) Auxin transport. In: Davies, P.J. (ed.) *Plant Hormones: Physiology, Biochemistry and Molecular Biology*, 2nd edn. Kluwer Academic Publishers, Dordrecht, The Netherlands, pp. 509–530.

Martens, H.J., Hansen, M. and Schulz, A. (2004) Caged probes: a novel tool in studying symplasmic transport in plant tissues. *Protoplasma* 223, 63–66.

Martínez-Ballesta, M.C., Alcaraz-López, C., Muries, B., Mota-Cadenas, C. and Carvajal, M. (2010) Physiological aspects of rootstock–scion interactions. *Scientia Horticulturae* 127, 112–118.

Maxwell, K. and Johnson, G.N. (2000) Chlorophyll fluorescence – a practical guide. *Journal of Experimental Botany* 51, 659–668.

Mikulic-Petkovsek, M., Slatnar, A., Stampar, F. and Veberic, R. (2010) The influence of organic/integrated production on the content of phenolic compounds in apple leaves and fruits in four different varieties over a 2-year period. *Journal of the Science of Food and Agriculture* 90, 2366–2378.

Milien, M., Renault-Spilmont, A.S., Cookson, S.J., Sarrazin, A. and Verdeil, J.L. (2012) Visualization of the 3D structure of the graft union of grapevine using X-ray tomography. *Scientia Horticulturae* 144, 130–140.

Miller, H. and Barnett, J.R. (1993) The structure and composition of bead-like projections on Sitka spruce callus cells formed during grafting and in culture. *Annals of Botany* 72, 441–448.

Mng'omba, S.A., du Toit, E.S. and Akinnifesi, F.K. (2008) The relationship between graft incompatibility and phenols in *Uapaca kirkiana* Muell Arg. *Scientia Horticulturae* 117, 212–218.

Moing, A., Salesses, G. and Saglio, P.H. (1987) Growth and the composition and transport of carbohydrate in compatible and incompatible peach plum grafts. *Tree Physiology* 3, 345–353.

Moore, R. (1983) Physiological aspects of graft formation. In: Moore, R. (ed.) *Vegetative Compatibility Responses in Plants*. Baylor Universtity Press, Waco, Texas, pp. 89–105.

Moore, R. (1991) Graft compatibilities in vitro. In: Bajaj, Y.P.S. (ed.) *Biotechnology in Agriculture and Forestry*, Vol. 17: *High-Tech and Micropropagation I*. Springer, Berlin, pp. 71–84.

Moore, R. and Walker, D.B. (1981) Study of vegetative compatibility-incompatibility in higher plants. I. A structural study of a compatible autograft in *Sedum telephoides* (Crassulaceae). *American Journal of Botany* 68, 820–830.

Mudge, K., Janick, J., Scofield, S. and Goldschmidt, E.E. (2009) A history of grafting. *Horticultural Reviews* 35, 437–493.

Mulkey, T.J., Kuzmanoff, K.M. and Evans, M.L. (1982) Promotion of growth and shift in the auxin dose/response relationship in maize roots treated with the ethylene biosynthesis inhibitors aminoethoxyvinylglycine and cobalt. *Plant Science Letters* 25, 43–48.

Musacchi, S., Pagliuca, G., Kindt, M., Piretti, M.V. and Sansavini, S. (2000) Flavonoids as markers for pear-quince graft incompatibility. *Journal of Applied Botany Angewandte Botanik* 74, 206–211.

Nocito, F.F., Espen, L., Fedeli, C., Lancilli, C., Musacchi, S., Serra, S., Sansavini, S., Cocucci, M. and Sacchi, G.A. (2010) Oxidative stress and senescence-like status of pear calli co-cultured on suspensions of incompatible quince microcalli. *Tree Physiology* 30, 450–458.

Penella, C., Nebauer, S. G., Lopéz-Galarza, S., SanBautista, A., Gorbe, E. and Calatayud, A. (2013). Evaluation for salt stress tolerance of pepper genotypes to be used as rootstocks. *Journal of Food, Agriculture and Environment* 11, 1101–1107.

Penella, C., Nebauer, S.G., Quiñones, A., San Bautista, A., López-Galarza, S. and Calatayud, A. (2015) Some rootstocks improve pepper tolerance to mild salinity through ionic regulation. *Plant Science* 230, 15–22.

Pina, A. and Errea, P. (2005) A review of new advances in mechanism of graft compatibility–incompatibility. *Scientia Horticulturae* 106, 1–11.

Pina, A. and Errea, P. (2008a) Differential induction of phenylalanine ammonia-lyase gene expression in response to in vitro callus unions of *Prunus* spp. *Journal of Plant Physiology* 165, 705–714.

Pina, A. and Errea, P. (2008b) Influence of graft incompatibility on gene expression and enzymatic activity of UDP-glucose pyrophosphorylase. *Plant Science* 174, 502–509.

Pina, A., Errea, P., Schulz, A. and Martens, H.J. (2009) Cell-to-cell transport through plasmodesmata in tree callus cultures. *Tree Physiology* 29, 809–818.

Pina, A., Errea, P. and Martens, H.J. (2012) Graft union formation and cell-to-cell communication via plasmodesmata in compatible and incompatible stem unions of *Prunus* spp. *Scientia Horticulturae* 143, 144–150.

Pizzi, A. and Cameron, F.A. (1986) Flavonoid tannins – structural wood components for drought-resistance mechanisms of plants. *Wood Science and Technology* 20, 119–124.

Rahman, A., Amakawa, T., Goto, N. and Tsurumi, S. (2001) Auxin is a positive regulator of ethylene-mediated response in the growth of Arabidopsis roots. *Plant and Cell Physiology* 42, 301–307.

Richardson, F.V.M., tSaoir, S.M. and Harvey, B.M.R. (1996) A study of the graft union in *in vitro* micrografted apple. *Plant Growth Regulation* 20, 7–23.

Rodrigues, C.A., Bender-Machado, L., Campos Diniz, A., Fachinello, J.C. and De Luces-Fortes, G.R. (2002) Peroxidases e fenóis totais em tecidos de porta-exertos de *Prunus* sp. nos períodos de crecimiento vegetativo e de donnéncia. *Ciencia Rural* 4, 559–564.

Santamour, F.S. (1992) Predicting graft incompatibility in woody plants. Combined Proceedings of the International Plant Propagators Society 42, 131–134.

Schöning, U. and Kollmann, R. (1997) Phoem translocation in regenerating *in vitro* heterografts of different compatibility. *Journal of Experimental Botany* 48, 289–295.

Shimomura, T. and Fujihara, K. (1977) Physiological study of graft union formation in cactus. II. Role of auxin on vascular connection between stock and scion. *Journal of the Japanese Society for Horticultural Science* 45, 397–406.

Singh, S., Cornilescu, C.C., Tyler, R.C., Cornilescu, G., Tonelli, M., Lee, M.S. and Markley, J.L. (2005) Solution structure of a late embryogenesis abundant protein (LEA14) from *Arabidopsis thaliana*, a cellular stress related protein. *Protein Science* 14, 2601–2609.

Slesak, I., Slesak, H., Libik, M. and Miszalski, Z. (2008) Antioxidant response system in the short-term post-wounding effect in *Mesembryanthemum crystallinum* leaves. *Journal of Plant Physiology* 165, 127–137.

Soumelidou, K., Battey, N.H., John, P. and Barnett, J.R. (1994) The anatomy of the developing bud union and its relationship to dwarfing in apple. *Annals of Botany* 74, 605–611.

Spiegelman, Z., Ham, B.-K., Zhang, Z., Toal, T.W., Brady, S.M., Zheng, Y., Fei, Z., Lucas, W.J. and Wolf, S. (2015) A tomato phloem-mobile protein regulates the shoot-to-root ratio by mediating the auxin response in distant organs. *Plant Journal* 83, 853–863.

Sterjiades, R., Dean, L.F.D., Gamble, G., Himmelsbach, D.S. and Eriksson, K.E.L. (1993) Extracellular laccases and peroxidases from sycamore maple (*Acer pseudoplatanus*) cell suspension cultures. Reactions with monolignols and lignin model compounds. *Planta* 190, 75–87.

Temperini, O., Calabrese, N., Temperini, A., Rouphael, Y., Tesid, R., Lenzi, A., Carito, A. and Colla, G. (2013) Grafting artichoke onto cardoon rootstocks: graft compatibility, yield and Verticillium wilt incidence. *Scientia Horticulturae* 149, 22–27.

Thomas, C.L., Bayer, E.M., Ritzenthaler, C., Fernandez-Calvino, L. and Maule, A.J. (2008) Specific targeting of a plasmodesmal protein affecting cell-to-cell communication. *PLoS Biology* 6, 7.

Tiedemann, R. (1989) Graft union development and symplastic phloem contact in the heterograft *Cucumis sativus* on *Cucurbita ficifolia*. *Journal of Plant Physiology* 134, 427–440.

Tiedemann, R. and Carsens-Behrens, U. (1994) Influence of grafting on the phloem protein patterns in Cucurbitaceae. I. Additional phloem exudate proteins in *Cucumis sativus* grafted on two *Curcubita* species. *Journal of Plant Physiology* 143, 189–194.

Treutter, D. (2006) Significance of flavonoids in plant resistance and enhancement of their biosynthesis. *Plant Biology* 7, 581–591.

Treutter, D. and Feucht, W. (1988) Accumulation of the flavonoid prunin in *P. avium/P. cerasus* grafts and its possible involvement in the process of compatibility. *Acta Horticulturae* 227, 74–78.

Treutter, D., Galensa, R., Feucht, W. and Schmid, P.P.S. (1985) Flavanone glucosides in callus and phloem of *Prunus avium*: identification and stimulation of their synthesis. *Physiologia Plantarum* 65, 95–101.

Treutter, D., Feucht, W. and Schmid, P.P.S. (1986) Polyphenols of the phloem in relation to incompatibility of interspecific *Prunus* graftings (*Prunus avium* L., *Prunus cerasus* L.). 1. Flavanones and flavanols above the graft union. *Gartenbauwissenschaftg* 51, 77–83.

Trinchera, A., Pandozy, G., Rinaldi, S., Crinò, P., Temperini, O. and Rea, E. (2013) Graft union formation in artichoke grafting onto wild and cultivated cardoon: an anatomical study. *Journal of Plant Physiology* 170, 1569–1578.

Turquois, N. and Malone, M. (1996) Non-destructive assessment of developing hydraulic connections in the graft union of tomato. *Journal of Experimental Botany* 47, 701–707.

Usenik, V., Krska, B., Vican, M. and Stampar, F. (2006) Early detection of graft incompatibility in apricot (*Prunus armeniaca* L.) using phenol analyses. *Scientia Horticulturae* 109, 332–338.

Vermerris, W. and Nicholson, R. (2006) Isolation and identification of phenolic compounds: a practical guide. In: *Phenolic Compound Biochemistry*. Springer, Dordrecht, The Netherlands, pp. 151–196.

Wang, Y. and Kollmann, R. (1996) Vascular differentiation in the graft union of *in vitro* grafts with different compatibility - structural and functional aspects. *Journal of Plant Physiology* 147, 521–533.

Wang, J., Jin, Z., Yin, H., Yan, B., Ren, Z.Z., Xu, J., Mu, C.J., Zhang, Y., Wang, M.Q. and Liu, H. (2014) Auxin redistribution and shifts in *PIN* gene expression during *Arabidopsis* grafting. *Russian Journal of Plant Physiology* 61, 688–696.

Webster, A.D. and Schmidt, H. (1996) Rootstocks for sweet and sour cherries. In: Webster, A.D. and Looney, N.E. (eds) *Cherries: Crop Physiology, Productions and Uses*. CAB International, Wallingford, UK, pp. 127–163.

Yang, S., Xiang, G., Zhang, S. and Lou, C. (1992) Electrical resistance as a measure of graft union. *Journal of Plant Physiology* 141, 98–104.

Yin, H., Yan, B., Sun, J., Jia, P., Zhang, Z., Yan, X., Chai, J., Ren, Z., Zheng, G. and Liu, H. (2012) Graft-union development: a delicate process that involves cell–cell communication between scion and stock for local auxin accumulation. *Journal of Experimental Botany* 63, 4219–4232.

Yoda, H., Hiroi, Y. and Sano, H. (2006) Polyamine oxidase is one of the key elements for oxidative burst to induce programmed cell death in tobacco cultured cells. *Plant Physiology* 142, 193–206.

Zarrouk, O., Gogorcena, Y. and Moreno, M.A. (2006) Graft compatibility between peach cultivars and *Prunus* rootstocks. *HortScience* 41, 1389–1394.

Zarrouk, O., Testillano, P.S., Risueño, M.C., Moreno, M.Á. and Gogorcena, Y. (2010) Changes in cell/tissue organization and peroxidase activity as markers for early detection of graft incompatibility in peach/plum combinations. *Journal of the American Society for Horticultural Science* 135, 9–17.

Zheng, B.S., Chu, H.L., Jin, S.H., Huang, Y.J., Wang, Z.J., Chen, M. and Huang, J.Q. (2010) cDNA-AFLP analysis of gene expression in hickory (*Carya cathayensis*) during graft process. *Tree Physiology* 30, 297–303.

6 Grafting as Agrotechnology for Reducing Disease Damage

Roni Cohen,[1*] Aviv Dombrovsky[2] and Frank J. Louws[3]

[1]*Newe Ya'ar Research Center, Agricultural Research Organisation, Ramat Yischai, Israel;* [2]*Volcani Center, Agricultural Research Organisation, Bet Dagan, Israel;* [3]*Center for Integrated Pest Management, North Carolina State University, Raleigh, North Carolina, USA*

6.1 Introduction

Grafted vegetables have been cultivated in eastern Asia for decades. Use of grafted vegetables in the western world, however, received more recent attention with the banning of the fumigant methyl bromide (Miguel *et al.*, 2004). One of the primary motives for using grafted plants is to avoid damage caused by soilborne pests and pathogens when genetic or chemical approaches for disease management are not available (Lee and Oda, 2003). The grafting of a susceptible scion on to a resistant rootstock provides a 'resistant' plant with no need for a prolonged breeding programme if adapted rootstocks are available. Furthermore, grafting plants allows a more rapid response to the appearance of new pathogen species or races, and overall provides a more rapid and flexible solution for controlling soilborne diseases than breeding new, resistant cultivars.

Several reviews have been published that address the integrated use of grafting techniques in vegetable production. The pioneers were Lee (1994) and Lee and Oda (2003), who provided comprehensive reviews addressing major topics associated with vegetable grafting, including grafting methods and procedures, grafting physiology, and production status and trends in the grafted vegetable industry. The adoption of grafting in the USA, including technological developments using robots, has been described in numerous publications (Jifon *et al.*, 2006; Kubota *et al.*, 2008; King *et al.*, 2008; Lee *et al.*, 2010). Cucurbit grafting, including the history, current status and benefits of vegetable grafting, has been well documented by Davis *et al.* (2008). Louws *et al.* (2010) summarized concurrent literature on the use of grafting technology to manage diseases and pests in vegetable production, and provided up-to-date information on disease cycles,

* ronico@volcani.agri.gov.il

complementary disease-management practices, and the effects of rootstocks on controlling different diseases of cucurbits and solanaceous crops (Tables 6.1 and 6.2). This chapter emphasizes additional plant-pathology-associated aspects of critical importance to long-term success in grafting systems.

6.2 First Step: Managing Diseases in the Nursery

Horticultural nurseries are exceedingly complex agricultural systems, making pest and pathogen management challenging (Werner *et al.*, 2002; Parke and Grünwald, 2012). Diseases can spread from the nursery to the field, and avoiding this risk can be much more challenging with grafted plants. A grafted plant is a

Table 6.1. Grafting in cucurbitaceous crops (watermelon, melon and cucumber): most important pathogens, and the countries and years in which grafting experiments were conducted. (Adapted from Louws *et al.*, 2010.)

Fungal pathogen	Watermelon	Melon	Cucumber
Fusarium oxysporum f. sp. *melonis*		Ital: 2000, 2002, 2007; Israel: 2002; Greece: 2005	
F. oxysporum f. sp. *niveum*	Korea: 2002; Turkey: 2003, 2007; Spain: 2004; Greece: 2007; Tunisia: 2007; Vietnam: 2009		
F. oxysporum f. sp. *lagenariae*	Japan: 1992, 2007; Korea: 2003		Japan: 1992, 2007
F. oxysporum f. sp. *cucumerinum*			Japan: 1971, 2008
F. oxysporum f. sp. *radicis-cucumerinum*			Greece: 2002; Israel: 2007; Spain: 2007
Fusarium solani f. sp. *cucurbitae*	Spain: 2000; Tunisia: 2007; 2008		
Monosporascus cannonballus	Spain: 2008; USA: 2008	Israel: 1999, 2005; Spain: 2007; Tunisia: 2009	
Verticillium dahliae	Greece: 2002, 2007	France: 1974; Greece: 2002	France: 1974; Greece: 2002
Phomopsis sclerotioides			England: 1969; France: 1974, 1994; Japan: 2009
Didymella bryoniae		Greece: 2000; Italy: 2007; Brazil: 2009	
Macrophomina phaseolina and *F. solani* (vine decline)		Israel: 2007	

Table 6.2. Grafting in solanaceous crops (tomato, pepper and aubergine): most important pathogens, countries and years in which grafting experiments were conducted. Adapted from Louws *et al.* (2010)

Pathogen	Tomato	Pepper	Aubergine
Fungi			
Verticillium dahliae (race 1)	Switzerland: 1976; France: 1979; Greece: 2002		USA: 1970; Switzerland: 1976; France: 1982; China: 1998; Greece: 2003; Cyprus: 2001; Italy: 2005; 2006; Brazil: 2009
Verticillium albo-atrum	UK: 1973		
Fusarium oxysporum f. sp. *lycopersici*	France: 1979; Korea: 1997; Taiwan: 2003; USA: 2008		
F. oxysporum f. sp. *radicis-lycopersici*	Canada: 1981		
F. oxysporum f. sp. *redolens*		Egypt: 2003	
F. oxysporum f. sp. *melongenae*			Indonesia: 2005
Sclerotium rolfsii	Taiwan: 2003; USA: 2010		
Pyrenochaeta lycopersici	France: 1974, 1979, 1998; Sweden: 2009		Cyprus: 2001; Italy: 2005
Colletotrichum coccodes	Italy: 2008, 2008		Cyprus: 2001
Rhizoctonia solani	Italy: 2007		
Oomycetes			
Phytophthora spp.	UK: 1968; Spain: 2001, 2007	France: 1974; Brazil: 2004; Spain: 2005, 2010; Italy: 2006; Taiwan: 2008, 2009; Tunisia: 2010	
Bacteria			
Ralstonia solanacearum	India: 1979; Brunei: 1982; Japan: 1993, 1996; Guadeloupe: 1994; Taiwan: 2003, 2008; USA: 2008; China: 2009	Taiwan: 2008, 2009	Italy: 2006
Nematodes			
Root-knot	Japan: 1993; Germany: 2002; Spain: 2002; 2008, 2008, 2009; Taiwan: 2003; USA: 2006, 2010; China: 2007; Cuba: 2009	Israel: 2004; Spain: 2005; Italy: 2006; USA: 2009	Cyprus: 2001; Italy: 2005, 2006; Brazil: 2009

Continued

Table 6.2. Continued.

Pathogen	Tomato	Pepper	Aubergine
Viruses			
Tomato spotted wilt virus	USA: 2008		
Pepino mosaic virus	Spain: 2004		
Weeds			
Broomrape	Israel: 2010		

combination of two plants, the rootstock and the scion. Each of them can potentially transmit diseases, and this is compounded by the 'risky' step of the grafting process. This process involves touching both plants and cutting them with knives, allowing vertical transmission of the pathogen across the scion–rootstock interface, as well as horizontal transmission from one plant to another on the contaminated knife. In addition, the resultant wound provides pathogens with an easy route for plant penetration. Finally, it takes the plants a long time to heal under the moist chamber conditions required for this process, and these conditions further enhance disease development. The typical agricultural crop starts with uniform propagation material – a seed. Disease transmission by seeds and the grafting procedure seem to be the two riskiest steps, and these require a systematic approach to disease-management tactics to minimize the risk of disease transmission from the nursery to the field. Producing disease-free seeds is the responsibility of seed companies, whereas creating a safe grafting procedure is a challenge for the nurseries. Sanitation and safe technology for plant cutting using tool sterilization protocols or new technologies that are being developed for this industry, such as laser-beam cutters, could dramatically improve grafted plant safety. Two examples of pathogens of major concern are highlighted here.

6.2.1 Tobamovirus management: grafted cucurbits and cucumber green mottle mosaic virus: an example of risk and a solution

The tobamovirus cucumber green mottle mosaic virus (CGMMV) was first described in the 1930s as cucumber viruses 3 and 4 (Ainsworth, 1935). Since then, the virus has spread among cucurbit crops worldwide, causing damage in the Far East (Ryu *et al.*, 2008), Europe, India and Israel (Antignus *et al.*, 1990). Recently, CGMMV has been discovered in Canada and the USA (Tian *et al.*, 2014).

Tobamoviruses can be transmitted efficiently by mechanical means, such as on workers' hands and tools (Kamenova and Adkins, 2004), and they remain viable in the soil, enabling transmission between growth cycles (Dornai *et al.*, 1993). Their viability in and on seeds (Reingold *et al.*, 2014) allows them to be easily transferred to distant geographical areas (Genda *et al.*, 2011). All of these modes of transmission have led to the uncontrollable spread of tobamoviruses and worldwide damage.

The grafting procedure involves at least two cutting actions, one of the rootstock and the other of the scion. In the case of a CGMMV-infected plant (rootstock or scion), the grafting procedure might expose many plants to contamination risk through the use of a contaminated knife. In recent study, the mechanical transmission rate using contaminated tools (shears or knife), followed by sequential cutting, displayed a maximum infection of five plants out of ten plants in total. No statistical differences were observed between the transmission rates with knife and shears, and the average transmission rate was 33%; no obvious pattern of infection was observed along the rows (Reingold *et al.*, 2016).

Nurseries play an important role in tobamovirus management and should be inspected regularly for tobamoviruses in commercial seed lots. They should regularly chlorinate the used seedling trays, benches, etc. and be aware that the use of high-pressure water may result in the spread of the virus.

Frequent replacement or disinfection of the grafting knife between each cut plant using stabilized chlorine (e.g. 1% Klor Bac™), 2% Virkon™ S or 1% Virocid® can contribute to reduced disease spread in the nursery especially when dealing with grafted plants. The actions at the nurseries contribute to minimizing the number of primary infected plants within the plots prior to the performance of agrotechnical activities in trellised plants, and therefore contribute to reducing secondary disease spread.

6.2.2 Bacterial canker management: grafted tomatoes and an old nemesis

Bacterial canker, caused by *Clavibacter michiganensis* subsp. *michiganensis*, is another example of a major risk to the vegetable grafting industry and is one of the most important diseases of tomato (Gleason *et al.*, 1993; Sen *et al.*, 2015). The pathogen is seedborne (de Leon *et al.*, 2011) and is transmitted mechanically by grafting knives, water movement, insects and human touch (Xu *et al.*, 2010; Sen *et al.*, 2015). It can colonize internal tissue or multiply on the phyllosphere without showing symptoms (Gitaitis *et al.*, 1991), and is rapidly transmitted from one plant to another as a result of standard plant-handling practices in nursery production systems (Chang *et al.*, 1991). It can cause major plant losses due to interplant spread early or late in the tomato production cycle. Successful management requires a systems-based approach from the seed production phase through the grafting phase and into the fruit production phase of the industry (Miller *et al.*, 2015; Sen *et al.*, 2015). Low infection/infestation rates in the seed, plus the capacity for the pathogen to rapidly multiply and spread, present the so-called 'one in 10,000' problem. Detection technologies are not sufficiently sensitive enough to detect low numbers of infected seeds but one seed in 10,000 can be enough to cause major epidemics in fruit production systems (Chang *et al.*, 1991; Gitaitis *et al.*, 1991). Despite extensive work to limit seed infections/infestations and manage the pathogen in all phases of production, major outbreaks are sporadic and continue to occur (Gleason *et al.*, 1993; Fanigliulo *et al.*, 2011), and are aggravated by grafting (Xu *et al.*, 2010). Seed producers and grafting nurseries typically follow the Good Seed and Plant Practices (GSPP, 2016) to minimize the risk associated with *C. michiganensis* subsp. *michiganensis*.

6.3 Disease Spread from the Nursery to the Field: the Example of Powdery Mildew of Watermelons

Powdery mildew of watermelon provides an example of disease spread from the nursery to the field – a new and unexpected situation demonstrating how introducing a grafted crop can create a new risk or enhance disease development where such a disease was unknown or did not pose a problem in the same non-grafted crop. In Israel, powdery mildew attacks plants relatively late in the season, after fruit set. The disease can cause severe damage in the north of the country, whereas in the Arava Valley in southern Israel, where early spring watermelons are grown, the disease in young plants is rare and insignificant. When grafted watermelons were first used in the Arava Valley, there was a severe epidemic of powdery mildew on very young transplants. An investigation of the situation revealed that the cotyledon of the *Cucurbita* rootstock, which is necessary for proper healing of the grafting site, is highly susceptible to the disease. The cotyledons had probably already been infected in the nursery, but the disease developed in the field. As a result, the grafted plants that were translocated to the field were essentially 'inoculum bombs'. In the following season, powdery mildew control was strictly applied in the nursery and the *Cucurbita* cotyledons were removed after healing and before plants were shipped to field. Powdery mildew also developed in US grafting operations, and one to two preventative applications of a biocontrol/biorational mix of products or efficacious fungicides was found to provide optimum management (Keinath and DuBose, 2012).

6.4 Intra- and Interspecific Grafting and their Relationship to Diseases

In contrast to long and expensive breeding programmes, grafting provides a fast and easy solution for acute pathological problems when rootstocks are available. Two types of grafting are used: the susceptible scion is grafted either on to a resistant rootstock of the same species (intraspecific grafting) or on to a close member of the same botanical family (interspecific grafting). In tomato, for example, there is a collection of rootstocks. These rootstocks contain different genes for resistance to fungal pathogens, bacteria, viruses and nematodes, and therefore intraspecific grafting is very common (Friedmann *et al.*, 1988; Augstin *et al.*, 2002; Paplomatas *et al.*, 2002). The tomato industry does use hybrids but rarely utilizes rootstocks from other species or another genus. *Solanum torvum* has been used as a rootstock for interspecific grafting with aubergine (Rahman *et al.*, 2002; Bletsos *et al.*, 2003), but in some countries, such as the USA, it is on the Federal Noxious Weed List and therefore is not used. In cucurbits, however, the two approaches (i.e. the use of resistant rootstock from the same species (muskmelons, watermelon) or other species (*Cucurbita* rootstocks)) have different purposes (Yetisir *et al.*, 2003; Cohen *et al.*, 2007; Thies *et al.*, 2015). Intraspecific grafting is used mainly to avoid damage caused by wilt pathogens such as *Fusarium oxysporum* f. sp. *melonis*, for which monogenic resistance genes in certain melon varieties used as rootstocks exist. This approach provides complete protection from the disease, with no reduction in fruit quality or quantity (Cohen *et al.*, 2002, 2007). *Cucurbita* rootstocks can also protect melons from Fusarium wilt, but in certain cases, the protection will only be partial, and some wilting of old vines may occur. Moreover, grafting on *Cucurbita* rootstocks may cause a reduction in fruit quality. The possibility of using melon rootstock to reduce Monosporascus wilt (root and stem rot diseases) was studied to avoid the undesirable effects of *Cucurbita* rootstock on fruit quality. Indeed, there are muskmelon accessions with different levels of quantitative resistance to the disease (Jifon *et al.*, 2006) that may serve as rootstocks. Nevertheless, it has not been possible to obtain a grafted plant that is reliably resistant to Monosporascus wilt and other soilborne pathogens under a variety of environmental conditions (R. Cohen and M. Edelstein, unpublished data). In fact, only *Cucurbita* rootstocks provide the required non-specific resistance to control several root and stem rot diseases for which dependable resistance is not available in melons or watermelons. Most of the watermelons grown in the Mediterranean basin are grafted on to *Cucurbita* rootstocks, which provide efficient protection against a wide range of soilborne pathogens. Nevertheless, grafting watermelon on watermelon rootstocks with certain resistances may eliminate the fruit-quality issues resulting from the use of *Cucurbita* rootstocks (Cohen *et al.*, 2014; Thies *et al.*, 2015).

To develop advanced selections of watermelon rootstocks, the response of exotic watermelon accessions to Fusarium wilt caused by *Fusarium oxysporun* f. sp. *niveum*, Fusarium crown rot caused by *Fusarium oxysporun* f. sp. *radicis-cucumerinum*, and infection by the nematodes *Meloidogyne javanica* and *Meloidogyne incognita* was evaluated in pot experiments. The response to *Macrophomina phaseolina* and *Monosporascus cannonballus* was evaluated under field conditions. The screened accessions exhibited various responses to the tested diseases, including resistance to Fusarium wilt, Fusarium crown rot and *M. phaseolina*. The findings suggest the possibility of breeding watermelon rootstocks with high levels of resistance to several diseases and no negative effect on fruit quality (Cohen *et al.*, 2014).

6.5 Biotic or Abiotic Stress? Different Responses of Grafted Plants to Environmental Conditions: the Case of 'Physiological Wilt'

When grafting was first introduced to western agriculture, the assumption was that the performance of the grafted plant, including the response to soilborne diseases, depended only on the contribution of the rootstock. Nurseries supplying grafted transplants to growers were convinced that finding a resistant and vigorous rootstock was all that was required, and that everything else needed for successful cultivation of grafted plants could be modified or adopted from the experience acquired in the Far East. The reality, however, turned out to be different. As with non-grafted plants or the introduction of new cultivars, the grafted plants must become adapted to the specific combination of the crop and environment. Not only does the grafted plant's response to the environment depend on the rootstock, but its total performance also depends to a great extent on the rootstock–scion interaction, and on the response of the scion to disease and to other environmental factors prevailing in the field (Cohen *et al.*, 2002, 2005).

The use of grafted plants should take into account the differences between the rootstocks and combinations compared with their non-grafted counterparts. For example, melon growers must take care of two plants simultaneously, the *Cucurbita* root system and the melon scion. Each of these components of the graft union has different needs, and different conditions may affect their responses to diseases. The response of the *Cucurbita* root system to high temperatures, for example, is a key factor in the performance of grafted plants in the hot season. The *Cucurbita* rootstock has a different response to soil temperature from other cucurbits that are usually grafted, such as watermelon, cucumber and melon. Grafted muskmelons that were transplanted into disease-free soil in the central Arava Valley in mid-July showed markedly suppressed growth in comparison with the normal development of non-grafted muskmelons. The hypothesis was that the *Cucurbita* roots are susceptible to high soil temperatures. Indeed, the problem was solved the following summer by growing the grafted melons on beds covered with yellow plastic, which reduced the soil temperature relative to transparent plastic (*see* Plate 11). However, the different response to soil temperature was useful for growing melons in cold soil, as some *Cucurbita* rootstocks are less susceptible to low temperatures. As a case in point, grafting is used for the very early melon crop in Europe when the soil is cold and non-grafted melons cannot be grown properly (Rivero *et al.*, 2003).

An example of an environmental stress (that was not related to pathogens) causing plant collapse occurred in the Arava Valley of southern Israel during early summer cropping. Trellised grafted muskmelons wilted at harvest time when the temperatures rose and the plants were loaded with fruit, in contrast to the non-grafted melons and grafted plants grown prostrate (Cohen *et al.*, 2007). The cause of this physiological disorder is still unknown. The assumption, however, is that it was related to a combination of environmental conditions, such as high soil temperature, water salinity and insufficient mineral uptake, which the grafted plants could not cope with. Interestingly, the same rootstock–scion combination grew normally in more moderate temperatures, indicating that the physiological collapse was related to the harsh conditions prevailing in the area.

Working with both Galia and Ananas type melons grafted on to several interspecific *Cucurbita* rootstocks and grown in a disease-free protected soil environment, Soteriou *et al.*, (2016) concluded that the expression of physiological incompatibility as plant collapse shortly before harvest was a one-time event not necessarily reflecting on the yield or fruit-quality parameters of the asymptomatic surviving plants. Negative rootstock effect on plant dry mass, loss of mesocarp firmness and attenuation of the first internode's diameter relative to that of the hypocotyl were proposed as plausible indices for rootstock–scion combinations sensitive to physiological incompatibility. Moreover, it was proposed that pronounced earliness of harvest maturity, evidenced as sensitive climacteric muskmelon scions, might relate to ethylene-mediated comprehensive acceleration of ripening, stressing rootstock–scion synergy to the point of collapse.

Rootstock breeding was assessed using a random crossing of *Cucurbita maxima* and *Cucurbita moschata* accessions (R. Cohen, unpublished data) and tested in the field using a 'trial and error' approach. Knowledge of successful rootstock–scion interactions is still limited, but we have some information on the requirements in certain areas, such as the need for tolerance to high soil temperatures or appropriate uptake of nutrients by the rootstock. Searching for an explanation for the above-described 'physiological wilt' demonstrated the need to select *Cucurbita* accessions that are resistant to certain conditions. Such selection will allow rational breeding based on knowledge of designated threats. The recent sequencing of the melon genome, together with advances in sequencing technologies, are enabling the profiling of differences in gene expression under diverse conditions across melon scions and the subsequent characterization of the cellular processes associated with the grafting process, under various stress conditions. Looking for phenotypic, physiological or molecular biomarkers for such traits can direct the search for suitable parents and facilitate rational breeding programmes.

6.6 Response of Grafted Plants to Nematodes

In experiments conducted in Israel, in most cases melons and cucumbers were more susceptible to nematodes than watermelons. Differences were found between the damages caused by the two nematode species evaluated. The severity of galling was always higher for *Meloidogyne incognita* than for *Meloidogyne javanica* in all cultivars tested. Two *Momordica charantia* accessions out of 12 tested exhibited no or very low root galling in response to *M. incognita* and *M. javanica*. However, attempts to use *Momordica* as rootstocks failed, as compatibility with the melon scion was poor. Commercial rootstocks used for watermelon, such as *Cucurbita moschata* × *Cucurbita maxima* F_1 hybrids and *Lagenaria siceraria*, are even more susceptible to root-knot nematodes than non-grafted watermelon cultivars (Edelstein *et al.*, 2010; Thies *et al.*, 2010). Due to the low susceptibility of watermelon cultivars to these nematodes, several attempts have been made to find sources of resistance in wild or exotic watermelon accessions (Hussey and Barker, 1973; Thies and Levi, 2003, 2007; Thies *et al.*, 2015). Some accessions from the wild-type watermelon *Citrullus lanatus* var. *citroides* were found to be moderately resistant to *M. incognita* and *Meloidogyne arenaria* and could be useful as rootstocks for watermelon or as germplasm for rootstock breeding (Thies and Levi, 2003, 2007; Thies *et al.*, 2015).

Susceptibility or resistance to nematodes can differ in certain geographical zones. *M. charantia*, which was resistant in Israel, is known to be very susceptible in Japan. Is this because it is a different *Momordica* or different nematodes? Significant variations in the genetic backgrounds of the evaluated organisms – both host and pest – are likely. The environmental conditions prevailing in a certain area, such as soil type and weather conditions, make it almost impossible to compare performances of biological systems in different regions.

Watermelon, which is the most resistant species among the evaluated cucurbits, can be used as a rootstock for commercial watermelon in nematode-infested fields. Will the level of resistance offered by the wild watermelon be sufficient to reduce nematode damage under field conditions? Although *Cucurbita* is defined as susceptible to nematodes, the susceptibility level and response of different *Cucurbita* rootstocks can vary. Thus, melon grafted on to *Cucurbita* rootstock seems to perform better than non-grafted melon (*see* Plate 12), even though the rootstock is relatively susceptibility to nematodes. The root system of grafted cucurbits and tomatoes seems to be capable of supporting the foliage, despite some nematode damage (Siguenza *et al.*, 2005). Nevertheless, the resistance level offered by the rootstocks does not seem to be able to offer a comprehensive solution, and therefore additional means of nematode control should be added to grafted plants to achieve a desirable level of control.

6.7 Commercial Rootstocks and Unknown Genetics

A challenge encountered by users of grafted plants is that limited proprietary information may be shared with regard to the pedigree of rootstocks or the genes that confer resistance. For example, bacterial wilt caused by *Ralstonia solanacearum* is a serious disease problem in south-eastern USA (Rivard *et al.*, 2012), as it is in many subtropical and tropical regions of the world. The pathogen persists in the soil for years, and available sources of resistance are not effective in all regions of the world, suggesting a differential host-genotype × pathogen-strain interaction, or a more complex host × pathogen × environment interaction that leads to the failure of host resistance (Wang *et al.*, 1998). The pathogen has a high level of diversity worldwide (Remenant *et al.*, 2010), and it is possible that local populations may adapt through mutation, selection or migration to increase new populations that overcome host resistance if the same host genes (typically quantitative trait loci (QTLs)) are continually deployed on a farm. Therefore, one strategy to limit the loss of efficacious resistance is to use rootstocks that harbour different genes for resistance. In south-eastern USA for example, four rootstocks have shown superior resistance to indigenous strains, conferring up to 100% plant-stand resistance, whereas non-grafted plants have 50–100% plant loss (*see* Plate 13). In preliminary experiences in North Carolina, each rootstock confers comparable yield benefits, and certain selections are being used in large areas of over 20 ha. These rootstocks have been developed by different commercial companies, but it is not clear if the rootstocks have the same or different pedigrees and therefore represent the same mechanisms of resistance or a diversity of mechanisms. If the rootstocks represent a diversity of genetic backgrounds, then using them in rotation in subsequent years may allow a more durable resistance than sole reliance on one rootstock. This problem is a specific issue where QTLs confer host resistance, unlike single genes for resistance that are well known and clearly identified by rootstocks suppliers, e.g. the *I*, *I-2* and *I-3* genes for resistance to Fusarium wilt that are progressively becoming well characterized (Catanzariti *et al.*, 2015).

6.8 Different Mechanisms Involved in Disease Resistance Induced by Grafting

A review by Guan *et al.* (2012) described the different mechanisms involved in disease resistance induced by grafting. Can the reduction in disease incidence or severity provided by grafting always be defined as resistance? Does a grafted plant become a resistant one? There is no simple answer to these questions, because the interaction with the pathogen is not limited to one plant species but to a combination of two species, the rootstock and the scion.

Different mechanisms contribute to the reduction in disease development. The first is the resistant nature of the rootstock, which can be monogenic, as with the resistance of melon and tomato rootstocks to Fusarium wilt, or polygenic, governed by QTLs, as is the case of bacterial wilt resistance in tomato rootstocks.

Among the other factors is that grafted plants can cope better with pathogens because of better uptake of water and nutrients, or they may have larger root systems that can compensate for the loss of active root area that might occur in non-grafted plants. Yet another mechanism is that the grafted combination may cause a shift of microorganism populations in the rhizosphere of the rootstock. A unique documentation of a mechanism was published from China where certain rootstocks grafted to aubergine had a different root exudate profile from the non-grafted aubergines and this exudate inhibited *Verticillium dahliae* mycelium, reducing disease incidence (Liu *et al.*, 2009).

Does a grafted plant become a resistant one? Soilborne pathogens seem to be able to penetrate the roots of resistant rootstock, as has been shown with Fusarium wilt (Cohen *et al.*, 2002), *M. phaseolina*, and a complex of *M. phaseolina* and *Rhizoctonia solani* (Fig. 6.1; Cohen *et al.*, 2012). In these examples, the plant became symptomatic, but the pathogen's rate of development and its ability to induce damage were significantly suppressed. Despite pathogen penetration, the plants were able to complete their life cycle and produce high-quality commercial yields.

One part of the complete picture is the role of induced resistance within the grafted plant. The possible mechanisms involved in inducing resistance, including signal transaction, RNA transcript movement and defence-related phytohormones, have been summarized by Guan *et al.* (2012). A clear demonstration of this phenomenon is presented in two different graft–pathogen systems. In the

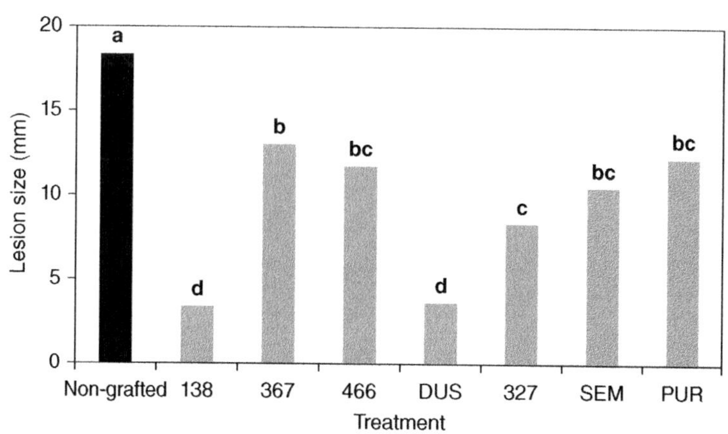

Fig. 6.1. Induction of resistance to *Rhizoctonia solani*. Graph showing lesion sizes on melon scions grafted on to different rootstocks in comparison with the lesion size on the non-grafted melon. Different lower-case letters indicate statistically significant differences. (R. Cohen, R. Grouch and D. Schwarz, unpublished data.)

first, hyphae of *R. solani* were placed on a melon scion grafted on to different *Cucurbita* rootstocks. The lesions that developed on the scion tissue in all cases were significantly smaller than those that developed on the non-grafted melon, serving as a control (Fig. 6.2). The second system (Fig. 6.6; Orgil, 2014) involves induced resistance to Fusarium crown rot pathogen (*F. oxysporum* f. sp. *radicis-cucumerinum*). Interestingly, in this case, the resistance induced by the resistant melon was greater than in the melon grafted on to the *Cucurbita* rootstock. In these two examples, only the scion was exposed to the pathogen and the outcome appeared to be modulated by the rootstock.

It seems that overall disease reduction in grafted plant derives from a combination of direct resistance of the rootstock and induced resistance of the scion conferred by the rootstock. In future research, it will be important to distinguish between the effects of the grafting process and rootstock–scion interactions. Changes in both the roots and shoots of grafted plants need to be elucidated to gain a complete understanding of rootstock–scion interactions and the contribution in induced systemic defence.

Perhaps the different mechanisms of resistance can also be exploited to act in an additive or even synergistic way to manage soilborne diseases. Application of a commercial compound (acibenzolar-*S*-methyl (ASM)) that induces systemic acquired resistance enhanced the control of bacterial wilt compared with the

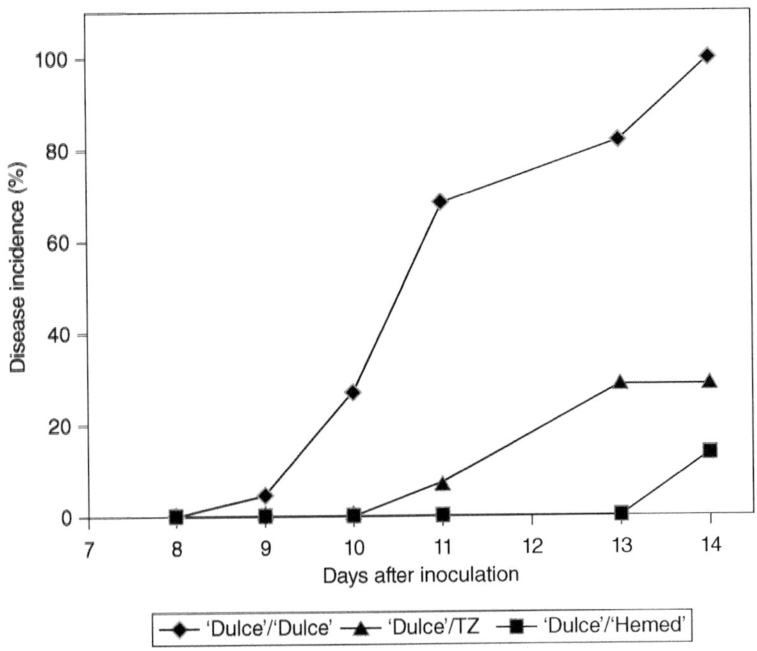

Fig. 6.2. Induction of resistance to Fusarium crown rot in grafted melons. The highest resistance was exhibited in the susceptible melon 'Dulce' grafted on to the resistant melon 'Hemed', followed by susceptible melon 'Dulce' grafted on to the *Cucurbita* rootstock 'TZ-148' (TZ), compared with the non-grafted 'Dulce' melon. Only the scion was inoculated by the stem puncher. (Adapted from Orgil, 2014.)

levels of control seen using tomato cultivars with modest resistance to this disease (Pradhanang *et al.*, 2005). In a complementary study, ASM did not decrease bacterial wilt incidence significantly when applied to grafted plants and compared with rootstocks that conferred a high level of resistance with no ASM (Kunwar *et al.*, 2014), but this combination of tactics is worthy of further study. Employing multiple mechanisms to manage bacterial wilt is likely to offer a more durable form of control than any one tactic alone.

6.9 Conclusions

Grafting of vegetables has expanded in most vegetable production regions of the world and provides a critical integrated pest management (IPM) tool to manage soilborne diseases. In addition to the direct effect that the rootstock confers on reducing disease incidence and severity, the grafting system introduces many complexities and opportunities related to disease management. Critical factors include considerations at each phase of production including seed health, management of diseases in the nursery, selection of the best genetics, unforeseen surprises, and an understanding and exploitation of mechanisms of resistance to limit disease severity. As the science and practice of IPM move forward, complemented with advances in genomics, an enhanced understanding of vegetable physiology and optimization of scion–rootstock interactions, grafting will become a more efficacious and precise tool to manage a large number of soilborne diseases of vegetables.

References

Ainsworth, G.C. (1935) Mosaic disease of cucumber, *Annals of Applied Biology* 22, 55–67.

Antignus, Y., Pearlsman, M., Benyoseph, R. and Cohen, S. (1990) Occurrence of a variant of cucumber green mottle mosaic virus in Israel. *Phytoparasitica* 18, 50–56.

Augstin, B., Graf, V. and Laun, N. (2002) Temperature influencing efficacy of grafted tomato cultivars against root-knot nematodes (*Meloidogyne arenaria*) and corky root (*Pyrenochaeta lycopersici*). *Zeitschrift für Pflanzenkrankheiten und Pflanzenschutz* 109, 371–383.

Bletsos, F., Thanassoulopoulos, C. and Roupakias, D. (2003) Effect of grafting on growth, yield, and Verticillium wilt of eggplant. *HortScience* 38, 83–186.

Catanzariti, A.M., Lim, G.T. and Jones, D.A. (2015) The tomato *I-3* gene: a novel gene for resistance to Fusarium wilt disease. *New Phytologist* 207, 106–118.

Chang, R.J., Ries, S.M. and Pataky, J.K. (1991) Dissemination of *Clavibacter michiganensis* subsp. *michiganensis* by practices used to produce tomato transplants. *Phytopathology* 81, 1276–1281.

Cohen, R., Horev, C., Burger, Y., Shriber, S., Hershenhorn, J., Katan, J. and Edelstein, M. (2002) Horticultural and pathological aspects of Fusarium wilt management using grafted melons. *HortScience* 37, 1069–1073.

Cohen, R., Burger, Y., Horev, C., Porat, A. and Edelstein, M. (2005) Performance of Galia type melons grafted onto *Cucurbita* rootstock in *Monosporascus cannonballus*-infested and non-infested soils. *Annals of Applied Biology* 146, 381–387.

Cohen, R., Burger, Y., Horev, C., Koren, A. and Edelstein, M. (2007) Introducing grafted cucurbits to modern agriculture: the Israeli experience. *Plant Disease* 91, 916–923.

Cohen, R., Omari, N., Porat, A. and Edelstein, M. (2012) Management of *Macrophomina phaseolina* in melons using grafting or fungicide soil application: pathological, horticultural and economical aspects. *Crop Protection* 35, 58–63.

Cohen, R., Tyutyunik, J., Fallik, E., Oka, Y., Tadmor, Y. and Edelstein, M. (2014) Phytopathological evaluation of exotic watermelon germplasm as a basis for rootstock breeding. *Scientia Horticulturae* 165, 203–210.

Davis, A.R., Perkins-Veazie, P., Sakata, Y., López-Galarza, S., Maroto, J.V., Lee, S.G., Huh, Y.C., Sun, Z., Miguel, A., King, S.R., Cohen, R. and Lee, J.M. (2008) Cucurbit grafting. *Critical Reviews in Plant Sciences* 27, 50–74.

de Leon, L., Siverio, F., Lopez, M.M. and Rodriguez, A. (2011) *Clavibacter michiganensis* subsp *michiganensis*, a seedborne tomato pathogen: healthy seeds are still the goal. *Plant Disease* 95, 1328–1338.

Dornai, D., Mingelgrin, U., Frenkel, H. and Barjoseph, M. (1993) Direct quantification of unadsorbed viruses in suspensions of adsorbing colloids with the enzyme-linked immunosorbent assay. *Applied Environmental Microbiology* 59, 3123–3125.

Edelstein, M., Oka, Y., Burger, Y., Eizenberg, H. and Cohen, R. (2010) Variation in the response of cucurbits to *Meloidogyne incognita* and *M. javanica*. *Israel Journal of Plant Sciences* 58, 77–84.

Fanigliulo, A., Viggiano, A., Piegari, G. and Crescenzi, A. (2011) Serious outbreak of *Clavibacter michiganensis* subsp. *michiganensis* on tomato in southern Italy. *Acta Horticulturae* 914, 43–46.

Friedmann, M., Lapidot, M., Cohen, S. and Pilowaky, M. (1988) A novel source of resistance to tomato yellow leaf curl virus exhibiting a symptomless reaction, *Journal of the American Society for Horticultural Science* 123, 1004–1007.

Genda, Y., Sato, K., Nunomura, O., Hirabayashi, T. and Tsuda, S. (2011) Immunolocalization of pepper mild mottle virus in developing seeds and seedlings of *Capsicum annuum*. *Journal of General Plant Pathology* 77, 201–208.

Gitaitis, R.D., Beaver, R.W. and Voloudakis, A.E. (1991) Detection of *Clavibacter michiganensis* subsp. *michiganensis* in symptomless tomato transplants. *Plant Disease* 75, 834–838.

Gleason, M.L., Gitaitis, R.D. and Ricker, M.D. (1993) Recent progress in understanding and controlling bacterial canker of tomato in eastern North America. *Plant Disease* 77, 1069–1076.

GSPP (2016) Hygiene in seed production and plant raising to prevent infection with pathogen. Good Seed and Plant Practices. Available at: http://www.gspp.eu (accessed 8 November 2016).

Guan, W., Zhao, X., Hassell, R. and Thies, J. (2012) Defense mechanisms involved in disease resistance of grafted vegetables. *HortScience* 47, 164–170.

Hussey, R.S. and Barker, K.R. (1973) A comparison of methods of collecting inocula of *Meloidogyne* spp., including a new technique. *Plant Disease Report* 57, 1025–1028.

Jifon, J., Crosby, K., Miller, M. and Leskovar, D. (2006) Physiological characteristics of grafted muskmelon grown in *Monosporascus cannonballus*-infested soil in south Texas. In: Holmes, G.J. (ed.) *Proceedings of Cucurbitaceae 2006*. Universal Press, Raleigh, North Carolina, pp. 23–30.

Kamenova, I. and Adkins, S. (2004) Transmission, in planta distribution, and management of *Hibiscus latent Fort Pierce virus*, a novel tobamovirus isolated from Florida hibiscus. *Plant Disease* 88, 674–679.

Keinath, A.P. and DuBose, V.B. (2012) Controlling powdery mildew on cucurbit rootstock seedlings in the greenhouse with fungicides and biofungicides. *Crop Protction* 42, 338–344.

King, S.R., Davis, A.R., Liu, W. and Levi, A. (2008) Grafting for disease resistance. *HortScience* 43, 1673–1676.

Kubota, C., McClure, M.A., Kokalis-Burelle, N., Bausher, M.G. and Rosskopf, E.N. (2008) Vegetable grafting: history, use, and current technology status in North America. *HortScience* 43, 1664–1669.

Kunwar, S., Paret, M.L., Jones, J.B., Ritchie, L., Freeman, J.H. and Olson, S.M. (2014) Management of bacterial wilt of tomato with a combination of grafting and a systemic acquired resistance inducer. *Phytopathology* 104, S64.

Lee, J.M. (1994) Cultivation of grafted vegetables. I. Current status, grafting methods, and benefits. *HortScience* 29, 235–239.

Lee, J.M. and Oda, M. (2003) Grafting of herbaceous vegetable and ornamental crops. *Horticultural Reviews* 28, 61–124.

Lee, J.M., Kubota, C., Tsao, S.J., Bie, Z., Hoyos, E., Morra, L. and Oda, M. (2010) Current status of vegetable grafting: diffusion, grafting techniques, automation. *Scientia Horticulturae* 127, 93–105.

Liu, N., Zhou, B., Zhao, X., Lu, B., Li, Y. and Hao, J. (2009) Grafting eggplant onto tomato rootstock to suppress *Verticillium dahliae* infection: the effect of root exudates. *HortScience* 44, 2058–2062.

Louws, F.J., Rivard, C.L. and Kubota, C. (2010) Grafting fruiting vegetables to manage soilborne pathogens, foliar pathogens, arthropods and weeds. *Scientia Horticulturae* 127, 127–146.

Miguel, A., Maroto, L.V., San Bautista, A., Baixauli, C., Cebolla, V., Pascual, B., López, S. and Guardiola, J.L. (2004) The grafting of triploid watermelon is an advantageous alternative to soil fumigation by methyl bromide for control of Fusarium wilt. *Scientia Horticulturae* 103, 9–17.

Miller, S.A., Ivey, M.L., Baysal-Gurel, F and Xu, X.L. (2015) A systems approach to tomato disease management. *Acta Horticulturae* 1069, 167–172.

Orgil, G. (2014) Root and stem rot in cucumber and melon: studying the biology of the disease and controlling it using grafted plants. MSc thesis, Robert H. Smith Faculty of Agriculture, Food, and Environment, Hebrew University of Jerusalem, Israel.

Paplomatas, E.J., Elena, K., Tsagkarakou, A. and Perdikaris, A. (2002) Control of Verticillium wilt of tomato and cucurbits through grafting of commercial varieties on resistant rootstocks. *Acta Horticulturae* 579, 445–448.

Parke, J.L. and Grünwald, N.J. (2012) A systems approach for management of pests and pathogens of nursery crops. *Plant Disease* 96, 1236–1244.

Pradhanang, P.M., Ji, P., Momol, M.T., Olson, S.M., Mayfield, J.L. and Jones, J.B. (2005) Application of acibenzolar-*S*-methyl enhances host resistance in tomato against *Ralstonia solanacearum*. *Plant Disease* 89, 989–993.

Rahman, M.A., Rashid, M.A., Hossain, M.M., Salam, M.A. and Masum, A. (2002) Grafting compatibility of cultivated eggplant varieties with wild *Solanum* species. *Pakistan Journal of Biological Sciences* 5, 755–757.

Reingold, V., Lachman, O., Blaosov, E. and Dombrovsky, A. (2014) Seed disinfection treatments do not sufficiently eliminate the infectivity of *Cucumber green mottle mosaic virus* (CGMMV) on cucurbit seeds. *Plant Pathology* 64, 245–255.

Reingold, V., Lachman, O., Belausov, E., Koren, A., Mor, N. and Dombrovsky, A. (2016) Epidemiological study of *Cucumber green mottle mosaic virus* (CGMMV) in greenhouses enables reduction of disease damage in cucurbit production. *Annals of Applied Biology* 168, 29–40.

Remenant, B., et al. (2010) Genomes of three tomato pathogens within the *Ralstonia solanacearum* species complex reveal significant evolutionary divergence. *BMC Genomics* 11, 379.

Rivard, C.L., O'Connell, S., Peet, M.M., Welker, R.M. and Louws, F.J. (2012) Grafting tomato to manage bacterial wilt caused by *Ralstonia solanacearum* in the southeastern United States. *Plant Disease* 96, 973–978.

Rivero, R.M., Ruiz, J.M. and Romero, L. (2003) Role of grafting in horticultural plants under stress conditions. *Food, Agriculture and Environment* 1, 70–74.

Ryu, K.H., Yoon, J.Y., Choi, G.S., Choi, S.K., Hong, J.S., Choi, J.K., Kim, W. and Lee, G.P. (2008) Molecular and biological diversities of *Cucumber green mottle mosaic virus* from cucurbitaceous crops in Korea. *Journal of Phytopathology* 156, 408–412.

Sen, Y., van der Wolf, J., Visser, R.G.F. and van Heusden, S. (2015) Bacterial canker of tomato: current knowledge of detection, management, resistance, and interactions. *Plant Disease* 99, 4–13.

Siguenza, C., Schochow, M., Turini, T. and Ploeg, A. (2005) Use of *Cucumis metuliferus* as a rootstock for melon to manage *Meloidogyne incognita*. *Journal of Nematology* 37, 276–280.

Soteriou, A.G., Papayiannis, L.C. and Kyriacou, M.C. (2016) Indexing melon physiological decline to fruit quality and vine morphometric parameters. *Scientia Horticulturae* 203, 207–215.

Thies, J.A. and Levi, A. (2003) Resistance of watermelon germplasm to the peanut root-knot nematode. *HortScience* 38, 1417–1421.

Thies, J.A. and Levi, A. (2007) Characterization of watermelon (*Citrullus lanatus* var. *citroides*) germplasm for resistance to root-knot nematodes. *HortScience* 42, 1530–1533.

Thies, J.A., Ariss, J.J., Hassell, R.L., Olson, S., Kousik, C.S. and Levi, A. (2010) Grafting for management of southern root-knot nematode, *Meloidogyne incognita*, in watermelon. *Plant Disease* 94, 1195–1199.

Thies, J.A., Buckner, S., Horry, M., Hassell, R. and Levi, A. (2015) Influence of *Citrullus lanatus* var. *citroides* rootstocks and their F_1 hybrids on yield and response to root knot nematode, *Meloidogyne incognita*, in grafted watermelon. *HortScience* 50, 9–12.

Tian, T., Posis, K., Maroon-Lango, C.J., Mavrodieva, V., Haymes, S., Pitman, T.L. and Falk, B.W. (2014) First report of *Cucumber green mottle mosaic virus* on melon in the United States. *Plant Disease* 98, 1163.

Wang, J.F., Hansen, P. and Barnes, J.A. (1998) Worldwide evaluation of an international set of resistance sources to bacterial wilt in tomato. In: Prior, P.H., Allen, C. and Elphinstone, J. (eds) *Bacterial Wilt Disease: Molecular and Ecological Aspects*. Springer, Berlin/Heidelberg, pp. 269–275.

Werner, N.A., Fulbright, D.W., Podolsky, R., Bell, J. and Hausbeck, M.K. (2002) Limiting populations and spread of *Clavibacter michiganensis* subsp. *michiganensis* on seedling tomatoes in the greenhouse. *Plant Disease* 86, 535–542.

Xu, X.L., Miller, S.A., Baysal-Gurel, F., Gartemann, K.H., Eichenlaub, R. and Rajashekara, G. (2010) Bioluminescence imaging of *Clavibacter michiganensis* subsp *michiganensis* infection of tomato seeds and plants. *Applied and Environmental Microbiology* 76, 3978–3988.

Yetisir, H., Sari, N. and Seral Yticel, S. (2003) Rootstock resistance to Fusarium wilt and effect on watermelon fruit yield and quality. *Phytoparasitica* 31, 163–169.

7 Grafting as a Tool for Tolerance of Abiotic Stress

Youssef Rouphael,[1] Jan Henk Venema,[2] Menahem Edelstein,[3] Dimitrios Savvas,[4] Giuseppe Colla,[5] Georgia Ntatsi,[4] Meni Ben-Hur,[6] Pradeep Kumar[7] and Dietmar Schwarz[8]*

[1]University of Naples Federico II, Portici, Italy; [2]University of Groningen, Groningen, The Netherlands; [3]Newe Ya'ar Research Center, Agricultural Research Organization, Ramat Yishai, Israel; [4]Agricultural University of Athens, Athens, Greece; [5]University of Tuscia, Viterbo, Italy; [6]Institute of Soil, Water and Environmental Sciences, Agricultural Research Organization, Bet Dagan, Israel; [7]ICAR-Central Arid Zone Research Institute, Jodhpur (Rajasthan), India; [8]Leibniz Institute for Vegetable and Ornamental Crops, Großbeeren, Germany

7.1 Introduction

Vegetable crops are cultivated mainly in the field but also intensively under protected structures of greenhouses year-round or for several months. If the conditions are not as comprehensively controlled as they are in highly sophisticated greenhouses, such plants will frequently be subjected to poor and extreme environmental conditions. As a result, farmers frequently encounter various problems, particularly abiotic stress, and do not attain the yield potential of their crops. It is estimated that 60–70% of the gap from the actual to the potential yield is attributable to abiotic factors, such as salinity, drought, non-optimal temperatures and nutritional deficiencies/toxicities (Peleg *et al.*, 2011). Even in modern high-tech greenhouses, imbalanced fertilization and high temperatures are possible. Moreover, due to climate change, these factors are expected to become aggravated, posing a major threat for productivity and food security in the future.

The use of grafted vegetable plants can minimize problems associated with successive cropping and abiotic stress (Colla *et al.*, 2010; Savvas *et al.*, 2010;

* schwarz@igzev.de

Schwarz *et al.*, 2010), and their enhanced vigour and root growth can provide yield benefits independently of mechanisms to tolerate abiotic stress conditions. Therefore, the application of grafting in crops of the Cucurbitaceae and Solanaceae is currently acknowledged worldwide; however, two main challenges remain when grafted plants are used as a tool to improve abiotic stress tolerance. One is the selection of the ideal rootstock–scion combination that matches the actual or expected abiotic stress factors. The other is to improve our understanding of the underlying tolerance mechanisms, which may support the breeding of rootstocks that broaden the abiotic stress tolerance (robustness) of vegetable crops. This chapter presents the current knowledge on these two issues organized by individual abiotic stress factors.

7.2 Temperature Stress

7.2.1 Diminishing temperature constraints for vegetable production

The optimal performance of vegetable crops is related to a specific temperature range that should be maintained during the entire production period. However, short chilling or heating episodes occur occasionally in many temperate regions. To protect plants from temperature stress, vegetable crops are cultivated in plastic or sophisticated climate-controlled greenhouses. The use of plastic greenhouses is common in southern Europe, but has the disadvantage that it is difficult to manage both temperature and humidity within these structures. However, the use of climate-controlled greenhouses under the temperate climatic conditions of central and northern Europe is an absolute prerequisite for commercial vegetable fruit production, as otherwise temperatures are too cool. Besides the high investment costs, heated-greenhouse vegetable production has the disadvantage that it requires a considerable energy input, which contributes significantly to the emission of CO_2.

Cucurbitaceae and Solanaceae, cultivated in the field or in unheated greenhouses, frequently have to face temperature conditions below or above their optimum. This occurs particularly at the start of the cultivation period when grown early in winter/spring or at the end of the production season when grown late in autumn/winter. Additionally, during the seasons, they have to cope with large temperature ranges and rapid diurnal variations in air and, to a lesser extent, soil temperatures (*see* Plate 15). Cultivation below or above the optimum temperature generates stress and impairs specific morphological, physiological and biochemical processes in both roots and shoots. Depending on the duration and severity of the temperature stress, these processes may lead to a reduction in crop development, growth, fruit yield and quality, or even complete crop loss (reviewed by Wang *et al.*, 2003; Venema *et al.*, 2005; van der Ploeg and Heuvelink, 2005; Schwarz *et al.*, 2010; Bita and Gerats, 2013). Below the optimum, three temperature zones are described: freeze (<0°C), chill (0–12°C) and suboptimum (>12°C to optimum). These ranges are only rough indications, as each aspect of growth, development and fruit formation has its own temperature optimum, which varies among and within species, as well as with plant age. Thus, the underlying effects of low-temperature stress are manifold and complex.

As vegetables originating from (sub)tropical areas are thermophilic and not freeze-tolerant, air temperatures below zero will kill such plants. They apparently lack the genetic information to allow cold acclimation or to become freeze-tolerant. Chilling can have a devastating effect on plant metabolism, as it disrupts cellular homeostasis and uncouples major physiological processes such as photosynthesis, thylakoid electron transport, the carbon reduction cycle and control of stomatal conductance (Allen and Ort, 2001; Suzuki and Mittler, 2006). Depending on the intensity and length of exposure, crops may not complete their life cycle and consequently provide no harvestable yield.

Suboptimal temperature starts above the threshold temperature at which a plant can successfully complete its life cycle (Greaves, 1996). Such temperatures impair cell membrane fluidity and permeability, which may result in ion leakage (Abbas, 2012), inhibition of intra- and extracellular water and nutrient movement (Salinas, 2002; Mahajan and Tuteja, 2005), inhibition of photosynthesis (Theocharis *et al.*, 2012), generation of reactive oxygen species (Gill and Tuteja, 2010), reduced water uptake and inhibition of stomatal closure (Wilkinson *et al.*, 2001) resulting in cellular dehydration. Suboptimal temperatures induce fluctuations in soluble sugar concentrations, which may involve changes in CO_2 assimilation, source–sink carbon partitioning and the activity of related enzymes, as well as in the expression of specific genes via stress- and sugar-signalling pathways that modulate whole-plant metabolism (Gupta and Knaur, 2005). During vegetative growth, this slows the rate of leaf initiation and leaf expansion. In the generative phase, this decreases fruit set as a result of poorer pollen quality and retards inflorescence appearance rate. The period between anthesis and maturity of the fruit is extended and fruit size increases (reviewed by Zhou *et al.*, 2004; van der Ploeg and Heuvelink, 2005).

At supra-optimal temperatures (>30°C), metabolic rates increase exponentially with temperature. As a consequence, a series of complex morphological, physiological, biochemical and molecular changes occur that adversely affect plant growth and productivity (reviewed by Wang *et al.*, 2003; Wahid *et al.*, 2007; Bita and Gerats, 2013; Jha *et al.*, 2014). If high-temperature stress coincides with high humidity, as frequently occurs in plastic greenhouses, the decrease in growth and yield becomes worse. The largest problem for tomato, particularly at temperatures above 32°C, is a reduction in the number of pollen grains and their viability. Moreover, pollen sterility, which limits fertilization and ultimately fruit development, is also problematic (Pressman *et al.*, 2002). Temperatures above 35°C limit the cultivation of Solanaceae rather than Cucurbitaceae because the supra-optimal temperature threshold of the latter is much higher (Bita and Gerats, 2013).

In addition to the breeding of more robust cultivars or the development of sophisticated cultivation systems that protect plants against harmful temperatures, the use of cold- or heat-tolerant rootstocks is now regarded as an additional tool to broaden the temperature range under which vegetable crops can produce optimally (Schwarz *et al.*, 2010). A rootstock with a more extensive root system with a broad temperature optimum can help the plant to withstand low root-zone temperatures (i.e. in the soil or substrates), which limit root water uptake, and may increase the tolerance of the scion to high air temperatures, which increase transpirational water loss. Grafting is advantageous, as not all cultivars require separate root optimization, allowing improved rooting at non-optimal temperatures of already-existing elite cultivars. Designing vigorous rootstocks for specific growth conditions is becoming a feasible target to cope with global climate change and the increased occurrence of temperature extremes (Gregory *et al.*, 2013). In the following section, we therefore focus on the role rootstocks may play in the improvement of vegetable crop performance under non-optimal temperatures.

7.2.2 Contribution of rootstocks to improved low- and high-temperature tolerance

Many grafting experiments have illustrated that rootstocks can alleviate the impact of chilling temperatures (Ahn *et al.*, 1999; Bloom *et al.*, 2004), suboptimal temperatures (Venema *et al.*, 2008, Li *et al.*, 2014a,b; Ntatsi *et al.*, 2014a,b) and supra-optimal (root-zone) temperatures (Rivero *et al.*, 2003a; Li *et al.*, 2014a). The underlying responses were associated with better root growth, increased root biomass partitioning and many other physiological mechanisms (reviewed by Schwarz *et al.*, 2010). This section discusses novel insights into how rootstocks have facilitated improved scion performance at non-optimal temperatures. The discussion focuses on two key aspects: root growth and rootstock–scion communication.

Temperature modulation of root growth and root system architecture

Root growth has an optimum temperature range, which is often slightly lower than that for shoot growth (Gosselin and Trudel, 1985). The optimal temperature range for root development in the following cultivars is: tomato 22–26 °C (Gosselin and Trudel, 1984; Rivero *et al.*, 2005), aubergine 22–30°C (Wu *et al.*, 2014), pepper 25–30°C (Airaki *et al.*, 2012), cucumber 19–23°C (Tachibana, 1982; Lee, 1994), melon 20–32°C (Korkmaz and Dufault, 2001) and watermelon 33–37°C (Rivero *et al.*, 2005). As well as this variation among vegetable species, the range may also vary within closely related species, as in the Curcurbitaceae, where figleaf gourd (*Cucurbita ficifolia* Bouché), originating from the highland regions of Latin America (from Mexico to Chile), has a 6°C lower optimum temperature for root growth than cucumber (Ahn *et al.*, 1999; Rivero *et al.*, 2003b).

The stunting effect of sub- and supra-optimal root-zone temperatures on root growth reduces the soil/substrate volume that roots may access for the uptake of water and nutrients. Root architecture also changes by decreasing: (i) the primary root length; (ii) the lateral root density (numbers of lateral roots per unit primary root length); and (iii) the angle under which lateral roots emerge from the primary root (McMichael *et al.*, 1993; Seiler, 1998; Nagel *et al.*, 2009). Roots subjected to supra-optimal temperatures started to initiate second- and third-order laterals (Pardales *et al.*, 1999) and showed an increased average root diameter (Qin *et al.*, 2007).

Rapid cell elongation in the elongation/differentiation zone is one of the key determinants of root growth (Takatsuka and Umeda, 2014), and its inhibition is the primary stunting effect of non-optimal temperatures (Pahlavanian and

Silk, 1988; Pardales *et al.*, 1992; Gladish and Rost, 1993). Root elongation rate is one of the most important characteristics determining nutrient uptake rate (Silberbush and Barber, 1983). An extensive study based on 29 genotypes from three Solanaceae crops (tomatoes, pepper and aubergine) under optimal conditions revealed a strong positive correlation between leaf area expansion rate and root elongation rate, as well as root branching density (Bui *et al.*, 2015). Among *Arabidopsis* accessions, variation in root elongation rate was equally determined by the mature cortical cell length (cell expansion) and the cell production rate within the root meristem (Beemster *et al.*, 2002). The latter was strongly correlated with A-type cyclin-dependent kinase (CDKA;1) activity. Thus, transferring *Arabidopsis* roots for 3 days from 22 to 16°C decreased the expression of genes (*CYCD1;1*, *CYCD4;2*, *CYCD6;1* and *E2Fb*) that positively regulate cell-cycle progression, while increasing the expression of two genes (*E2Fc* and *RBR*) that negatively regulate cell proliferation (Zhu *et al.*, 2015). Temperatures below the optimum temperature for root development repressed the division potential of root meristematic cells by reducing both the meristem size and cell number (Zhu *et al.*, 2015). Inhibition of the division potential of root meristematic cells at a suboptimal temperature (16°C) was ascribed to decreased auxin levels in the root apex, which was associated with inhibited expression of the PIN-FORMED auxin transport proteins PIN1, -3 and -7, and auxin biosynthesis-related genes. Root elongation of several *Arabidopsis* cytokinin signalling mutants was much less sensitive to low temperature measured in terms of a reduction in root length, meristem cell number, PIN1/3 transcripts and auxin levels, suggesting that cytokinin signalling is involved in reduced auxin accumulation at a suboptimal temperature. Subjecting *Arabidopsis* roots to 4°C inhibited the root basipetal (shootwards) auxin transport by selectively blocking the intracellular cycling of PIN2 and asymmetric redistribution of PIN3. This diminished the root's ability to form an auxin gradient, which collectively led to reduced cell elongation (Shibasaki *et al.*, 2009).

The limited information that is available in the literature on the underlying inhibitory effects of high-temperature stress on root elongation excludes the involvement of altered indole-3-acetic acid (IAA) transport or levels (Gladish *et al.*, 2000), but supports the involvement of increased ethylene levels (Qin *et al.*, 2007).

In contrast to *Arabidopsis*, the mechanisms that underlie the strong decrease in root elongation rate at non-optimal temperatures have barely been studied in thermophilic vegetable crops. Therefore, Illumina RNA sequencing was used to compare genome-wide root transcriptomes at optimal (23°C) and suboptimal (15°C) temperatures in two tomato species (Solanum lycopersicum L. 'Moneymaker' and *Solanum habrochaites* S. Knapp & D.M. Spooner LA1777), in which root elongation rates showed a different susceptibility to low temperature (Zamir and Gadish, 1987). Significant differences in gene expression between both species were detected for genes involved in: (i) the response to auxin (e.g. auxin-responsive protein, auxin-repressed protein, auxin-induced SAUR-like protein); (ii) ethylene synthesis (1-aminocyclopropane-1-carboxylate oxidases); (iii) lateral root formation (cyclin-dependent kinase inhibitor 7); (iv) root morphogenesis; (v) cell-wall synthesis (xyloglucan endotransglucosylase/hydrolase 2, β-galactosidase, glucosyltransferase, chitinase-like protein); and (vi) ethylene-responsive transcription factors (P. Heidari and J.H. Venema, unpublished results).

Further elaboration of these expression profiles is ongoing and may provide candidate genes and pathways that can be used by breeders to select low-temperature-tolerant rootstocks.

Rootstock factors that mediate improved scion performance at non-optimal temperatures

Fast cooling of roots from 20 to 5°C (within 2 h) at ambient air temperature (22°C) induced wilting in a cultivated tomato (*S. lycopersicum* 'T5') but not in a high-altitude accession of the wild tomato (*S. habrochaites* LA1778) (Truco *et al.*, 2000; Bloom *et al.*, 2004). During root cooling, root hydraulic conductance declined similarly in both species, but the more cold-tolerant *S. habrochaites* rapidly closed its stomata, thereby maintaining water potential and shoot turgor, in contrast to the cold-sensitive *S. lycopersicum* (Bloom *et al.*, 2004). A major quantitative trait locus (QTL) named *shoot-turgor maintenance9* (*stm9*) under root chilling was detected and fine-mapped on chromosome 9 in an interspecific backcross population derived from *S. lycopersicum* 'T5' and *S. habrochaites* LA1778 (Truco *et al.*, 2000; Goodstal *et al.*, 2005). Using high-resolution mapping, the QTL *stm9* was located in a 0.32 cM region in the *S. lycopersicum* genome sequence that is gene rich and contains representatives of gene families that have been associated with abiotic stress tolerance (Arms *et al.*, 2015). Introgression lines from two other interspecific backcross populations containing alleles of the high-altitude accessions *S. habrochaites* LA1777 and *Solanum lycopersicoides* Dunal LA2951 at the chromosome 9 region collinear to the QTL *stm9* region in *S. habrochaites* LA1778 were tested in reciprocal grafting experiments with their cultivated parents for their susceptibility to root chilling-induced water stress. Shoot turgor maintenance conferred by this introgression during root chilling was root based: tomato scions grafted on to a chill-induced wilting-tolerant introgression line rootstock maintained shoot turgor during root chilling regardless of the scion genotype (Easlon *et al.*, 2013). The presence of an antitranspirant signal in the xylem sap was confirmed by feeding detached leaves with xylem sap (collected from chilled roots of wilting or non-wilting introgression lines and measuring their transpiration rate; H.M. Easlon and A.J. Bloom, unpublished results). The higher antitranspirant activity of the xylem sap of non-wilting introgression lines could not be explained by increased abscisic acid (ABA) concentration or pH (*see* Chapter 4, this volume). Despite extensive xylem sap profiling, the root-derived signal that induces stomatal closure in root-chilled *S. habrochaites*, *S. lycopersicoides* and introgression lines containing the QTL *stm9* remains unknown.

Comparative hormone profiling of non-grafted *S. lycopersicum* 'Moneymaker' and *S. habrochaites* LA1777 demonstrated that the stronger growth inhibition of 'Moneymaker' at a suboptimal root-zone temperature (15°C) was associated mainly with a stronger decrease in auxin (IAA) and cytokinin concentrations, in combination with a more pronounced increase in ABA concentration in the leaves (J.H. Venema, unpublished results). Remarkably, the 15°C root-zone temperature treatment hardly changed the root hormone profiles of either species. The only root hormone levels that were differently affected by the 15°C root-zone temperature treatment compared with the 25°C treatment were ABA and jasmonic acid, which increased significantly in 'Moneymaker' and in LA1777.

The involvement of ABA in the growth inhibition of tomato at a root-zone temperature of 15°C was studied using reciprocal grafts with *notabilis*, a null mutant in the ABA biosynthesis gene *LeNCED1* (Ntatsi *et al.*, 2014a). All rootstock–scion combinations, however, showed a similar growth inhibition and increase in leaf ABA level in response to a 15°C root-zone temperature. From this experiment, it was concluded that *LeNCED1* is not involved in leaf ABA accumulation induced by suboptimal root temperatures. In another study with the same rootstock–scion combinations, however, shoot growth and leaf ABA concentrations under complete suboptimal temperature conditions (air and root zone) were inhibited by the *LeNCED1* mutation. The decrease in leaf ABA concentration was only obvious in grafts with *notabilis* as the scion, whereas the genotype of the rootstock had no additional effect on the leaf ABA status. The observed differences in both studies may be because, under complete suboptimal growth conditions, as in the latter study, the leaves accumulated higher levels of soluble sugars, which may trigger ABA biosynthesis differently. Under complete suboptimal growth conditions, ABA exerts a protective role on tomato scion growth, presumably by decreasing the endogenous ethylene levels (Ntatsi, 2013).

Scion acclimation to low temperatures might be governed not only directly by the upward transfer of rootstock-derived signal molecules such as hormones, ions, small RNAs or proteins (*see* Chapter 4, this volume), but also indirectly by triggering transcriptional changes in the scion. A comparative Agilent microarray transcriptome analysis in the cultivated tomato 'Kommeet' grafted either on to a cold-sensitive (*S. lycopersicum* 'Moneymaker') or a more cold-tolerant rootstock (*S. habrochaites* LA 1777) revealed that reducing the root-zone temperature from 25 to 15°C for 30 days induced significant transcriptional changes in the rootstock and to a lesser extent in the scion (Ntatsi, 2013). When the transcriptomes of the two graft combinations were compared in both leaves and roots, the number of differentially expressed genes in response to the suboptimal root-zone temperature (15°C) was much larger in grafts with 'Moneymaker' (361 and 2036 transcripts, respectively) than in grafts with LA 1777 (0 and 1509 transcripts, respectively) as rootstock. This large difference indicated that a suboptimal root-zone temperature was more stressful for tomato cultivars grafted on to 'Moneymaker' than on to LA 1777, which was supported by results obtained in previous grafting experiments with these rootstocks (Venema *et al.*, 2008; Ntatsi *et al.*, 2014b). 'Cell-wall genes associated with cellulose synthesis through cellulose synthase' were significantly enriched in LA 1777 at a suboptimal root-zone temperature (15°C) in contrast to 'Moneymaker' (Ntatsi, 2013). This result corroborates the faster root growth of this species compared with 'Moneymaker' at 15°C, as well as the RNA sequencing data discussed above, which also pointed to interspecific differences in the expression of genes involved in cell-wall metabolism at 15°C. Further evaluation of this microarray data may identify additional genes involved in the physiological or molecular mechanisms by which the LA1777 rootstock improves tolerance to suboptimal root-zone temperatures.

In leaves of cucumber scions grafted on to cold-tolerant figleaf gourd (*C. ficifolia*) and heat-tolerant luffa (*Luffa cylindrica* (L.) M. Roem.) rootstocks, a subset of stress-responsive genes involved in signal transduction (*MAPK1* and *RBOH*), transcriptional regulation (*MYB* and *MYC*), protein protection (*HSP45.9* and *HSP70*), antioxidant response (*Cu/Zn-SOD*, *cAPX* and *GR*) and photosynthesis (*RBCL*, *RBCS*, *RCA*, and *FBPase*) were upregulated at suboptimal (18/13°C) and supra-optimal (36/31°C) day/night temperatures, respectively (Li *et al.*, 2014b). This upregulation was accompanied by increased activity of the encoded enzymes and reduced redox homeostasis in the leaves. The enhanced heat tolerance of cucumber scions grafted on to luffa rootstocks was associated with increased accumulation of ABA and apoplastic hydrogen peroxide (H_2O_2), *RBOH* (respiratory burst oxidase homologue) transcripts and heat-shock protein 70 (HSP70) expression, and a decrease in oxidative stress (Li *et al.*, 2014a). Increased leaf ABA accumulation in scions grafted on to a luffa rootstock was attributed to increased root-to-shoot transport of ABA and/or increased foliar ABA, depending on whether the root and/or the scion were heat stressed, respectively. The increased ABA biosynthesis that triggered the expression of HSP70 was induced by increased H_2O_2 generation via upregulation of the NADPH oxidase transcript encoded by *RBOH*. These data suggest that an ABA-dependent H_2O_2-driven mechanism underlies a systemic response by which luffa rootstocks increase the heat tolerance of cucumber scions.

In tomato, high-temperature tolerance was conferred to non-transgenic scions by grafting on to a rootstock in which the fatty acid desaturase 7 (*LeFAD7*) gene was silenced (Nakamura *et al.*, 2016). Silencing of the *LeFAD7* gene in the rootstock repressed *LeFAD7* expression in the scion via graft transmission of a small interfering RNA. This resulted in decreased amounts of unsaturated fatty acids, which increased the high-temperature tolerance. This study is an example how graft transmission of RNA silencing from rootstocks to non-transgenic scions may induce temperature tolerance (*see* Chapter 4, this volume).

7.2.3 Rootstock selection for improved temperature-stress tolerance

The selection and breeding of suitable rootstocks offers a powerful tool to sustain and expand the cultivation of vegetable crops under non-optimal temperatures, as well as many other suboptimal growth conditions (Gregory *et al.*, 2013). To breed rootstocks that broaden the temperature optimum of vegetable crops via grafting, it is essential to improve our understanding of the regulatory processes and underlying genetic components that regulate root growth, plasticity of the root system architecture (RSA) and root functioning (e.g. root hydraulic conductance). Root growth studies in young chilling-sensitive maize plants indicated that investment in lateral root formation improves plant performance in the suboptimal temperature range, whereas stimulation of axillary root growth is needed in the supra-optimal temperature range (Hund *et al.*, 2007, 2008). These RSA changes are of adaptive value to improve uptake capacity and allowing root proliferation in lower soil layers to enhance water uptake, respectively. Year-round production systems, during which roots experience a broad range of temperatures, makes rootstock selection based on particular RSA traits challenging.

Root growth regulation, and its response to changing environmental conditions such as temperature, is a highly complicated process that is controlled at many different levels by complex actions of gene networks in both time and space (Kuijken *et al.*, 2015). Advances in this area, mainly derived from work in *Arabidopsis*, were reviewed recently (Wachsman *et al.*, 2015; Slovak *et al.*, 2016). It will be important to gain more knowledge of: (i) the natural variation in root development, root biomass partitioning and RSA that exists within vegetable crops (cultivars, old landraces and related wild accessions) under optimal and non-optimal temperatures; (ii) the physiological and molecular basis of this variation; and (iii) the root-derived signals by which the rootstock improves scion performance under adverse temperatures (Warschefsky *et al.*, 2016). Extensive research on these topics may deliver physiological and molecular indicators for temperature-stress tolerance, which can be incorporated by breeders into their current selection pipeline for low- and high-temperature-tolerant rootstocks.

In recent decades, several QTLs, genes and molecular pathways have been discovered that control root traits with agricultural relevance (reviewed by de Dorlodot *et al.*, 2007; Rogers and Benfey, 2015). Much progress in our understanding of control mechanisms involved in root development is expected in the next 5–10 years, due to the increasing number of highly efficient root phenotyping platforms, the use of genome-wide association studies for root traits, the increasing availability of functional genomics resources for roots, and the development of smart root model systems. Grafting experiments are an excellent tool to advance our understanding of the adaptive value of genetically determined differences in root growth and RSA at the level of crop performance, marketable yield and fruit quality in targeted root environments and growth conditions. This will improve the efficiency of future breeding programmes aimed at the selection of rootstocks that broaden the temperature optimum of vegetable crops.

7.2.4 Cold- and heat-tolerant Cucurbitaceae and Solanaceae rootstocks

During the 1960s, the use of unheated plastic greenhouses for horticultural production became common worldwide to extend the growing period. This was accompanied by renewed interest in grafting to enhance fruit yield and earliness during the cold seasons in Japan and Korea, when low soil temperatures may seriously affect or even kill seedlings (Ahn *et al.*, 1999; Lee and Chung, 2005). Since then, rootstocks have been commonly used in Cucurbitaceae (cucumber, squash, melon and watermelon) in unheated greenhouses and in the open field. The successful use of rootstocks for Solanaceae until the end of the 1980s was restrained by inefficient grafting techniques and poor germination of rootstock seed. Due to the introduction of heated cultivation and soil-less culture, the need to use rootstocks disappeared. The introduction of Japanese grafting techniques (*see* Chapter 1, this volume) to Europe allowed cheaper and more successful grafting activities. Extensive research by private seed companies led to a drastically improved germination capacity of tomato rootstock seed and the development of new vigorous rootstock hybrids, which boosted fruit production. For Solanaceae, cultivated in unheated greenhouses, rootstocks may improve the tolerance to low soil temperatures and high air temperatures. In climate-controlled greenhouses, such rootstocks are regarded as an additional tool to improve the energy efficiency.

For Cucurbitaceae, figleaf gourd (*C. ficifolia*), bur cucumber (*Sicos angulatus* L.) and squash are used commercially as rootstocks under field conditions with low soil temperatures (Table 7.1). Watermelons grafted on to a 'Shin-tosa'-type interspecific *Curcubita maxima* Duch. × *Curcubita moschata* Duch. squash hybrid resulted in low-temperature tolerance and higher yields compared with non- and self-grafted melons (Okimura *et al.* 1986; Gao *et al.*, 2009). Figleaf gourd is unique among cucurbit species, having an optimal root temperature of approximately 15°C, which is 6°C lower than that of cucumber roots (Tachibana, 1982; Lee, 1994; Ahn *et al.*, 1999; Rivero *et al.*, 2003b). Several studies have demonstrated that these two rootstocks improved vegetative growth and early yield at suboptimal temperatures (e.g. den Nijs, 1980; Tachibana, 1982; Bulder *et al.*, 1991; Zhou *et al.*, 2007), particularly when only the roots were subjected to chilling temperatures below 8°C (Ahn *et al.*, 1999).

Cucumber is the Curcurbitaceae crop that is most susceptible to high-temperature stress. Grafting cucumbers on to luffa (*L. cylindrica*) rootstocks increased the tolerance to 40°C air and/or root-zone temperature (Li *et al.*, 2014a,b, 2015a). However, there is only one commercial rootstock variety available for cucumber that improves heat tolerance, i.e. SQ60 F_1.

In comparison with the Curcurbitaceae, the availability of commercial rootstock varieties that may improve low-temperature tolerance in the Solanaceae is still low. Nine commercial rootstock varieties from different seed companies were tested under low-temperature and low-light conditions, however, none of these rootstocks improved the yield or fruit-quality parameters considered (Riga, 2015). In contrast, several old tomato cultivars such as 'Gartenfreude' and 'Radio' demonstrated advanced flowering and increased yield in a screening of 29 potential tomato rootstocks performed under energy-saving growth conditions with a low night temperature (18/7°C, day/night) (Zijlstra and den Nijs, 1987). All cultivars tested as rootstocks varied largely in root size, earliness, truss and shoot fresh mass, and fruit production. Wild tomato species (*Solanum* spp.), specifically accessions originating from high altitudes with large diurnal variations in day/night temperatures (Tachibana, 1982; Venema *et al.*, 2005), have great potential to improve low-temperature tolerance. Comparative analyses of physiological responses to chilling and suboptimal temperature stress conducted during recent decades have provided some explanations for the tolerance or sensitivity to chilling in tomato, using *S. habrochaites* (LA1777 and LA1778) as model species in grafting experiments (Bloom *et al.*, 2004; Venema *et al.*, 2008; Easlon *et al.*, 2013; Ntatsi *et al.*, 2014a). These cold-tolerant high-altitude accessions are therefore of particular interest as potential germplasm sources to widen the genetic variation for low-temperature tolerance of commercial tomato rootstocks, which are interspecific hybrids between a vigorous cultivated tomato and a wild tomato species, mostly *S. habrochaites* accessions (Keatinge *et al.*, 2014). Nevertheless, there are only a few commercial tomato rootstock varieties available and recommended by seed companies that improve cold tolerance. Two of them i.e. 'Actimino' (E 28.33465) and 'Fortamino', are useful in unheated soil but totally unsuitable for heated soil-less cultures. For aubergines, several cold-tolerant rootstocks are available based on the low-temperature tolerance of *Solanum torvum* Sw.

Table 7.1. Cucurbitaceae and Solanaceae genotypes, wild accessions and commercial varieties with improved cold (CT) and/or heat (HT) tolerance when used as rootstocks. References are for studies where these genotypes were successfully tested under low- or high-temperature stress, or refer to rootstock catalogues on the websites of seed companies.

Species	Rootstock genotypes/F_1/ accession/variety	Tolerance	Reference(s)
Cucumber	Figleaf gourd (*Cucurbita ficifolia*)	CT	Gao *et al.* (2009); Zhou *et al.* (2007, 2009); (Li *et al.* (2014b)
	Bur cucumber (*Sicos angulatus*)	CT	Zhang S.P. *et al.* (2008)
	Luffa (*Luffa cylindrica*)	HT	Li *et al.* (2014a,b)
	'Kilameki', 'Tielizhen' (*Cucurbita moschata* × *C. moschata*)[a]	CT	Li *et al.* (2015b)
	'Azman RZ F_1', 'Kobalt RZ F_1'	CT	Rijk Zwaan (2016)
	Bottle gourd (*Lagenaria siceraria*)	CT	Bulder *et al.* (1991)
	'Ancora F_1' (*C. moschata*)	CT	Takii Seed (2016)
	'SQ60 F_1'	CT, HT	Meridiem Seeds (2016)
Watermelon	'Shintosa'-type (interspecific squash hybrid, *Curcubita maxima* × *C. moschata*)	CT	Okimura *et al.* (1986); Davis *et al.* (2008); Gao *et al.* (2009)
	'Shintosa F1'	CT, HT	Mikado Kyowa Seeds (2016)
	'Shintosa F90'	CT	Semillas Fitó (2016)
	C. moschata[a]	CT	Sun *et al.* (2009)
	C. ficifolia[a]	CT	Sun *et al.* (2009)
	'Bond Z F_1' (*Lagenaria siceraria*)	CT	Mikado Kyowa Seeds (2016)
	'Tetsukabuto F_1" (*C. maxima* × *C. moschata*)	CT	Takii Seed (2016)
	'Hércules F_1', 'Ulises F_1'	CT	Ramiro Arnedo Semillas (2016)
Tomato	*Solanum habrochaites* LA 1777 and LA 1778 (synonym *Lycopersicon hirsutum*)[a]	CT	Venema *et al.* (2008); Ntatsi *et al.* (2014b); Easlon *et al.* (2013)
	Solanum lycopersicoides LA 2951	CT	Easlon *et al.* (2013)
	'Gartenfreude', 'Alfresco', 'Radio', 'Sonatine', 'Ohio205040'	CT	Zijlstra and den Nijs (1987)
	Recombinant inbred lines from backcrossed progeny of *Solanum lycopersicum* 'T5' × *S. habrochaites* LA1778 and *S. lycopersicoides* LA2951[a]	CT	Bloom *et al.* (2004); Easlon *et al.* (2013)
	'Armada F_1'	CT	Takii Seed (2016)
	'Actimino' (E 28.33465), 'Fortamino'	CT	Enza Zaden (2016)
	'Top Bental', 'Top Gun', 'Top Pittam'	CT, HT	Top Seeds Ibérica (2016)
	'VI046103' (EG195), 'VI045276' (EG203)	HT	AVRDC (2016)
	'Arnold'	CT, HT	Syngenta (2016)

Continued

Table 7.1. Continued.

Species	Rootstock genotypes/F_1/ accession/variety	Tolerance	Reference(s)
Aubergine	'Torvum vigour'	CT	Okimura *et al.* (1986)
	'Hiranasu' (Japan)[a]	CT	Gao *et al.* (2009)
	'Javah F_1', 'Astro F_1'	CT	Takii Seed (2016)
	'*Solanum torvum* F_1'	CT	Semillas Fitó (2016)
	'Beo F_1' (*Solanum melongena* × *Solanum aethiopicum*)	CT	Meridiem Seeds (2016)

[a]Only tested during the vegetable stage: no fruit production data available, except in Ntatsi *et al.* (2014b).

Testing tomato rootstocks under hot and wet conditions in the tropics revealed 'VI043614' (Hawaii 7996) as a candidate rootstock that can be grown successfully under heat. However, this was related to its resistance to bacterial and Fusarium wilts, which often occur under hot and wet conditions (Black *et al.*, 2003). As aubergines are better adapted to hot arid climates than tomatoes, they were suggested as a rootstock to improve the heat tolerance of tomato. Grafting tomato on to rootstocks from *Solanum melongena* L. 'Nianmaoquie' prolonged growth and increased yield by up to 10% (Black *et al.*, 2003; Wang *et al.*, 2007). This led to the recommendation to use the aubergine varieties 'VI046103' (EG195) and 'VI045276' (EG203) as rootsocks for tomato production under hot and wet conditions (Keatinge *et al.*, 2014). Only one seed company, Top Seeds Ibérica, recommends rootstocks with heat tolerance: 'Top Bental', 'Top Gun' and 'Top Pittam'. Among the different species of chilli pepper that were tested as rootstocks (*Capsicum chacoense* Hunz., *Capsicum baccatum* L., *Capsicum frutescens* L. and *Capsicum annuum* L.), *C. annuum* ('Toom-1' and '9852-54') resulted in the highest yields under high-temperature conditions and is recommended as a rootstock by the World Vegetable Center (Palada and Wu, 2008).

7.3 Salinity Stress

The presence of sodium chloride (NaCl) in soil or water is one of the most critical environmental stresses limiting vegetable productivity worldwide, especially in arid and semi-arid regions (Colla *et al.*, 2010). Plant growth inhibition and yield reduction are typical responses of annual fruit crops to salt stress, and the magnitude of growth and productivity losses may change in relation to the plant phenological stage and the magnitude of the stress (i.e. time of exposure and concentration of salt) (Colla *et al.*, 2012).

Enhancement of plant growth and yield has been demonstrated in many grafting combinations of greenhouse and open-field annual fruit crops grown under saline conditions (Colla *et al.*, 2010). Grafting the salt-sensitive tomato 'Moneymaker' on to tolerant rootstocks such as 'Radja' and 'Pera' under saline conditions (50 mM NaCl) mitigated the detrimental effect of saline water by increasing the fruit yield compared with self-grafted plants (Santa-Cruz *et al.*, 2002; Martínez-Rodríguez *et al.*, 2008). Similarly, Estañ *et al.* (2005) reported an improvement in salt tolerance of the commercial tomato 'Jaguar' grafted on to several *S. lycopersicum* rootstocks ('Radja', 'Pera', 'Volgogradskij' and hybrid Volgogradskij × Pera) under increasing NaCl concentrations (0, 25, 50 and 75 mM). The same authors reported that the beneficial effect on salt tolerance caused by the rootstocks was more evident under moderate and severe salt stress (50 and 75 mM) than under 25 mM, indicating that the tolerance induced by the rootstock in the shoot was related more to ionic than to osmotic stress caused by NaCl toxicity. The different marketable yield responses among grafting combinations were attributed to the diverse abilities of the rootstocks selected to regulate the transport of toxic saline ions. The agronomical performance of the aubergine 'Suqiqie' was also enhanced under salinity conditions when the *S. torvum* 'Torvum vigor' was used as rootstock (Wei *et al.*, 2007). Similar to tomato and aubergine, Penella *et al.* (2015) reported increased fruit yield of the commercial pepper 'Adige' under salinity (40 mM NaCl) when grafted on to accessions *Capsicum chinense* Jacq. ECU-973 and *C. baccatum* var. *pendulum* (Willd.) Eshbaugh BOL-58, whereas no positive effect was recorded when grafted on to *C. annuum* 'Serrano'. The salt tolerance of the selected rootstocks was achieved due to a lower accumulation of Na^+ and Cl^- in the shoot, but also ion selectivity maintenance, in particular Na^+/K^+ ratio discrimination (Penella *et al.*, 2015). In a similar experiment, Penella *et al.* (2016) observed, under natural salinity conditions, a larger amount of marketable fruit (+ 75%) and lower blossom end rot incidence (– 31%) in commercial 'Adige' grafted on to a salt-tolerant pepper accession (A25) when compared with non-grafted plants.

The ability of rootstocks to confer salinity tolerance in comparison with self-grafted and non-grafted plants has been also reported in several cucurbit crops (*see* Plate 16). For instance, grafting watermelon 'Fantasy' on to pumpkin rootstock 'Strong Tosa' increased biomass production and leaf area under salt stress conditions compared with non-grafted plants (Goreta *et al.*, 2008). Moreover, other experiments have demonstrated that improving growth performance of melon under salinity could be achieved by grafting on to *Cucurbita* ('P360') hybrid rootstock (Rouphael *et al.*, 2012). Similarly, Romero *et al.* (1997) found that melons grafted on to rootstocks consisting of three hybrids of squash were more tolerant to salinity and gave higher marketable yields than non-grafted plants. However, these findings were not confirmed by Edelstein *et al.* (2005) and Colla *et al.* (2006a), who showed a similar sensitivity to salinity between grafted and non-grafted plants. Edelstein *et al.* (2005) and Colla *et al.* (2006a) indicated that the higher crop productivity between grafted and non-grafted plants could be attributed to grafting per se. Similar positive effects of grafting were also recorded in cucumber. The cultivar 'Jinchun No. 2' grafted on to figleaf gourd (*C. ficifolia*) and *Lagenaria siceraria* Standl. had higher marketable yield and also lower unmarketable production under two different salt levels (30 and 60 mM NaCl) compared with self-grafted plants (Huang *et al.*, 2009b). Moreover, in the study of Huang *et al.* (2013), six combinations of cucumber (non-grafted cucumber and pumpkin, self-grafted cucumber and pumpkin, cucumber grafted on to pumpkin, and pumpkin grafted on to cucumber) were exposed to a saline nutrient solution (91 mM NaCl). Pumpkin rootstock had a higher ability for Na^+ exclusion, which resulted in lower translocation of Na^+ to the aerial parts, thus increasing the salt tolerance of the cucumber. The authors concluded that low Na^+ translocation from the rootstock to scion is driven primarily by the rootstock. Colla *et al.* (2012, 2013b) investigated whether grafting would enhance the salt tolerance of vegetables, in particular cucumber, to salts other than NaCl, such as Na_2SO_4 and $CaCl_2$. These studies were performed with equimolar concentrations of Na^+ and Cl^- in order to evaluate the ion effects of the three salinity sources. The greater biomass and yield reduction in NaCl than in Na_2SO_4 (Colla *et al.*, 2012) and in $CaCl_2$ than in NaCl (Colla *et al.*, 2013b) was attributed mainly to the toxic effects of Cl^- rather than to the Na^+ and SO_4^{2-} because the tested rootstocks showed a limited ability to restrict shoot accumulation of Cl^-. Thus, Cl^- continues passing to the leaves and becomes the more significant toxic component of the saline solution.

The improvement in crop performance of grafted cucumber, watermelon, melon, aubergine and tomato plants has been often been attributed to Na^+ or Cl^- exclusion (Bai *et al.*, 2005; Edelstein *et al.*, 2005; Colla *et al.*, 2006a,b; Goreta *et al.*, 2008; Zhu and Guo, 2009; Yetisir and Uygur, 2010) or Na^+ and Cl^- exclusion (Romero *et al.*, 1997; Santa-Cruz *et al.*, 2002; Estañ *et al.*, 2005; Wei *et al.*, 2007; Martínez-Rodríguez *et al.*, 2008; Huang *et al.*, 2009b) (Table 7.2). Moreover, nutrient (e.g. K^+) uptake and translocation to the shoot is often improved in salt-tolerant grafting combinations, resulting in a reduction of nutrient imbalances and deficiencies caused by salinity. The Na^+ and/or Cl^- exclusion in grafted vegetables under saline conditions has been associated with the morphological root system characteristics, in particular total root length, root diameter, total root surface, and the number of root hairs and their length (Colla *et al.*, 2010). The root characteristics of the rootstock play a primary role in the mechanisms of salt tolerance in grafted plants. In tomatoes, He *et al.* (2009) recorded that root dry mass was significantly reduced under severe salt stress concentrations (100 and 150 mM NaCl), but the reduction was smaller in grafted compared with non-grafted plants. In fact, the partitioning of biomass between shoot and root showed that grafted plants tended to accumulate more biomass in the root, thus leading to a higher root-to-shoot ratio (Colla *et al.*, 2010). Both enzymatic and non-enzymatic antioxidants were also found to contribute to the salinity tolerance in grafted vegetables (Colla *et al.*, 2010). Salinity stress enhances the production of reactive oxygen species in plant tissues, causing oxidative damage. In several salt-tolerant grafting combinations of cucurbits and solanaceous crops, the antioxidant activity of enzymatic (superoxide dismutase, catalase, peroxidase, ascorbate peroxidase, dehydroascorbate reductase, glutathione reductase) and non-enzymatic (ascorbic acid, glutathione) systems were often higher in comparison with non-grafted or self-grafted plants, leading to a reduction of oxidative damage in plant tissues (Colla *et al.*, 2010). Rootstocks can also increase the salinity tolerance of a scion by changing the hormonal balance and especially ABA, cytokinins and polyamines (Colla *et al.*, 2010).

Table 7.2. Na^+ and Cl^- exclusion and/or inclusion in aerial parts of grafting combinations grown under saline conditions

Scion	Rootstock	Ion exclusion and/ or inclusion in aerial part of grafted plants	Reference
Cucumis sativus 'Jinyou No.3'	*Cucurbita moschata*	Na^+ exclusion	Chen and Wang (2008)
C. sativus 'Jinyou No.3'	*Cucurbita ficifoli*	Na^+ exclusion	Chen and Wang (2008)
C. sativus 'Prince'	*Cucurbita maxima* × *C. moschata* 'Ercole hybrid F_1 No. 6001'	Na^+ exclusion	El-Wanis *et al.* (2012)
C. sativus 'Prince'	*Lagenaria sicaria* var. Calabsha	Na^+ exclusion	El-Wanis *et al.* (2012)
C. sativus 'Zaoduojia'	*C. moschata* 'Chaojiquanwang'	Na^+ exclusion and Cl^- inclusion	Zhu *et al.* (2008)
C. sativus 'Jinchun No. 2'	*L. siceraria* 'Chaofeng 8848'	Na^+ and Cl^- exclusion	Huang *et al.* (2009a)
C. sativus 'Jinchun No. 2'	*L. siceraria* 'Chaofengkangshengwang'	Na^+ and Cl^- exclusion	Huang *et al.* (2009b)
C. sativus 'Jinchun No. 2'	*C. ficifolia*	Na^+ and Cl^- exclusion	Huang *et al.* (2009b)
C. sativus 'Jinyu No. 1'	*C. moschata* 'Chaojiquanwang'	Na^+ exclusion, similar Cl^-	Zhen *et al.* (2010)
C. sativus 'Akito'	*C. maxima* × *C. moschata* 'PS1313'	Na^+ exclusion and Cl^- inclusion	Colla *et al.* (2012)
C. sativus 'Ekron'	*C. sativus* 'Affyne'	Na^+ exclusion and Cl^- inclusion	Colla *et al.* (2013b)
C. sativus 'Jinchun No. 2'	*C. moschata* 'Chaojiquanwang'	Na^+ exclusion	Lei *et al.* (2014)
C. sativus 'Akito'	*C. maxima* × *C. moschata* 'PS1313'	Na^+ exclusion, similar Cl^-	Rouphael *et al.* (2012)
C. sativus '722 Virginia RZ.'	*C. moschata* 'Mammoth F1'	Na^+ inclusion	Petropoulos *et al.* (2011)
Citrullus lanatus 'Fantasy'	*C. maxima* × *C. moschata* 'Strongtosa'	Na^+ exclusion	Goreta *et al.* (2008)
C. lanatus 'Tex'	*C. maxima* × *C. moschata* 'Ercole'	Na^+ exclusion and Cl^- inclusion	Colla *et al.* (2006a)
C. lanatus 'Tex'	*L. siceraria* 'Macis'	Na^+ exclusion and Cl^- inclusion	Colla *et al.* (2006a)
C. lanatus 'Crimson Tide'	*L. siceraria* SKP and Berecik (landraces)	Na^+ exclusion	Yetisir and Uygur (2010)
C. lanatus 'Crimson Tide'	*C. maxima* (landrace)	Na^+ exclusion	Yetisir and Uygur (2010)
Cucumis melo 'Cyrano'	*C. maxima* × *C. moschata* 'P360'	Na^+ exclusion and Cl^- inclusion	Colla *et al.* (2006b)
C. melo 'Arava'	*C. maxima* × *C. moschata* 'TZ-148'	Na^+ exclusion, similar Cl^-	Edelstein *et al.* (2005)

Continued

Table 7.2. Continued.

Scion	Rootstock	Ion exclusion and/ or inclusion in aerial part of grafted plants	Reference
C. melo 'Arava'	*C. maxima* × *C. moschata* 'TZ-148'	Na^+ exclusion, similar Cl^-	Edelstein *et al.* (2011)
C. melo 'Resisto', 'Arava'	*C. maxima* × *C. Moschata* 'Shintoza'	Na^+ and Cl^- exclusion	Romero *et al.* (1997)
C. melo var. *cantalupensis* 'London' and 'Brennus'	*C. maxima* × *C. moschate* 'RS841'	Na^+ exclusion	Orsini *et al.* (2013)
C. melo 'Cyrano'	*C. maxima* × *C. moschata* 'P360'	Na^+ exclusion, similar Cl^-	Rouphael *et al.* (2012)
Solamun melongena 'Suqiqie'	*Solanum torvum* 'Torvum vigour'	Na^+ and Cl^- exclusion	Wei *et al.* (2007)
Solamun lycopersicum 'Hezuo903'	*S. lycopersicum* 'Zhezhen No. 1'	Similar Na^+ and Cl^-	He *et al.* (2009)
S. lycopersicum 'Boludo F1'	*S. lycopersicum* × *Solanum cheesmaniae* recombinant inbred lines	Similar Na^+	Albacete *et al.* (2009)
S. lycopersicum 'Fanny', 'Goldmar'	*S. lycopersicum* 'AR-9704'	Na^+ and Cl^- exclusion	Fernández-García *et al.* (2002)
S. lycopersicum 'Fanny'	*S. lycopersicum* 'AR-9704'	Na^+ and Cl^- exclusion	Fernández-García *et al.* (2004)
S. lycopersicum 'Moneymaker'	*S. lycopersicum* 'Kyndia'	Na^+ and Cl^- exclusion	Santa-Cruz *et al.* (2002)
S. lycopersicum 'Jaguar'	*S. lycopersicum* 'Radja' and 'Pera'	Na^+ and Cl^- exclusion	Estañ *et al.* (2005)
S. lycopersicum 'Moneymaker'	*S. lycopersicum* 'Radja' and 'Pera'	Na^+ and Cl^- exclusion	Martínez-Rodríguez *et al.* (2008)
S. lycopersicum 'Jaguar'	*S. lycopersicum* 'Radja' and 'Volgogradsjik'	Na^+ and Cl^- exclusion	Cuartero *et al.* (2006)
S. lycopersicum 'Cuore di Bue'	*S. lycopersicum* × *Solanum habrochaites* 'Arnold'	Na^+ accumulation in older leaves	di Gioia *et al.* (2013)
S. lycopersicum 'Farida'	*S. lycopersicum* × *S. habrochaites*	Na^+ and Cl^- exclusion	Wahb-Allah (2014)
S. lycopersicum 'Big Dena'	*S. lycopersicum* × *S. habrochaites* 'Maxifort'	Na^+ exclusion	Semiz and Suarez (2015)
Capsicum annum 'Adige'	*Capsicum chinense* 'ECU-973'; *Capsicum baccatum* var. *pendulum* 'BOL-58'	Na^+ and Cl^- exclusion	Penella *et al.* (2015)
C. annum 'Adige'	*C. annum* (wild accession, code A-25)	Similar Na^+ and Cl^-	Penella *et al.* (2015)

7.4 Nutrient Stress

7.4.1 Excessive nutrient availability

An increasingly encountered problem in soils used for intensive vegetable production worldwide is the occurrence of several nutrient ions at excessive concentrations due to overfertilization, prolonged use of organic fertilizers composed of sewage sludge or animal manure, or irrigation with wastewater. Direct toxicity to plants, for example on the lipid composition of membranes, may be caused by manganese (Mn), zinc (Zn), copper (Cu) and boron (B) if their concentrations in the root zone are excessively high (Savvas *et al.*, 2010). Excessive external concentrations of the other micronutrients except for phosphorus (P) may cause salinity stress, resulting in water stress due to reductions in leaf water and leaf osmotic potentials (Neocleous and Savvas, 2015). The impact of grafting on salinity stress is addressed above. This section will briefly review the responses of grafted plants to excessively high external concentrations of Mn, Zn, Cu and B and the possibilities of grafting on to selected rootstocks as a means to alleviate toxicity caused by these nutrients.

The ability of plant roots to control nutrient uptake depends on both the root structure and the uptake mechanisms at the biochemical level in root cells (Savvas *et al.*, 2010). Many rootstocks used to graft *Solanum* and *Cucurbita* spp. of vegetables are capable of increasing the uptake rates of some nutrients because they are characterized by more vigorous root systems than those of elite varieties (Lee, 1994; Zhang G.W. *et al.*, 2008; Huang *et al.*, 2010). Nevertheless, the mechanisms controlling nutrient uptake in the root may also be influenced by the shoot through alterations in the transport of some metabolites, such as organic acids, amino acids and carbohydrates, which are initiated by hormonal messengers (Zhou *et al.*, 2007). Indeed, the uptake of nutrients by grafted plants may be governed not only by the rootstock genotype (Jensen *et al.*, 2003) but also by the impact of the scion–rootstock compatibility and interactions affecting water relations, mineral absorption and fruit development, as well as fruit quality (Martínez-Ballesta *et al.*, 2010). Thus, the efficiency of a rootstock to confer tolerance to nutrient stress seems also to depend on the scion genotype (Guimarães *et al.* 2009).

The B deficiency and toxicity thresholds (*see* section 7.5) are separated by a narrow range of non-harmful concentrations (Mortvedt *et al.*, 1991). Therefore, B can readily occur at toxic levels in semi-arid regions where B is enriched through natural deposition due to poor drainage and/or geothermal activity (Landi *et al.*, 2013). As reported by Edelstein *et al.* (2005, 2007), increases in tissue B concentrations and decreases in growth and fruit yield brought about by too high external B concentrations were mitigated by grafting melon on to the commercial rootstock 'TZ-148' (*C. maxima* × *C. moschata*). However, Gimeno *et al.* (2012) found that tolerance to excess B in grafted citrus plants was not related to the leaf B level, which indicates that physiological, biochemical and anatomical alterations brought about by the rootstock may influence B tolerance. Furthermore, Romani (2014) did not find any interaction between grafting and external B concentration affecting yield or tissue B levels in tomato grafted on to the commercial tomato rootstock 'Beaufort', 'Energy' and 'Caramba'. Based on the currently available information, it seems that grafting on to suitable rootstocks may alleviate or even prevent growth and yield decreases associated with B toxicity in some grafted vegetables. However, whether grafting alleviates B toxicity by restricting B uptake is still open to debate.

As shown by Rouphael *et al.* (2008), grafting on to the rootstock 'Shintoza' (*C. maxima* × *C. moschata*) may mitigate Cu toxicity in cucumber exposed to excessively high Cu levels in the root zone, presumably by restricting the uptake and translocation of Cu to the shoot. Savvas *et al.* (2009) also found lower Cu levels in roots and shoots of tomato 'Belladonna' when grafted on to 'He-Man' (*S. lycopersicum* × *S. habrochaites*) compared with self-grafted 'Belladona' plants. According to Zhang *et al.* (2010), excess Cu, as well as inhibiting growth, photosynthesis and pigment synthesis of grafted and non-grafted cucumber transplants, significantly increased the accumulation of Cu in the roots of grafted cucumber plants compared with non-grafted plants, thereby reducing Cu concentration in the leaves. Moreover, the activities of protective antioxidant enzymes (e.g. superoxide dismutase, peroxidase, catalase, glutathione reductase) and the contents of ascorbate and glutathione in leaves of grafted plants were significantly higher than those in non-grafted plants under Cu stress.

Our current knowledge about the interactions between grafting and excessively high external Zn or Mn levels is rather limited. The only information published to date regarding the impacts of grafting and rootstock genotype on Mn uptake under conditions of excessive Mn levels in the root zone is that reported by Savvas *et al.* (2009), who showed that grafting tomato 'Belladona' on to the rootstock 'He-Man' decreased the tolerance of the scion to high external Mn concentrations in the root zone. This response seemed to be partly associated with the appreciably lower ability of plants grafted on to 'He-Man' to transport Mg to the leaves, which resulted in lower foliar Mg/Mn ratios in comparison with self-grafted plants. With respect to Zn, the only relevant information refers to the impact of grafting on Zn uptake under normal external Zn concentrations. In an experiment with cucumber, grafting reduced the Zn concentration in mature leaves regardless of the rootstock genotype (Savvas *et al.*, 2012). However, grafting tomato on to 'He-Man' rootstock had no effect on the tissue Zn levels when the external Zn levels were at normal levels (Savvas *et al.*, 2009).

7.4.2 Deficient nutrient availability

The 'low-input' vegetable production approach requires a significant reduction in fertilizer inputs and the selection of vegetable genotypes characterized by high nutrient-use efficiency. This strategy could lead to low environmental impact on agro-systems while maintaining high crop performance (Lynch, 1998; Hirsch and Sussman, 1999; Sorgonà *et al.*, 2006). Grafting high-yielding genotypes on to specific rootstocks is considered an effective tool to improve vegetable tolerance to stress caused by reduced nutrient availability in the root zone, and consequently to improve nutrient-use efficiency by the scion (Colla *et al.*, 2010, 2011). The enhanced nutrient stress tolerance of grafted vegetables has often been associated with a more vigorous root system in terms of root length, density and number of root hairs, which can acquire nutrients more efficiently (Savvas *et al.*, 2010).

Several studies have shown that grafted vegetable plants are more efficient in the uptake and translocation of macro- and micronutrients to the shoot than non-grafted plants (e.g. Savvas *et al.*, 2010). As a result, grafted plants may sustain growth and yield potential even when the availability of some nutrients in the root zone is marginal. For instance, the melon (*Cucumis melo* L.) 'Melina' and 'Gallicum' grafted on to *C. maxima* 'Kamel' commercial rootstocks were more efficient in translocating nitrogen (N) to the shoot in comparison with non-grafted melon plants (Ruiz *et al.*, 1997). Similar results were observed when watermelon was grafted on to rootstocks of the commercial *Cucurbita pepo* L. 'Brava', 'Kamel' and 'Shintoza' (Pulgar *et al.*, 2000). Phosphorus uptake and translocation were also affected when fruit vegetables were grafted on to selected rootstocks. Aubergine (*S. melongena*) grafted on to tomato 'Beaufort' was characterized by higher P contents in the shoot in comparison to non-grafted plants (Leonardi and Giuffrida, 2006). Higher concentrations of P in the aerial part were also observed in grafted melon (Salehi *et al.*, 2010) and mini-watermelon plants (Colla *et al.*, 2010).

Improved K uptake and translocation have been observed in several grafted vegetables, particularly melon (Qi *et al.*, 2006), watermelon (Rouphael *et al.*, 2008; Huang *et al.*, 2013), cucumber (Zhu *et al.*, 2008) and aubergine (Leonardi and Giuffrida, 2006); this was mostly associated with a more vigorous root system of the rootstock. Albacete *et al.* (2009) suggested that the higher K uptake of tomato rootstocks is controlled by cytokinins synthesized in roots.

Several studies have addressed the influence of grafting and rootstock–scion combinations on uptake and translocation of calcium (Ca) and magnesium (Mg). Fernández-García *et al.* (2004) reported a significant increase in leaf Ca contents when tomato 'Fanny' and 'Goldmar' were grafted on to the commercial rootstock 'AR-9704'. Similarly, Leonardi and Giuffrida (2006) found significantly increased leaf Ca concentrations in tomato and aubergine grafted on to the rootstock 'Beaufort', 'PG3' and 'Energy'. The effect of grafting on Mg is still contradictory and depends mainly on the rootstock adopted. The Mg level in leaf tissues increased when aubergine and mini-watermelon were grafted on to 'Energy' or 'PS1313' rootstocks, respectively (Leonardi and Giuffrida, 2006; Rouphael *et al.*, 2008), whereas an opposite trend was recorded in tomato grafted on to the commercial rootstock 'Beaufort', He-Man' and 'Resistar' (Savvas *et al.*, 2009).

Research regarding the impact of the rootstock genotype on micronutrient uptake by grafted vegetables in general has not shown any significant differences or decreases compared with self-grafted or non-grafted plants (Rouphael *et al.*, 2008; Savvas *et al.*, 2009; Huang *et al.*, 2010; Colla *et al.*, 2012, 2013a). However, in some exceptional cases, iron (Fe), Cu and/or Mn uptake increased when specific rootstocks were used (Colla *et al.*, 2010, 2011; Huang *et al.*, 2010).

In most plant nutrition studies with grafted vegetables, enhanced macro- and micronutrient uptake and translocation were observed under conditions of optimal nutrient status in the root zone. The impact of grafting on nutrient uptake by vegetable plants challenged with low nutrient concentrations in the root zone has been addressed in only a few reports to date (Colla *et al.*, 2010, 2011; Huang *et al.*, 2013; Schwarz *et al.*, 2013). Colla *et al.* (2010) observed that melon plants ('Proteo') grafted on to two *C. melo* rootstocks 'Dinero' and 'Jador', and on to the *Cucurbita* hybrid 'P360' needed 5.7, 5.2 and 6.1 mM NO_3^-, respectively, to reach half-maximal shoot dry mass, whereas plants grafted on to the *Cucurbita* hybrid

'PS1313' rootstock and the non-grafted plants needed 9.1 and 13.1 mM NO_3^-, respectively. Moreover, Colla *et al.* (2010) found no significant differences among treatments in nitrate reductase activity at high nitrate concentration (10 mM). However, when the external nitrate level was low (2.5 mM), the nitrate reductase activity in the leaves of all grafted plants was higher than in the non-grafted plants. These results indicate that, under low-nitrate conditions, grafted plants potentially have greater uptake efficiency compared with non-grafted plants (*see* Plate 17). According to Djidonou *et al.* (2013, 2015), grafted tomato plants had a greater N crop nutritional requirement than non-grafted plants together with an increase in predicted marketable yield. Indeed, 'Florida 47' tomato plants grafted on to 'Beaufort' and 'Multifort' rootstocks showed an average increase of 27 and 30% in total and marketable fruit yields, respectively, and had greater potential for yield improvement with increasing N rates compared with non-grafted plants.

With respect to potassium (K), Huang *et al.* (2013) found that watermelon grafted on to specific rootstocks could mitigate the negative effect of low K availability in the root zone. This was ascribed to a higher efficiency of grafted watermelon plants to take up K and translocate it to the above-ground biomass, rather than to utilize K. In line with previous research, Schwarz *et al.* (2013) demonstrated that the tomato rootstock 'Maxifort' and 'Brigeor' improved root growth and yield, and increased fruit titratable acidity and carotenoid content, even with a moderately deficient K supply, when used to graft tomato 'Piccolino', but not when 'Classy' was the scion. The authors indicated that the alleviation depended on the rootstock–scion combinations.

Interestingly, some rootstocks increase vigour under low-K supply compared with standard nutrition (Martínez-Andújar, C., Albacete Moreno, A.A., Martínez Pérez, A., Asins, M.J., Pérez Pérez, J.M. and Pérez-Alfocea, F., unpublished data). Indeed, some recombinant inbred lines derived from a cross between two species closely related to cultivated tomato (*S. lycopersicum* var. *cerasiforme* (Dunal) D.M.Spooner, G.J.Anderson & R.K.Jansen × *Solanum pimpinellifolium* L.) are able to convert the elite 'Boludo F1' into a nutrient-efficient phenotype with better K, S and Ca assimilation and micronutrient (Zn, Mn, Fe and B) uptake and transport to the shoot. The ethylene precursor 1-aminocyclopropane-1-carboxylic acid (ACC) and its interactions with other hormones (cytokinins, jasmonic acid and ABA) seem to play a pivotal role in rootstock-mediated nutrition and growth in tomato plants growing under low-K supply. In the meantime this article is published by Martínez-Andújar, *et al.* (2016). This kind of rootstock can provide benefits such as a reduction in fertilizer application and improved productivity under reduced fertilization.

7.5 Stress Induced by Metalloids and Heavy Metals

7.5.1 Boron

Although the metalloid B is an essential nutrient for plants, excessively high concentrations can be toxic to many plants (Ben-Gal and Shani, 2002; Goldberg *et al.*, 2003; Yau and Ryan, 2008), as described earlier in this chapter. Problems of B toxicity have been widely documented, particularly in dry regions such as western Asia, North Africa, South Australia and western USA (Yau and Ryan, 2008). Bingham *et al.* (1985) indicated that wheat, barley and sorghum are, respectively, sensitive, moderately tolerant and very tolerant to B. Francois (1984) found a threshold toxic value of 5.7 mg l^{-1} B in the soil solution for tomato plants; their productivity rate decreased by 3.7% for each additional 1.1 mg l^{-1} B. Symptoms of B toxicity have been found in cereals and legumes, such as lucerne, faba beans, chickpeas and lentils, manifested as chlorosis or necrosis of the tips and margins of the plant leaves, with these symptoms being more intense in older plants (Yau and Ryan, 2008). Crops may be exposed to B via two main sources: (i) growing in soils containing high levels of native B, usually accompanied by high salinity; and (ii) irrigation with water containing high levels of B such as treated sewage water, in which B is a common microelement that is not decreased by water treatment (Tsadilas, 1997; Ben-Hur, 2004).

Boron toxicity exerts various negative effects on vascular plants, including decreases in shoot and root growth (Nable *et al.*, 1997; Edelstein *et al.*, 2005), yields (Eaton, 1944; Goldberg *et al.*, 2003), chlorophyll content and photosynthetic rates (Nable *et al.*, 1997). Nevertheless, the physiological mechanisms underlying the reduction in plant yield by excess B are not fully clear. Reid *et al.* (2004) suggested three main mechanisms responsible for B toxicity: (i) alterations in cell-wall structure; (ii) binding of B to the ribose of ATP, NADH or NADPH, thus disrupting metabolic activity; and (iii) binding of B to the ribose of RNA, thus interfering with protein synthesis and cell division.

Grafting B-sensitive plants on to tolerant rootstocks can increase the tolerance of the grafted plants to B toxicity (Edelstein *et al.*, 2005, 2007). Edelstein *et al.* (2005) found that grafting melon 'Arava' plants on to commercial pumpkin *Cucurbita* 'TZ-148' rootstocks decreased B accumulation in the leaves of the grafted plants compared with non-grafted melon under irrigation with two water qualities (Fig. 7.1). They suggested that the root system of 'TZ-148' absorbed less B than that of 'Arava' plants (Edelstein *et al.*, 2005).

In another experiment, Edelstein *et al.* (2007) studied the effects of five different B concentrations ranging from 0.1 to 10.4 mg l^{-1} on vegetative growth, fruit yield and B uptake of non-grafted 'Arava' melon and the same melon grafted on to pumpkin 'TZ-148' rootstock. B concentrations in the plants were found to increase linearly with the B in the irrigation water; the highest concentrations (150–2224 mg kg^{-1}) were in old leaves, concentrations in the roots were intermediate (47–282 mg kg^{-1}) and the lowest concentrations (10–100 mg kg^{-1}) were in the fruits. As leaves have the highest transpiration rates in the plant, Edelstein *et al.* (2007) suggested that transpiration is the major cause for the high B accumulation in the leaves. Within the same B supply level, the B concentrations were significantly lower in grafted than in non-grafted plants. Shoot and root dry mass and fruit yield decreased linearly as the B concentration in the irrigation water increased, but the non-grafted plants were more sensitive to excessively high B levels than the grafted ones. Edelstein *et al.* (2007) also found that the root system of the non-grafted plants was more sensitive to B concentration than that of the grafted plants. The more pronounced reduction in the melon versus the 'TZ-148' root system with increasing B concentration in the irrigation water resulted in a more

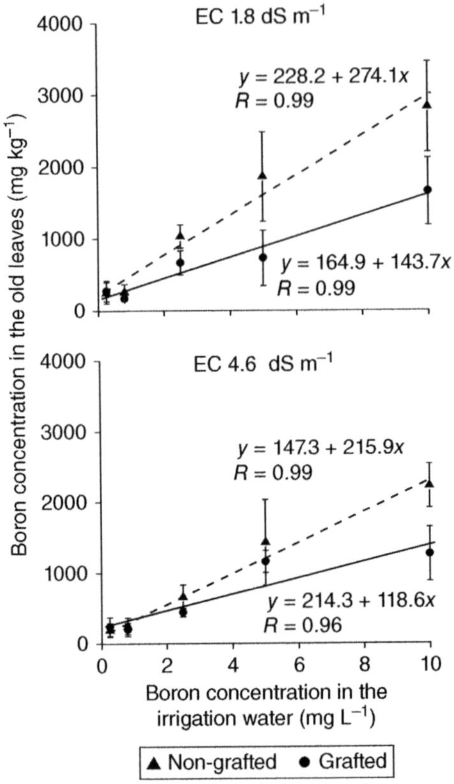

Fig. 7.1. Boron concentrations in old leaves of grafted and non-grafted melon plants as functions of boron concentration in irrigation water at two salinity levels. (Adapted from Edelstein *et al.*, 2005.)

pronounced decrease in the former's capacity to absorb water and nutrients, which in turn sharply decreased shoot dry mass and fruit yield of the non-grafted plants.

To understand better the mechanisms governing the tolerance of grafted plants to high B concentrations, Edelstein *et al.* (2011) exposed six melon/pumpkin combinations (melons and pumpkins were grown as non-grafted, self-grafted, melons grafted on pumpkins or pumpkins grafted on melons) to 0, 5 and 10 mg l^{-1} B in the root system. Shoot and root dry mass of all plant types decreased as the B concentrations increased, but the reduction was significantly more pronounced in plants with melon root systems than in those with pumpkin roots. This indicated that pumpkins are more tolerant to B than melons, and can impart this tolerance to the plants that are grafted on to them. The concentration of B in the xylem sap exudate was significantly lower in plants with pumpkin root systems than in those with melon roots, indicating that the pumpkin roots blocked some of the B uptake from the growth medium. Taken together, it was concluded that melon plants, which are sensitive to high B concentration, are less affected by B toxicity when they are grafted on B-tolerant rootstocks such as 'TZ-148'.

7.5.2 Heavy metals

Heavy metals are categorized as environmental pollutants due to their toxic effects on plants, animals and humans. Heavy metals interfere with the physiological functions of plants, such as photosynthesis, gas exchange and nutrient absorption, and thereby causing reductions in plant growth, dry matter accumulation and yield. Heavy metals also affect the levels of antioxidants in plants and reduce the nutritive value of the produce.

Heavy metals can be divided into two main groups. The first group comprises heavy metals such as cadmium (Cd), nickel (Ni), arsenic (As), lead (Pb) and mercury (Hg), which are toxic to plants and animals, and whose absorption by plants can increase the potential entry of contaminants into the food chain (An *et al.*, 2004; Gratao *et al.*, 2005; Clemens, 2006; Shao *et al.*, 2010). For example, Cd at >5 mg kg^{-1} in leaves is toxic to most plants (White and Brown, 2010), and a survey conducted in Japan indicated that ~7% of total aubergine (*S. melongena*) production has fruits that contain Cd at concentrations exceeding the internationally accepted limit for fruiting vegetables (Arao *et al.*, 2008). Similarly, Ni above a threshold level (5–10 mg kg^{-1} of dry mass) was found to disturb nutrient homeostasis and affect various physiological and biochemical processes in plants, resulting in decreased plant growth and fruit yield (Chen *et al.*, 2009; Kumar *et al.*, 2015b). The toxic heavy metals on agricultural lands can originate from various sources, including industry, soil fertilization, irrigation with treated wastewater and application of sewage sludge. The second group comprises micronutrients, such as B, Cu and Zn, which, in excess, can be toxic to plants and damaging to human health.

Recent studies (e.g. Rouphael *et al.*, 2008; Mori *et al.*, 2009; Savvas *et al.*, 2010; Lux *et al.*, 2011; Colla *et al.*, 2013a; Kumar *et al.*, 2015a,b,c) have shown that some vegetable rootstocks may limit heavy-metal uptake and even their translocation to the shoots. The effects of grafting melon 'Arava' on to commercial *Cucurbita* 'TZ-148' rootstock on concentrations of heavy metals in their vegetative shoots and fruit were studied under field conditions (Edelstein and Ben-Hur, 2012). This study was conducted in clay soil at an experimental station in Akko, northern Israel, where the field plots were irrigated with secondary treated sewage water or fresh water for 4 years. The irrigation-water qualities are presented in Table 7.3. The contents of strontium (Sr), Mn, chromium (Cr), titanium (Ti), Pb, Ni and Cd in the vegetative shoots and fruit of the grafted and non-grafted melon plants are presented in Fig. 7.2. The concentrations of most of the heavy metals studied in the vegetative shoots were one order of magnitude higher than in the fruit. These results provide a good indication that transpiration is the major cause of the accumulation of the various heavy metals in plant leaves, as the greatest transpiration occurs via the vegetative shoots. The concentrations of Sr, Mn, Cr, Ti, Pb, Ni and Cd in the shoots and fruit were significantly lower in the grafted versus non-grafted plants (Fig. 7.2), indicating that grafting of vegetables may decrease the content of heavy metals in plant tissues. Similar results were demonstrated by Savvas *et al.* (2013), who showed a significant reduction of Ni and Cd content in edible fruit of cucumber grafted on to rootstock 'Power' relative to non-grafted plants.

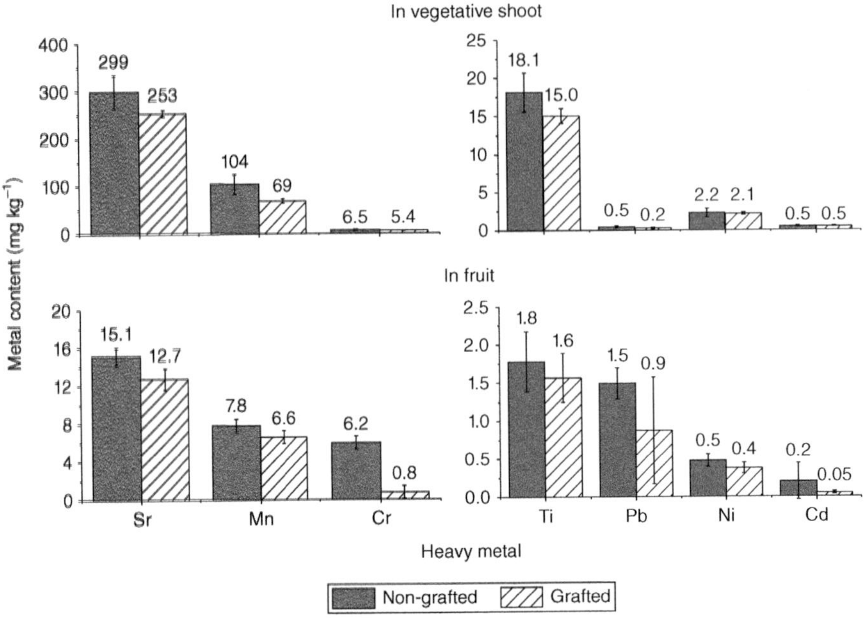

Fig. 7.2. Microelement concentrations in the shoots and fruit from grafted and non-grafted melon plants irrigated with secondary effluent water. Vertical bars indicate standard error. (Adapted from Edelstein and Ben-Hur, 2012.)

Table 7.3. Values of pH, electrical conductivity and sodium adsorption ratio, and concentrations of HCO_3^-, Cl^-, dissolved organic matter, and macro- and micronutrients in secondary treated wastewater and fresh water used in Akko, northern Israel. (M. Edelstein and M. Ben-Hur, unpublished data.)

Measurement	Treated wastewater (mean ± SD)	Fresh water
pH	7.5 ± 0.2	7.7
Electrical conductivity	1.3 ± 0.1	0.8
Sodium adsorption ratio	2.5 ± 0.2	0.3
HCO_3^- (mmol l^{-1})	11.4 ± 0.3	7.5
Cl^- (mmol l^{-1})	3.7 ± 0.3	1.6
Dissolved organic matter (mg l^{-1})	130 ± 45	~0
Macronutrients (mg l^{-1})		
NH_4-N	34.4 ± 8.2	~0
NO_3-N	6.5 ± 0.5	~0
PO_4-P	6.4 ± 0.4	~0
K	0.6 ± 0.1	0.1
Micronutrients (μg l^{-1})		
Fe	106 ± 4.2	9.3
Mn	72 ± 2.1	3.7
Zn	119 ± 8.5	19.5
Cu	5.2 ± 0.2	~0
Mo	0.8 ± 0.1	~0

SD, standard deviation.

In a series of experiments, Kumar *et al.* (2015a,b,c) exposed tomato plants (grafted, self-grafted or non-grafted) to different concentrations of heavy metals (0, 25 and 50 μM Cd and Ni) in soil-less culture medium, using quartziferous sand. Tomato 'Ikram', regardless of its grafting status (non-grafted, self-grafted or grafted on to tomato 'Maxifort' and 'Unifort' or aubergine 'Black Beauty' rootstocks), was invariably adversely affected by the presence of excess Cd and Ni (25 and 50 μM) in the growing medium, as evidenced by a marked reduction in shoot and root biomass, and fruit yield.

The Cd content in the aerial parts of plants grafted on to 'Maxifort' rootstock was significantly lower than that in non-grafted or self-grafted plants, indicating that 'Maxifort' rootstocks are better able to limit Cd transfer from the roots to aerial parts (Kumar *et al.*, 2015c). Moreover, 'Black Beauty' rootstock showed some incompatibility with 'Ikram' scion, as reflected by the lower transfer of nutritional elements to the aerial parts, including the toxic heavy metals, followed by a decline in plant physiological and biochemical activities and, finally, poor growth and yield of 'Black Beauty'-grafted plants.

Kumar *et al.* (2015a) explored the tolerance mechanism of 'Maxifort'-grafted plants at the physiological, metabolomic and molecular levels by exposing them to moderate Cd stress (25 μM). They found that plants grafted on to 'Maxifort' rootstock could effectively mitigate the adverse effects of Cd stress by maintaining better physiological and metabolic plant activities, as reflected by improved plant nutritional status, photosynthetic pigments, photochemical activity of photosystem II (F_v/F_m), antioxidant enzyme activity (catalase, ascorbate peroxidase), proline content and contents of metabolites linked to oxidative stress (i.e. phytochelatin and fructans (inulins)). These metabolites helped neutralize the level of Cd-induced oxidative stress by decreasing the level of reactive oxygen species, lipid peroxidation (malondialdehyde content) and electrolyte leakage. In addition, there was less interference of Cd, in particular of root-to-shoot translocation of nutrients, and therefore better nutrient homeostasis in the shoots of the grafted plants.

Rootstock-mediated tolerance mechanisms were investigated at the molecular level in self-grafted and 'Maxifort'-grafted tomatoes by analysing the relative expressions of two important candidate genes (*LeNRAMP3* and *LeFER*) (Kumar *et al.*, 2015a). *LeNRAMP3* expression was upregulated in the leaves of 'Maxifort'-grafted plants, and this was linked to the availability of certain nutritional elements in the leaves. As the function of NRAMP (natural resistance-associated macrophage protein) in mobilizing Fe, Mn and Zn has already been established in several studies (Bereczky *et al.*, 2003; Ishimaru *et al.*, 2012; Yang *et al.*, 2013), this result confirmed the unique characteristic of 'Maxifort' rootstock in delivering a tolerance response to the shoots of tomato 'Ikram' compared with self-grafted tomato plants.

Similarly, Kumar *et al.* (2015b) found that plants grafted on to the tomato rootstocks 'Unifort' and especially 'Maxifort' could effectively mitigate the adverse effects of Ni by maintaining a better leaf nutritional status, higher chlorophyll content and a higher F_v/F_m ratio. Moreover, the 'Maxifort'-grafted plants had increased antioxidant enzyme activity (ascorbate peroxidase, guaiacol peroxidase) with decreased oxidative stress (malondialdehyde content) and electrolyte leakage.

Finally, Ni content was also lower in the leaves of 'Maxifort'-grafted versus self-grafted plants. In contrast, Stazi et al. (2016) reported that the use of 'Maxifort' rootstock increased As accumulation in tomato fruits. The above findings indicate that the behavior of a rootstock on the uptake and translocation of heavy metals to the scion can vary depending on the type of heavy metal, grafting combination and growing conditions.

Rouphael *et al.* (2008) studied the effects of grafting cucumber ('Akito') on to the commercial rootstock 'Shintoza' (*C. maxima* × *C. moschata*) on Cu uptake and translocation to the shoot, and on plant biomass and yield, by exposing the plants to different external concentrations of this micronutrient. Grafting of cucumber mitigated the adverse effects of excess Cu concentration in the nutrient solution by restricting its uptake and translocation to the shoot (Rouphael *et al.*, 2008). The authors concluded that the reduction of Cu uptake in the grafted plants was associated with the higher ability of the 'Shintoza' rootstock to exclude it.

In some cucurbits (*Cucumis sativus* L., *Lagenaria siceraria* (Mol.) Standl. *Luffa cylindrica* L., *Momordica charantia* L.) that were exposed to Hg and Pb stress, a high level of peroxidase activity was observed in all plant tissues compared with the control plants (Khan and Chaudhry, 2006). This activity was ascribed to Hg- and Pb-induced impairment of biochemical processes, which severely stress the plant, thereby leading to initiation of antioxidative defence reactions.

The mechanisms responsible for decreasing the uptake and accumulation of heavy metals in grafted plant tissues are still not completely understood (Savvas *et al.*, 2010). To elucidate them further, Arao *et al.* (2008) studied the effects of grafting aubergine on to itself, *S. torvum* or *Solanum paniculatum* var. *integrifolium* Dunal rootstocks on Cd absorption and translocation to the shoot and fruit, and on plant growth and yield, in Cd-polluted and unpolluted soils. Grafting aubergine (*S. melongena*) on to *S. torvum* reduced the Cd content in the leaves and stems by 67–73% compared with self-grafting or grafting on to *S. integrifolium* in the Cd-polluted and unpolluted soils. Moreover, Cd concentration in xylem sap collected from the stems of *S. torvum* was 22% of that in the stems of *S. melongena*, indicating appreciable restriction of Cd translocation from the root to the shoot in the aubergine grafted on to *S. torvum*. However, Mori *et al.* (2009) found similar Cd concentrations in the roots of *S. melongena* and *S. torvum* when these plants were exposed to identical external Cd levels. Savvas *et al.* (2010) concluded from these results that *S. torvum* specifically restricts the translocation of Cd to the shoots but not its absorption by the root system. Mori *et al.* (2009) suggested that restriction of Cd translocation to the fruit of aubergine grafted on to *S. torvum* results from a process occurring during xylem loading. In contrast, Yamaguchi *et al.* (2010) enlisted molecular mechanisms to explain the reduced Cd absorption by *S. torvum* rootstock, namely, dehydration-related transcription factors and aquaporin isoforms as potential constituents of Cd-induced biochemical impediments. Other results have shown that the rootstock significantly affects gene expression in the scion, thereby indicating that some signals transported from the root to the shoot may also influence Cd uptake and translocation (Si *et al.*, 2010). Savvas *et al.* (2013) studied the impact of grafting on Ni and Cd contents in a greenhouse experiment. Cucumber 'Creta', which was non-grafted, self-grafted or grafted on to four different *Cucurbita* rootstocks, was exposed to high Cd and Ni concentrations

(10 and 50 μmol l^{-1}, respectively). The concentrations of Ni and Cd in the edible cucumber fruit were efficiently restricted by grafting the cucumber plants on to the various rootstocks.

Thus, it can be concluded that grafted plants accumulate fewer heavy metals or B in the shoot tissues in comparison with non-grafted plants exposed to the same concentrations. This difference is ascribed either to a higher absorption selectivity or to a higher capacity of the rootstock to retain these toxic microelements. Therefore, grafting can serve as a useful tool to decrease the contents of toxic microelements in the edible parts of vegetables, thus preventing their entry into the food chain.

7.6 Stress by Adverse Soil pH

Approximately 30–40% of the world's total land area consists of alkaline and acid soils, especially arable land and land in semi-arid climates, where food production is critical (Troeh and Thompson, 2005; Seguel *et al.*, 2013; Rouphael *et al.*, 2015). Alkalinity and acidity in soils and/or irrigation water may adversely affect vegetable crop productivity. For instance, alkaline soils and water are characterized by high concentrations of bicarbonate (HCO_3^-) and calcium carbonate ($CaCO_3$), high pH (pH > 8–9) and low bioavailability of macro- and micronutrients (Marschner, 1995). HCO_3^- interacts with the uptake and translocation of mono- and bivalent cations such as K^+, Ca^{2+} and Mg^{2+} (Pissaloux *et al.*, 1995), and also interferes with the availability of Fe, an important micronutrient, causing chlorosis, which may lead to serious yield losses (Savvas *et al.*, 2010). Acid soils (pH <5) favour the solubilization of aluminium (Al), which is generally present in harmless forms such as oxides and aluminosilicates, into the toxic trivalent cation Al^{3+} (Seguel *et al.*, 2013), which is released into the soil solution to levels that affect root elongation, growth and crop productivity (Sivaguru *et al.*, 2013).

The impact of grafting on adverse soil pH conditions is addressed in this section, where we will briefly review the responses of grafted plants to excessively high and low external pH and the possibilities of grafting on to selected rootstocks as a means of improving alkalinity, acidity and Al toxicity tolerance in annual fruit crops.

Colla *et al.* (2010) investigated the mechanisms underlying the effects of grafting on physiological and biochemical responses to alkalinity. The authors studied the effectiveness of grafting in terms of leaf gas exchange, root Fe(III)-chelate reductase, mineral composition and root exudates under a nutrient solution pH of 6.0 and 8.1, using either non-grafted watermelon plants (*Citrullus lanatus* (Thunb.) Matsum. and Nakai 'Ingrid') or plants grafted on to two pumpkin (*C. maxima* × *C. moschata*) rootstocks ('P360' and'PS1313') and two bottle gourd (*L. siceraria*) rootstocks ('Macis' and 'Argentario'). The leaf chlorosis symptoms and shoot biomass reduction in the non-grafted watermelon plants and the plants grafted on to bottle gourd rootstocks were, in general, more pronounced than those in plants grafted on to the two pumpkin rootstocks. Under alkaline stress conditions, plants grafted on to 'P360' and 'PS1313' were capable of maintaining higher net assimilation of CO_2, and had a greater capacity to accumulate P, Mg and especially Fe in the shoot tissue in comparison with the non-grafted plants and those grafted on to 'Macis' and 'Argentario'. The authors also reported that watermelon grafted on to pumpkin rootstocks exuded more citric and malic acids (189% and 91%, respectively) than those grafted on to *L. siceraria* and non-grafted plants, especially under alkalinity stress conditions. These findings support the hypothesis that the uptake and translocation of P, Mg and Fe from the nutrient solution by 'P360' and 'Ps1313' was facilitated by exudation of organic acids from roots (Gent *et al.*, 2005).

The better nutritional status recorded in grafted plants under adverse pH conditions, has been also demonstrated by Borgognone *et al.* (2013), who reported highest Ca, Fe, Zn and Cu concentrations in leaf tissues of tomato (*S. lycopersicum*) 'Moneymaker' grafted on to 'Maxifort' (*S. lycopersicum* × *S. habrochaites*) than in self-grafted plants under five pH levels in the nutrient solution (pH 3.5, 4.5, 5.5, 6.5 or 7.5). Furthermore, Mohsenian *et al.* (2012) and Mohsenian and Roosta (2015) found that certain grafting combinations affected plant growth parameters, mineral uptake and translocation, soluble sugars content, pigment synthesis and maximum quantum use efficiency of photosystem II in the dark-adapted state (F_v/F_m). In the former two experiments, the authors studied the vegetative growth and physiological response of alkaline-stressed tomato 'Red Stone' non-grafted, or grafted on to five rootstocks of aubergine (*S. melongena* 'Long Purple'), datura (*Datura patula* L.), orange nightshade (*Solanum luteum* Mill.), local Iranian tobacco (*Nicotiana tabacum* L.) and field tomato 'Cal.jn3', exposed to 0, 5 and 10 mM $NaHCO_3$ in the nutrient solution. Overall, the tomato plants grafted on to datura rootstock and exposed to excess external $NaHCO_3$ were capable of maintaining better vegetative growth in comparison with non-grafted plants and plants grafted with other rootstocks. The better crop performance recorded with the rootstock datura was attributed to the higher soluble sugar contents in leaves, photosynthetic pigments and F_v/F_m ratio. The authors concluded that appropriate selection of scion × rootstock combinations could improve the high pH tolerance of tomato plants under $NaHCO_3$ stress conditions.

The effectiveness of grafting to improve alkalinity tolerance has also been observed in other important vegetable crops such as cucumber (*Cucumis sativus* L.) (Roosta and Karimi, 2012). The authors investigated the effect of three grafting combinations: non-grafted or grafted plants on to *C. moschata* and *C. maxima* under alkali-stress conditions (10 mM sodium bicarbonate). They report that rootstocks appear to play an important role in alleviating the effect of alkalinity stress, in particular *C. moschata*, which provided better osmotic adjustment under stress and thus exhibited the highest shoot length, leaf number, and fresh and dry biomass.

To the best of our knowledge, the impact of grafting on the rhizosphere low pH and Al toxicity has not been addressed in any peer-reviewed publication to date, and thus research in this field is currently urgently needed (Savvas *et al.*, 2010). Based on the currently available information coming from Italian and German research groups, it seems that grafting on to suitable rootstocks may alleviate growth and yield decreases associated with acidity and Al toxicity in some grafted vegetables such as tomato and cucumber (Rouphael *et al.*, 2016). However, the mechanisms underlying possible interactions between the rootstock genotype and Al toxicity are still unclear, and the question as to whether grafting alleviates Al toxicity by restricting metal uptake by roots and/or its translocation to the shoots, or by facilitating metal detoxification processes by regulating the level of antioxidants and certain metabolite concentrations in plants, is still unknown.

7.7 Drought and Flood Stresses

7.7.1 Drought

Drought is one of the most common abiotic stresses that affects plant development and growth, and has become the primary cause for reductions in crop productivity (Passioura, 2007a). Drought is reported to disturb physiological, biochemical and metabolic processes, leading to alterations in anatomical structures, growth inhibition and consequently yield reduction (Rouphael *et al.*, 2012).

Most vegetable crops are rather shallow-rooted and are therefore susceptible to water-limited conditions, even for short periods. Although these crops are typically irrigated, water shortage may be a problem due to increasing limitations in water resources, which will negatively affect crop productivity. Thus, using drought-tolerant cultivars might effectively contribute to maximizing productivity and promoting sustainable practices (Schwarz *et al.*, 2010; Nilsen *et al.*, 2014). Numerous attempts have been made to enhance drought tolerance of vegetable crops by traditional breeding and genetic engineering, but commercial success has been very limited (Gong *et al.*, 2010; Choi *et al.*, 2011; Ziaf *et al.*, 2011). The traits associated with drought tolerance in xerophytes typically concern survival during drought through adaptation mechanisms that limit growth rates (Passioura, 2007b). However, in an agricultural context, 'drought tolerance' is defined in terms of yield in relation to a limiting water supply. An additional difficulty in breeding drought-tolerant commercial cultivars is due to the complexity of multigenic traits: complicated genetic control of traits such as photosynthetic responses to drought may need several decades of research in plant engineering for the development of commercial cultivars (Nilsen *et al.*, 2014). A sustainable tool to improve drought tolerance in vegetables would be to graft sensitive vegetable cultivars on to specific rootstocks that are able to enhance water conservation and water-use effciency, leading to improved plant growth and crop yield stability. Nilsen *et al.* (2014) found that certain scion–rootstock combinations affected plant growth, diffusion conductance to CO_2 and photosynthetic responses under mild drought. They studied the vegetative growth and photosynthetic parameters of drought-stressed tomato (*S. lycopersicum*) 'BHN 602' non-grafted or grafted on to two hybrid rootstocks, 'Cheong Gang' and 'JjakKkung'. The shoot biomass of 'BHN 602' grafted on to 'JjakKkung' was significantly lower in comparison with 'BHN 602' grafted on to 'Cheong Gang' and non-grafted plants. However, the 'JjakKkung' rootstock improved also water conservation and increased the tolerance of the photosynthetic apparatus to mild drought. The different response between these two rootstocks suggested that a root signal improved water conservation in 'JjakKkung' in comparison with 'Cheong Gang'. Grafting trials with ABA-deficient tomato mutants, *sitiens* and *flacca*, and their near-isogenic wild-type parent demonstrated that there is a chemical signal produced by the roots that controls stomatal conductance (Holbrook *et al.*, 2002). The authors concluded that the strategy for water conservation by the 'JjakKkung' rootstock was a combination of growth reduction (i.e. reducing leaf area) and an ability to maintain mean midday CO_2-saturated photosynthesis at lower water potentials.

Recently, it has also been demonstrated that agronomic water-use efficiency (defined as fruit yield per water use) in tomato can be improved by exploiting rootstock-mediated hormonal regulation of leaf biomass (Cantero-Navarro *et al.*, 2016). Interestingly, some recombinant inbred lines, grown as grafted rootstocks with a common scion with low biomass and water use, improved fruit yield and water-use efficiency by up to 40% compared with self-grafted plants. Although other hormonal factors appear implicated in this response, xylem ACC concentration seems an important root-derived trait that inhibits leaf growth but does not limit fruit yield.

It is well known that drought depresses nutrient uptake, particularly N, by root and translocation to the shoot due to a restricted transpiration rate as well as reduced membrane permeability, leading to growth and yield reduction (Rouphael *et al.*, 2012). In a recent study, Sánchez-Rodriguez *et al.* (2013) investigated the mechanisms underlying the effects of grafting on enzymes involved in N metabolism in response to moderate water stress, focusing on the generation of ammonium ions (NH_4^+) through photorespiration. The authors studied the effectiveness of grafting in terms of N metabolism under moderate water stress (50% of field capacity) using different combinations of grafted scion–rootstock (reciprocal grafting), self-grafted and non-grafted tomato plants, 'Zarina' (drought tolerant) and 'Josefina' (drought-sensitive). The 'Josefina'–'Zarina' (scion–rootstock) combination showed improved N uptake and NO_3^- assimilation in the scion, which resulted in a favourable leaf relative water content and growth response to moderate water stress. However, use of the 'Zarina'–'Josefina' combination exhibited better growth under stress conditions due to an improved photorespiration cycle. In a similar experiment, using the same grafting combinations, Sánchez-Rodriguez *et al.* (2014) demonstrated that the use of 'Zarina' as rootstock improved the ionome, with increases in N, P and K concentration and uptake fluxes, and an increase in Fe and Cu concentration and uptake under water stress. The better nutritional status of the 'Josefina'–'Zarina' combination could be attributed to the vigorous root system of the rootstocks (lateral and vertical development), which resulted in enhanced uptake of water and minerals (Rouphael *et al.*, 2008). Sánchez-Rodriguez *et al.* (2016), using the same grafting combinations (non-grafted, self-grafted and reciprocal grafting), also analysed the changes in antioxidant enzymes in tomato fruits of non-grafted and grafted plants under well watered and water-stress conditions. The authors demonstrated that polyamines, small aliphatic polycations, mitigated drought-induced oxidative damage (H_2O_2 content) by enhancing the antioxidant enzymes activities (catalase and superoxide dismutase), as well as reducing membrane lipid peroxidation, estimated by measuring malondialdehyde in the tomato fruit harvest in 'Zarina' non-grafted and 'Zarina'–'Josefina' plants. The authors concluded that polyamines, especially spermine, were associated with the reduction of tissue damage in tomato fruits, which may also contribute to enhance dehydration tolerance (Sánchez-Rodriguez *et al.*, 2016). Ibrahim *et al.* (2014) also showed that it is possible to increase tomato production and improve the water-use efficiency when plants are grafted on to vigorous rootstocks. For example, grafting tomato 'Faridah' on to the commercial 'Unifort' (*S. lycopersicum* × *S. habrochaites*) under water-stress conditions (80% of crop evapotranspiration) resulted in a 16.7% saving in irrigation water, with a slight reduction in yield (0.7–1.3%) compared with non-grafted tomato plants.

Balancing stomata-dependent CO_2 assimilation and transpiration, which is a key challenge for increasing water-use efficiency, was also possible by grafting cucumber (*C. sativus* 'Jinyan No. 4') on to luffa (*L. cylindrica* 'Xiangfei No. 236') rootstock (Liu *et al.*, 2016). The authors showed that luffa roots have developed an improved ability to sense the changes in root-zone moisture, delivering modest levels of ABA from the roots to the aerial parts to improve water-use efficiency under water-stress conditions. Such inherent drought-tolerance mechanisms could be greatly exploited to improve horticultural production, especially in arid and semi-arid areas.

Similarly to tomato and cucumber, the growth performance of pepper (*C. annuum*) 'Verset' was improved under moderate water-stress conditions when grafted on to selected rootstocks, in comparison with non-grafted plants (Penella *et al.*, 2014). Two experiments were designed. In the first, the behaviour of 18 greenhouse pepper genotypes during vegetative (2 months) and reproductive (5 months) stages under water stress was evaluated based on photosynthetic parameters: net CO_2 fixation rate, stomatal conductance of water vapour and maximum quantum yield of photosystem II. The aim of this screening experiment was to select tolerant genotypes that might be used as rootstocks for pepper cultivation. The genotypes 'Atlante', 'C-40', 'Serrano', 'PI-152225', 'ECU-973', 'BOL-58' and 'NuMex Conquistador' exhibited the highest tolerance to water stress, as they were able to maintain the net photosynthetic rate under stress conditions (Penella *et al.*, 2014). The second experiment aimed to confirm whether the use of selected rootstocks to graft pepper 'Verset' could improve crop performance. It was found that 'Verset' plants grafted on to the rootstocks 'Atlante', 'PI-152225' and 'ECU-973' showed higher marketable yields when compared with non-grafted 'Verset' plants (Penella *et al.*, 2014).

Enhancement of marketable yield (by 60%) and water-use efficiency (7–10%) due to grafting under conditions of deficit irrigation was reported by Rouphael *et al.* (2008) for mini-watermelon (*C. lanatus* 'Ingrid') scions grafted on to the commercial rootstock 'PS 1313' (*C. maxima* × *C. moschata*). The higher crop performance recorded with grafting was attributed mainly to higher concentrations of N, K and Mg in the leaves and higher net CO_2 assimilation. To the best of our knowledge, there are currently no additional reports in the literature on the effectiveness of grafting to improve drought tolerance in aubergine or melon. Futher research on species-specific responses to drought is urgently needed.

7.7.2 Flooding and waterlogging

Flooding and waterlogging are frequent and extensive abiotic stresses occurring during the hot summer period in the lowland tropics (Schwarz *et al.*, 2010). As a consequence of erratic and unpredictable rainfall associated with the climate change scenario, the probability of facing flooding in vegetable crops is increasing worldwide, with negative consequences on plant growth and productivity. The different mechanisms involved in plant tolerance to flooding include: (i) the capacity of tolerant plants to leak O_2 from their root surfaces and to oxidize the reduced forms of Fe and Mn, consequently protecting themselves from excessive uptake of these ions that might cause toxicity (Ashraf and Arfan, 2005); and (ii) developing aerenchyma, which allows exchange of gases from the shoots to the roots (Armstrong, 1982).

An effective and sustainable tool to mitigate the consequences of flooding would be to graft non-tolerant vegetable cultivars on to tolerant ones (*see* Plate 18). Grafting bitter melon (*M. charantia* 'New Known You # 3') on to luffa (*L. cylindria* 'Cylinder #2') improved flooding tolerance (Liao and Lin, 1996). The higher flooding tolerance of grafted plants was attributed mainly to less reduction in CO_2 exchange rate, stomatal conductance, transpiration rate and ribulose-1,5-bisphosphate carboxylase/oxygenase (RuBisCO) activity, in comparison with those recorded in non-grafted bitter melon plants (Liao and Lin, 1996).

When grafting watermelon 'Crimson Tide' on to *L. siceraria* SKP (landrace), the decreases imposed by flooding in chlorophyll content, leaf CO_2 exchange rate, stomatal conductance and transpiration rate were less pronounced compared with those measured in non-grafted watermelon (Yetisir *et al.*, 2006). The authors also showed that flooding decreased the dry mass of plants at the end of the growing cycle by about 230% and 80% in non-grafted and grafted watermelons, respectively. Moreover, adventitious roots and aerenchyma were observed only in the grafted watermelon (Yetisir *et al.*, 2006).

In a recent study carried out in India, Bhatt *et al.* (2015) showed that flooding tolerance of the flooding-sensitive tomato 'ArkaRakshak' could be improved by grafting this cultivar on to aubergine rootstocks. Among the four aubergine rootstocks used, the grafting combination 'ArkaRakshak'–'ArkaNeelkanth' followed by 'ArkaRakshak'–'BPLH-1' and 'ArkaRakshak'–'MattuGulla' exhibited the highest level of tolerance in comparison with that recorded in non-grafted and self-grafted plants. The authors concluded that concentration changes in sugars such as glucose and sucrose, and the formation of adventitious roots, were associated with the higher tolerance of grafted plants to flood stress (Bhatt *et al.*, 2015).

7.8 Conclusions

Grafting of annual fruit crops is expected to expand in the coming decades as an alternative means to overcome abiotic stresses such as suboptimal and/or supra-optimal temperatures, drought, salinity, nutrient stress, heavy-metal contamination, adverse soil pH and flooding. Grafting on to tolerant and suitable rootstocks enables greater adaptability of elite scion varieties to adverse environments. The responses of grafted plants to abiotic stress conditions depend on the scion as well as the rootstock genotype and the scion × rootstock interactions. The anatomical, physiological and agronomical processes implicated in the abiotic stress tolerance of grafted plants have received much attention. However, the molecular aspects involved remain relatively unknown. 'Omic' approaches, such as genomics, metabolomics, xylomics and proteomics, are needed to clarify the functions of many genes, proteins and metabolite networks involved in root-to-shoot interactions of grafted plants in response to unfavourable soil and environmental conditions. Unravelling the root-based mechanisms responsible for the higher scion performance of grafted plants will not only contribute to a better understanding of their mode of action but also improve their breeding efficiency.

Acknowledgements

The authors of each section are: J.H. Venema, G. Ntatsi and D. Schwarz (section 7.2); Y. Rouphael and G. Colla (section 7.3); D. Savvas, G. Ntatsi and Y. Rouphael (section 7.4); M. Edelstein, G. Colla, P. Kumar and M. Ben-Hur (section 7.5); Y. Rouphael, G. Colla and D. Schwarz (section 7.6); and Y. Rouphael, G. Ntatsi and D. Savvas (section 7.7).

References

Abbas, S.M. (2012) Effects of low temperature and selenium application on growth and the physiological changes in sorghum seedlings. *Journal of Stress Physiology and Biochemistry* 8, 268–286.

Ahn, S.J., Im, Y.J., Chung, G.C., Cho, B.H. and Suh, S.R. (1999) Physiological responses of grafted-cucumber leaves and rootstock roots affected by low root temperature. *Scientia Horticulturae* 81, 397–408.

Airaki, M., Leterrier, M., Mateos, R.M., Valderrama, R., Chaki, M., Barroso, J.B., Del Río, L.A., Palma, J.M. and Corpas, F.J. (2012) Metabolism of reactive oxygen species and reactive nitrogen species in pepper (*Capsicum annuum* L.) plants under low temperature stress. *Plant, Cell and Environment* 35, 281–295.

Albacete, A., Martínez-Andújar, C., Ghanem, M.E., Acosta, M., Sánchez-Bravo, J., Asins, M.J., Cuartero, J., Lutts, S., Dodd, I.C. and Pérez-Alfocea, F. (2009) Rootstock-mediated changes in xylem ionic and hormonal status are correlated with delayed leaf senescence, and increased leaf area and crop productivity in salinized tomato. *Plant, Cell and Environment* 32, 928–938.

Allen, D.J. and Ort, D.R. (2001) Impacts of chilling temperatures on photosynthesis in warm-climate plants. *Trends in Plant Science* 6, 36–42.

An, Y., Kim, Y., Kwon, T. and Jeong, S. (2004) Combined effect of copper, cadmium, and lead upon *Cucumis sativus* growth and bioaccumulation. *Science of the Total Environment* 326, 85–93.

Arao, T., Takeda, H. and Nishihara, E. (2008) Reduction of cadmium translocation from roots to shoots in eggplant (*Solanum melongena*) by grafting onto *Solanum torvum* rootstocks. *Soil Science and Plant Nutrition* 54, 555–559.

Arms, E.A., Bloom, A.J. and St Clair, D.A. (2015) High-resolution mapping of a major QTL from wild tomato *Solanum habrochaites* that influences water relations under root chilling. *Theoretical and Applied Genetics* 12, 1713–1725.

Armstrong, W. (1982) Waterlogged soils. In: Etherington, J.R. (ed.) *Environment and Plant Ecology*. Wiley, Chichester, UK, pp. 290–330.

Ashraf, M. and Arfan, M. (2005) Gas exchange characteristics and water relations in two cultivars of *Hibiscus esculentus* under waterlogging. *Biologia Plantarum* 49, 459–462.

AVRDC (2016) World Vegetable Center seed catalogue. Available at: http://avrdc.org/seed/ (accessed 29 June 2016).

Bai, L.P., Zhou, B.L., Li, N., Huo, S.F. and Fu, Y.W. (2005) The ion absorption and transportation of grafted eggplants (*Solamun melongena* L.) under NaCl stress. *Plant Physiology Communications* 41, 767–769.

Beemster, G.T.S., de Vusser, K., de Tavernier, E., de Bock, K. and Inze, D. (2002) Variation in growth rate between Arabidopsis ecotypes is correlated with cell division and A-type cyclin-dependent kinase activity. *Plant Physiology* 129, 854–864.

Ben-Gal, A. and Shani, U. (2002) Yield, transpiration and growth of tomatoes under combined excess boron and salinity stress. *Plant and Soil* 247, 211–221.

Ben-Hur, M. (2004) Sewage water treatment and reuse in Israel. In: F. Zereini (ed.) *Water in the Middle East and in North Africa: Resources, Protection and Management*. Springer, New York, pp. 167–180.

Bereczky, Z.,Wang, H.Y., Schubert, V., Ganal, M. and Bauer, P. (2003) Differential regulation of *nramp* and *irt* metal transporter genes in wild type and iron uptake mutants of tomato. *Journal of Biological Chemistry* 278, 24697–24704.

Bhatt, R.M., Upreti, K.K., Divya, M.H., Bhat, S., Pavithra, C.B. and Sadashiva, A.T. (2015) Interspecific grafting to enhance physiological resilience to flooding stress in tomato (*Solanum lycopersicum* L.). *Scientia Horticulturae* 182, 8–17.

Bingham, E.T., Strong, J.E., Rhoads, J.D. and Keren, R. (1985) An application of the Maas–Hoffman salinity response model for boron toxicity. *Soil Science Society of American Journal* 49, 672–674.

Bita, C.E. and Tom Gerats, T. (2013) Plant tolerance to high temperature in a changing environment: scientific fundamentals and production of heat stress-tolerant crops. *Frontiers in Plant Science* 4, 273.

Black, L.L., Wu, D.L., Wang, J.F., Kalb, T., Abbass, D. and Chen, J.H. (2003) Grafting tomatoes for production in the hot-wet season. AVRDC International Cooperators' Guide. AVRDC publication no. 03-551. Available at: http://www.ces.ncsu.edu/fletcher/programs/ncorganic/research/grafting.pdf (accessed 30 April 2016).

Bloom, A.J., Zwieniecki, M.A., Passioura, J.B., Randall, L.B., Holbrook, N.M. and St. Clair, D.A. (2004) Water relations under root chilling in a sensitive and tolerant tomato species. *Plant, Cell and Environment* 27, 971–979.

Borgognone, D., Colla, G., Rouphael, Y., Cardarelli, M., Rea, E. and Schwarz, D. (2013) Effect of nitrogen form and nutrient solution pH on growth and mineral composition of self-grafted and grafted tomatoes. *Scientia Horticulturae* 149, 61–69.

Bui, H.H., Serra, V. and Pagès, L. (2015) Root system development and architecture in various genotypes of the Solanaceae family. *Botany* 93, 465–474.

Bulder, H.A.M., den Nijs, A.P.M., Speek, E.J., van Hasselt, P.R. and Kuiper, P.J.C. (1991) The effect of low root temperature on growth and lipid composition of low temperature tolerant rootstock genotypes for cucumber. *Journal of Plant Physiology* 138, 661–666.

Cantero-Navarro, E., Romero-Aranda, R., Fernández-Muos, R., Martínez-Andújar, C., Perez-Alfocea, F. and Albacete, A. (2016) Improving agronomic water use efficiency in tomato by rootstock-mediated hormonal regulation of leaf biomass. *Plant Science* 251, 90–100.

Chen, G. and Wang, R. (2008) Effects of salinity on growth and concentrations of sodium, potassium, and calcium in grafted cucumber seedlings. *Acta Horticulturae* 771, 217–224.

Chen, C., Huang, D. and Liu, J. (2009) Functions and toxicity of nickel in plants: recent advances and future prospects. *Clean* 37, 304–313.

Choi, J.Y., Seo, Y.S., Kim, S.J., Kim, W.T. and Shin, J.S. (2011) Constitutive expression of CaXTH3, a hot pepper xyloglucan endotransglucosylase/hydrolase, enhanced tolerance to salt and drought stresses without phenotypic defects in tomato plants (*Solanum lycopersicum* cv. Dotaerang). *Plant Cell Reports* 30, 867–877.

Clemens, S. (2006) Toxic metal accumulation, responses to mechanisms of tolerance in plants. *Biochimie* 88, 1707–1717.

Colla, G., Rouphael, Y., Cardarelli, M., Massa, D., Salerno, A. and Rea, E. (2006a) Yield, fruit quality and mineral composition of grafted melon plants grown under saline conditions. *Journal of Horticultural Science and Biotechnology* 81, 146–152.

Colla, G., Rouphael, Y., Cardarelli, M. and Rea, E. (2006b) Effect of salinity on yield, fruit quality, leaf gas exchange, and mineral composition of grafted watermelon plants. *HortScience* 41, 622–627.

Colla, G., Suãrez, C.M.C., Cardarelli, M. and Rouphael, Y. (2010) Improving nitrogen use efficiency in melon by grafting. *HortScience* 45, 559–565.

Colla, G., Rouphael, Y., Mirabelli, C. and Cardarelli, M. (2011) Nitrogen-use efficiency traits of mini-watermelon in response to grafting and nitrogen-fertilization doses. *Journal of Plant Nutrition and Soil Science* 174, 933–941.

Colla, G., Rouphael, Y., Rea, E. and Cardarelli, M. (2012) Grafting cucumber plants enhance tolerance to sodium chloride and sodium sulphate. *Scientia Horticulturae* 135, 177–185.

Colla, G., Kumar, P., Cardarelli, M. and Rouphael, Y. (2013a) Grafting: an effective tool for abiotic stress alleviation in vegetables. In: Chanda, K.L., Singh, A.K., Singh, S.K. and Dhillon, W.S. (eds) *Horticulture for Food and Environment Security*. Westville Publishing House, New Delhi, India, pp. 15–28.

Colla, G., Rouphael, Y., Jawad, R., Kumar, P., Rea, E. and Cardarelli, M. (2013b) The effectiveness of grafting to improve NaCl and $CaCl_2$ tolerance in cucumber. *Scientia Horticulturae* 164, 380–391.

Cuartero, J., Bolarin, M.C., Asins, M.J. and Moreno, V. (2006) Increasing salt tolerance in the tomato. *Journal of Experimental Botany* 57, 1045–1058.

Davis, A.R., Perkins-Veazie, P., Dakata, Y., Lopez-Galarza, S., Maroto, J.V., Lee, S.G., Hyh, Y.-C., Sun, Z., Miguel, A., King, S.R., Cohen, R. and Lee, J.M. (2008) Cucurbit grafting. *Critical Reviews in Plant Sciences* 27, 50–74.

de Dorlodot, S., Forster, B., Pagès, L., Price, A., Tuberosa, R. and Draye, X. (2007) Root system architecture: opportunities and constraints for genetic improvement of crops. *Trends in Plant Science* 12, 474–481.

den Nijs, A.P.M. (1980) The effect of grafting on growth and early production of cucumbers at low temperature. *Acta Horticulturae* 118, 57–63.

di Gioia, F., Signore, A., Serio, F. and Santamaria, P. (2013) Grafting improves tomato salinity tolerance through sodium partitioning within the shoot. *HortScience* 48, 855–862.

Djidonou, D., Zhao, X., Simonne, E.H., Koch, K.E. and Erickson, J.E. (2013) Yield, water-, and nitrogen-use efficiency in field-grown, grafted tomatoes. *HortScience* 48, 485–492.

Djidonou, D., Lopiano, K., Zhao, X., Simonne, E.H., Erickson, J.E. and Koch, K.E. (2015) Estimating nitrogen nutritional crop requirements of grafted tomatoes under field conditions. *Scientia Horticulturae* 182, 18–26.

Easlon, H.S., Asensio, J.S.R., St Clair, D.A. and Bloom, A.J. (2013) Chilling-induces water stress: variation in shoot turgor maintenance among wild tomato species from diverse habitats. *American Journal of Botany* 100, 1991–1999.

Eaton, F.M. (1944) Deficiency, toxicity and accumulation of boron in plants. *Journal of Agricultural Research* 69, 237–277.

Edelstein, M. and Ben-Hur, M. (2012) Use of grafting to mitigate chemical stresses in vegetables under arid and semiarid conditions. *Advances in Environmental Research* 20, 163–179.

Edelstein, M., Ben-Hur, M., Cohen, R., Burger, Y. and Ravina, I. (2005) Boron and salinity effects on grafted and non-grafted melon plants. *Plant Soil* 269, 273–284.

Edelstein, M., Ben-Hur, M. and Plaut, Z. (2007) Grafted melons irrigated with fresh or effluent water tolerate excess boron. *Journal of the American Society for Horticultural Science* 132, 484–491.

Edelstein, M., Plaut, Z. and Ben-Hur, M. (2011) Mechanism responsible for restricted boron concentration in plant shoots grafted on pumpkin rootstocks. *Israel Journal of Plant Sciences* 59, 207–215.

El-Wanis, A., Mona, M., Abdel-Baky, M.H. and Salman, S.R. (2012) Effect of grafting and salt stress on the growth, yield and quality of cucumber grown in NFT system. *Journal of Applied Sciences Research* 8, 5059–5067.

Enza Zaden (2016) Vegetable seed catalogue. Available at: http://www.enzazaden.co.uk/products-and-services/our-products/Rootstock (accessed 21 November 2016).

Estañ, M.T., Martinez-Rodriguez, M.M., Perez-Alfocea, F., Flowers, T.J. and Bolarin, M.C. (2005) Grafting raises the salt tolerance of tomato through limiting the transport of sodium and chloride to the shoot. *Journal of Experimental Botany* 56, 703–712.

Fernández-García, N., Martínez, V. and Carvajal, M. (2004) Effect of salinity on growth, mineral composition, and water relations of grafted tomato plants. *Journal of Plant Nutrition and Soil Science* 167, 616–622.

Francois, L.E. (1984) Effect of excess boron on tomato yield, fruit size and vegetative growth. *Journal of American Society for Horticultural Science* 109, 322–324.

Gao, J.J., Qin, A.G. and Yu, X.C. (2009) Effect of grafting on cucumber leaf SOD and CAT gene expression and activities under low temperature stress. *Chinese Journal of Applied Ecology* 20, 213–217.

Gent, M.P.N., Parrish, Z.D. and White, J.C. (2005) Nutrient uptake among subspecies of *Cucurbita pepo* L. is related to exudation of citric acid. *Journal of American Society for Horticultural Science* 130, 782–788.

Gill, S.S. and Tuteja, N. (2010) Polyamines and abiotic stress tolerance in plants. *Plant Signaling and Behavior* 5, 26–33.

Gimeno, V., Simón, I., Nieves, M., Martínez, V., Cámara-Zapata, J.M., García, A.L. and García-Sánchez, F. (2012) The physiological and nutritional responses to an excess of boron by Verna lemon trees that were grafted on four contrasting rootstocks. *Trees* 26, 1513–1526.

Gladish, D. and Rost, T.L. (1993) The effects of temperature on primary root-growth dynamics and lateral root distribution in garden pea (*Pisum sativum* cv. "Alaska"). *Environmental and Experimental Botany* 33, 2, 243–258.

Gladish, D.K., Sutter, E.G. and Rost, T.L. (2000) The role of free indole-3-acetic acid (IAA) levels, IAA transport, and sucrose transport in the high temperature inhibition of primary root development in pea (*Pisum sativum* L. cv. Alaska). *Journal of Plant Growth Regulation* 19, 347–358.

Goldberg, S., Shouse, P.J., Lesch, S.M., Grieve, C.M., Poss, J.A., Forster, H.S. and Suarez, D.L. (2003) Effect of high boron application on boron content and growth of melons. *Plant and Soil* 256, 403–411.

Gong, P.J., Zhang, J.H., Li, H.X., Yang, C.X., Zhang, C.J., Wang, T., Fei, Z. and Ye, Z. (2010) Transcriptional profiles of drought responsive genes in modulating transcription signal transduction, and biochemical pathways in tomato. *Journal of Experimental Botany* 61, 3563–3575.

Goodstal, F.J., Kohler, G.R., Randall, L.B., Bloom, A.J. and St Clair, D.A. (2005) A major QTL introgressed from wild *Lycopersicon hirsutum* confers chilling tolerance to cultivated tomato (*Lycopersicon esculentum*). *Theoretical and Applied Genetics* 111, 898–905.

Goreta, S., Bucevic-Popovic, V., Selak, G.V., Pavela-Vrancic, M. and Perica, S. (2008) Vegetative growth, superoxide dismutase activity and ion concentration of salt-stressed watermelon as influenced by rootstock. *Journal of Agricultural Science* 146, 695–704.

Gosselin, A. and Trudel, M.J. (1984) Interactions between air and root temperature and light levels on growth, development and photosynthesis of *Lycopersicon esculentum* Mill. cultivar 'Vendor'. *Scientia Horticulturae* 23, 313–321.

Gosselin, A. and Trudel, M.J. (1985) Influence of root-zone temperature on growth, development and yield of cucumber plants cv. Toska. *Plant and Soil* 85, 327–336.

Gratao, P.L., Polle, A., Lea, P.J. and Azevedo, R.A. (2005) Making the life of heavy metal-stressed plants a little easier. *Functional Plant Biology* 32, 481–494.

Greaves, J.A. (1996) Improving suboptimal temperature tolerance in maize – the search for variation. *Journal of Experimental Botany* 47, 307–323.

Gregory, P.J., Atkinson, C.J., Bengough, A.G., Else, M.A., Fernández-Fernández, F. and Harrison, R.J. (2013) Contributions of roots and rootstocks to sustainable, intensified crop production. *Journal of Experimental Botany* 64, 1209–1222.

Guimarães, M.D.A., Gustin, J.L. and Salt, D.E. (2009) Reciprocal grafting separates the roles of the root and shoot in zinc hyperaccumulation in *Thlaspi caerulescens*. *New Phytologist* 184, 323–329.

Gupta, A.K. and Knaur, N. (2005) Sugar signalling and gene expression in relation to carbohydrate metabolism under abiotic stresses in plants. *Journal of Biosciences* 30, 5, 761–776.

He, Y., Zhu, Z., Yang, J. and Zhu, B. (2009) Grafting increases the salt tolerance of tomato by improvement of photosynthesis and enhancement of antioxidant enzymes activity. *Environmental and Experimental Botany* 66, 270–278.

Hirsch, R. and Sussman, M. (1999) Improving nutrient capture from soil by genetic manipulation of crop plants. *Trends in Biotechnology* 17, 356–361.

Holbrook, N.M., Shashidhar, V.R., James, R.A. and Munns, R. (2002) Stomata control in tomato with AB- deficient roots: response of grafted plants to soil drying. *Journal of Experimental Botany* 53, 1503–1514.

Huang, Y, Zhu, J., Zhen, A. and Bie, Z. (2009a) Organic and inorganic solutes accumulation in the leaves and roots of grafted and ungrafted cucumber plants in response to NaCl stress. *Journal of Food Agriculture and Environment* 7, 703–708.

Huang, Y., Tang, R., Cao, Q. and Bie, Z. (2009b) Improving the fruit yield and quality of cucumber by grafting onto the salt tolerant rootstock under NaCl stress. *Scientia Horticulturae* 122, 26–31.

Huang, Y., Bie, Z., He, S., Hua, B., Zhen, A. and Liu, Z. (2010) Improving cucumber tolerance to major nutrients induced salinity by grafting onto *Cucurbita ficifolia*. *Environmental and Experimental Botany* 69, 32–38.

Huang, Y., Li, J., Hua, B., Liu, Z., Fan, M. and Bie, Z. (2013) Grafting onto different rootstocks as a means to improve watermelon tolerance to low potassium stress. *Scientia Horticulturae* 149, 80–85.

Hund, A., Richner, W., Soldati, A., Fracheboud, Y. and Stamp, P. (2007) Root morphology and photosynthetic performance of maize inbred lines at low temperature. *European Journal of Agronomy* 27, 52–61.

Hund, A., Fracheboud, Y., Soldati, A. and Stamp, P. (2008) Cold tolerance of maize seedlings as determined by root morphology and photosynthetic traits. *European Journal of Agronomy* 28, 178–185.

Ibrahim, A., Wahb-Allah, M., Abdel-Razzak, H. and Alsadon, A. (2014) Growth, yield, quality and water use efficiency of grafted tomato plants grown in greenhouse under different irrigation levels. *Life Science Journal* 11, 118–126.

Ishimaru, Y., Takahashi, R., Bashir, K., Shimo, H., Senoura, T., Sugimoto, K., Ono, K., Yano, M., Ishikawa, S., Arao, T., Hiromi, N. and Nishizawa, N.K. (2012) Characterizing the role of rice NRAMP5 in manganese, iron and cadmium transport. *Scientific Reports* 2, 286.

Jensen, P.J., Rytter, J., Detwiler, E.A., Travis, J.W. and McNellis, T.W. (2003) Rootstock effects on gene expression patterns in apple tree scions. *Plant Molecular Biology* 493, 394–51.

Jha, U.C., Bohra, A.S. and Narendra, P. (2014) Heat stress in crop plants: its nature, impacts and integrated breeding strategies to improve heat tolerance. *Plant Breeding* 133, 679–701.

Keatinge, J.D.H., Lin, L.-J., Ebert, A.W., Chen, W.Y., Hughes, J.d'A., Luther, G.C., Wang, J.-F. and Ravishankar, M. (2014) Overcoming biotic and abiotic stresses in the Solanaceae through grafting: current status and future perspectives. *Biological Agriculture and Horticulture* 30, 272–287.

Khan, A.S. and Chaudhry, N.Y. (2006) Peroxidase activity (POX) in some cucurbits under the stress of mercury and lead. *Journal of Food, Agriculture and Environment* 4, 274–276.

Korkmaz, A. and Dufault, R. J. (2001) Developmental consequences of cold temperature stress at transplanting on seedling and field growth and yield. II. Muskmelon. *Journal of American Society and Horticultural Science* 126, 410–413.

Kuijken, R.C.P., van Eeuwijk, F.A., Marcelis, L.F.M. and Bouwmeester, H.J. (2015) Root phenotyping: from component trait in the lab to breeding. *Journal of Experimental Botany* 66, 5389–5401.

Kumar, P., Lucini, L., Rouphael, Y., Cardarelli, M., Kalunke, R.M. and Colla, G. (2015a) Insight into the role of grafting and arbuscular mycorrhiza on cadmium stress tolerance in tomato. *Frontiers in Plant Science* 6, 477.

Kumar, P., Rouphael, Y., Cardarelli, M. and Colla, G. (2015b) Effect of nickel and grafting combination on yield, fruit quality, antioxidative enzyme activities, lipid peroxidation, and mineral composition of tomato. *Journal of Plant Nutrition and Soil Science* 178, 848–860.

Kumar, P., Edelstein, M., Cardarelli, M., Ferri, E. and Colla, G. (2015c) Grafting affects growth, yield, nutrient uptake, and partitioning under cadmium stress in tomato. *HortScience* 50, 1654–1661.

Landi, M., Remorini, M., Pardossi, A. and Guidi, L. (2013) Boron excess affects photosynthesis and antioxidant apparatus of greenhouse *Cucurbita pepo* and *Cucumis sativus*. *Journal of Plant Research* 126, 775–786.

Lee, J.-M. (1994) Cultivation of grafted vegetables I. Current status, grafting methods, and benefits. *Hortscience* 29, 235–239.

Lee, S.H. and Chung, G.C. (2005) Sensitivity of root system to low temperature appears to be associated with the hydraulic properties through aquaporin activity. *Scientia Horticulturae* 105, 1–11.

Lei, B., Huang, Y., Xie, J.J., Liu, Z.X., Zhen, A., Fan, M.L. and Bie, Z.L. (2014) Increased cucumber salt tolerance by grafting on pumpkin rootstock and after application of calcium. *Biologia Plantarum* 58, 179–184.

Leonardi, C. and Giuffrida, F. (2006) Variation of plant growth and macronutrient uptake in grafted tomatoes and eggplants on three different rootstocks. *European Journal of Horticultural Science* 71, 97–101.

Liao, C.T. and Lin, C.H. (1996) Photosynthetic response of grafted bitter melon seedling to flood stress. *Environmental and Experimental Botany* 36, 167–172.

Li, H., Liu, S.-H., Yi, C.-Y., Wang, F., Zhou, J., Xia, X.-J., Shi, K., Zhou, Y.-H. and Yu, J.-Q. (2014a) Hydrogen peroxide mediates abscisic acid-induced HSP70 accumulation and heat tolerance in grafted cucumber plants. *Plant, Cell and Environment* 37, 2768–2780.

Li, H., Wang, F., Chen, X.-J., Shi, K., Xia, X.-J., Considine, M.J., Yu, J.-Q. and Zhou, Y.-H. (2014b) The sub/supra-optimal temperature-induced inhibition of photosynthesis and oxidative damage in cucumber leaves are alleviated by grafting onto figleaf gourd/luffa rootstocks. *Physiologia Plantarum* 152, 571–584.

Li, H., Ahammed, G.J., Zhou, G.N., Xia, X.J., Zhou, J., Shi, K., Yu, J.Q. and Zhou, Y.H. (2015a) Unraveling main limiting sites of photosynthesis under below- and above-ground heat stress in cucumber and the alleviatory role of *Luffa* rootstock. *Frontiers in Plant Science* 7, 746.

Li, Y., Tian, X., Wei, M., Shi, Q., Yang, F. and Wang, X. (2015b) Mechanisms of tolerance differences in cucumber seedlings grafted on rootstocks with different tolerance to low temperature and weak light stresses. *Turkish Journal of Botany* 39, 606–614.

Liu, S.S., Li, H., Lv, X.Z., Ahammed, G.J., Xia, X.J., Zhou, J., Shi, K., Asami, T., Yu, J.Q. and Zhou, Y.H. (2016) Grafting cucumber onto luffa improves drought tolerance by increasing ABA biosynthesis and sensitivity. *Scientific Reports* 6, 20212.

Lux, A., Martinka, M., Vaculík, M. and White, P.J. (2011) Root response to cadmium in the rhizosphere: a review. *Journal of Experimental Botany* 62, 21–37.

Lynch, J.P. (1998) The role of nutrient-efficient crops in modern agriculture. *Journal of Crop Production* 1, 241–264.

Mahajan, S. and Tuteja, N. (2005) Cold, salinity and drought stresses: an overview. *Archives Biochemistry and Biophysics* 444, 139–158.

Marschner, H. (1995) Functions of mineral nutrients: micronutrients. In: Marschner, H. (ed.) *Mineral Nutrition of Higher Plants*, 2nd edn. Academic Press Limited, San Diego, California, pp. 313–396.

Martínez-Andújar, C., Albacete, A., Martínez Pérez, A., Pérez Pérez, J.M., Asins, M.J., and Pérez-Alfocea, F. (2016) Root-to-shoot hormonal communication in contrasting rootstocks suggests an important role for the ethylene precursor aminocyclopropane-1-carboxylic acid in mediating plant growth under low-potassium nutrition in tomato. *Frontiers in Plant Science* 7, 1782.

Martínez-Ballesta, M.C., Alcaraz-López, C., Muries, B., Mota-Cadenas, C. and Carvajal, M. (2010) Physiological aspects of rootstock–scion interactions. *Scientia Horticulturae* 127, 112–118.

Martínez-Rodríguez, M.M., Estañ, M.T., Moyano, E., García-Abellan, J.O., Flores, F.B., Campos, J.F., Al-Azzawi, M.J., Flowers, T.J. and Bolarín, M.C. (2008) The effectiveness of grafting to improve salt tolerance in tomato when an 'excluder' genotype is used as scion. *Environmental and Experimental Botany* 63, 392–401.

McMichael, B.L. and Quisenberry, J.E. (1993) The impact of the soil environment on the growth of root systems. *Environmental and Experimental Botany* 33, 53–61.

Meridiem Seeds (2016a) Rootstock catalogue. Available at: http://www.meridiemseeds.com/en/products/rootstock (accessed 29 June 2016).

Mikado Kyowa Seeds (2016b) Seed catalogue. Available at: http://en.mikadokyowa.com/seed/rootstock/ (accessed 29 June 2016).

Mohsenian, Y. and Roosta, H.R. (2015) Effects of grafting on alkali stress in tomato plants: datura rootstock improve alkalinity tolerance of tomato plants. *Journal of Plant Nutrition* 38, 51–72.

Mohsenian, Y., Roosta, H.R., Karimi, H.R. and Esmaeilzade, M. (2012) Investigation of the ameliorating effects of eggplant, datura, orange nightshade, local Iranian tobacco, and field tomato as rootstocks on alkali stress in tomato plants. *Photosynthetica* 50, 411–421.

Mori, S., Uraguchi, S., Ishikawa, S. and Arao, T. (2009) Xylem loading process is a critical factor in determining Cd accumulation in the shoots of *Solanum melongena* and *Solanum torvum*. *Environmental and Experimental Botany* 67, 127–132.

Mortvedt, J.J., Cox, F.R., Shuman, L.M. and Welch, R.M. (1991) *Micronutrients in Agriculture*, 2nd edn. SSSA Book Series 4. Soil Science Society of America, Madison, Wiscosin.

Nable, R.O., Banuelos, G.S. and Paull, J.G. (1997) Boron toxicity. *Plant Soil* 193, 181–198.

Nagel, K., Kastenholz, B., Jahnke, S., van Dusschoten, D., Aach, T. and Muhlich, M. (2009) Temperature responses of roots: impact on growth, root system architecture and implications for phenotyping. *Functional Plant Biology* 36, 947–959.

Nakamura, S., Hondo, K., Kawara, T., Okazaki, Y., Saito, K., Kobayashi, K., Yaeno, T., Yamaola, N. and Nishiguchi, M. (2016) Conferring high-temperature tolerance to nontransgenic tomato scions using graft transmission of RNA silencing of the fatty acid desaturase. *Plant Biotechnology Journal* 14, 783–790.

Neocleous, D. and Savvas, D. (2015) Effect of different macronutrient cation ratios on macronutrient and water uptake by melon (*Cucumis melo*) grown in recirculating nutrient solution. *Journal of Plant Nutrition and Soil Science* 178, 320–332.

Nilsen, E.T., Freeman, J., Grene, R. and Tokuhisa, J. (2014) A rootstock provides water conservation for a grafted commercial tomato (*Solanum lycopersicum* L.) line in response to mild-drought conditions: a focus on vegetative growth and photosynthetic parameters. *PLoS ONE* 9, e115380.

Ntatsi, G. (2013) Impact of grafting on growth, yield, quality, and physiological characteristics of tomato grown under sub-optimal temperature conditions. PhD thesis, Agricultural University of Athens, Athens, Greece.

Ntatsi, G., Savvas, D., Huntenburg, K., Drüge, U., Hincha, D., Zuther, E. and Schwarz, D. (2014a) A study on ABA involvement in the response of tomato to suboptimal root temperature using reciprocal grafts with notabilis, a null mutant in the ABA-biosynthesis gene *LeNCED1*. *Environmental and Experimental Botany* 97, 11–21.

Ntatsi, G., Savvas, D., Klaering, H.-P. and Schwarz, D. (2014b) Growth, yield, and metabolic responses of temperature-stressed tomato to grafting onto rootstocks differing in cold tolerance. *Journal of American Society for Horticultural Science* 139, 230–243.

Okimura, M., Matso, S., Arai, K. and Okitsu, S. (1986) Influence of soil temperature on the growth of fruit vegetable grafted on different stocks. *Bulletin of the Vegetable and Ornamental Crops Research Station Japan* C9, 43–58.

Orsini, F., Sanoubar, R., Oztekin, G.B., Kappel, N., Tepecik, M., Quacquarelli, C., Tuzel, Y., Bona, S. and Gianquinto, G. (2013) Improved stomatal regulation and ion partitioning boosts salt tolerance in grafted melon. *Functional Plant Biology* 40, 628–636.

Pahlavanian, A.M. and Silk, W.K. (1988) Effect of temperature on spatial and temporal aspects of growth in the primary maize root. *Plant Physiol*ogy 87, 529–532.

Palada, M.C. and Wu, D.L. (2008) Evaluation of chili rootstocks for grafted sweet pepper production during the hot-wet and hot-dry seasons in Taiwan. *Acta Horticulturae* 767, 167–174.

Pardales, J., Kono, Y. and Yamauchi, A. (1992) Epidermal cell elongation in sorghum seminal roots exposed to high root-zone temperature. *Plant Science* 81, 143–146.

Pardales, J.R.J., Banoc, D.M., Yamauchi, A., Iijima, M. and Kono, Y. (1999) Root system development of cassava and sweetpotato during early growth stage as affected by high root zone temperature. *Plant Production Science* 2, 247–251.

Passioura, J.B. (2007a) The drought environment: physical, biological and agricultural perspectives. *Journal of Experimental Botany* 58, 113–117.

Passioura, J.B. (2007b) Drought and drought tolerance. In: Belhassen, E. (ed.) *Drought Tolerance in Higher Plants: Genetical, Physiological and Molecular Biological Analysis*. Springer Science Business Media B.V., Dordrecht, The Netherlands, pp. 1–5.

Peleg, Z., Reguera, M., Tumimbang, E., Walia, H. and Blumwald, E. (2011) Cytokinin-mediated source/sink modifications improve drought tolerance and increase grain yield in rice under water-stress. *Plant Biotechnology Journal* 9, 747–758.

Penella, C., Nebauer, S.G., Bautista, A.S., López-Galarza, S. and Calatayud, A. (2014) Rootstock alleviates PEG-induced water stress in grafted pepper seedlings: physiological responses. *Journal of Plant Physiology* 171, 842–851.

Penella, C., Nebauer, S.G., Quiñones, A., Lopéz-Galarza, S., SanBautista, A. and Calatayud, A.C. (2015) Some rootstocks improve pepper tolerance to mild salinity through ionic regulation. *Plant Science* 230, 12–22.

Penella, C., Landi, M., Guidi, L., Nebauer, S.G., Pellegrini, E., Bautista, A.S., Remorini, D., Nali, C., Lopez-Galarza, S. and Calatayud, A. (2016) Salt-tolerant rootstock increases yield of pepper under salinity through maintenance of photosynthetic performance and sinks strength. *Journal of Plant Physiology* 193, 1–11.

Petropoulos, S.A., Olympios, C., Ntatsi, G., Akoumianakis, C. and Passam, H. (2011) Effects of grafting and salinity on agronomic characteristics and ion uptake by cucumber plants grown in a closed hydroponic system. *Acta Horticulturae* 952, 637–643.

Pissaloux, A., Morarad, P. and Bertoni, G. (1995) Alkalinity-bicarbonate calcium effects on iron chlorosis in white lupine in soilless culture. In: Abadía, J. (ed.) *Development in Plant and Soil Science. Iron Nutrition in Soils and Plants. Seventh International Symposium on Iron Nutrition and Interactions in Plants*, Vol. 59. Zaragoza, Spain, 27 June–2 July 1993. Kluwer Academic Publishers, Dordrecht, The Netherlands, pp. 127–133.

Pressman, E., Peet, M.M. and Pharr, D.M. (2002) The effect of heat stress on tomato pollen characteristics is associated with changes in carbohydrate concentration in the developing anthers. *Annals of Botany* 90, 631–639.

Pulgar, G., Villora, G., Moreno, D.A. and Romero, L. (2000) Improving the mineral nutrition in grafted watermelon plants: nitrogen metabolism. *Biologia Plantarum* 43, 607–609.

Qi, H.Y., Liu, Y.F., Li, D. and Li, T.L. (2006) Effects of grafting on nutrient absorption, hormone content in xylem exudation and yield of melon (*Cucumis melo* L.). *Plant Physiology Communications* 42, 199–202.

Qin, L., He, J., Lee, S.K. and Dodd, I.C. (2007) An assessment of the role of ethylene in mediating lettuce (*Lactuca sativa*) root growth at high temperatures. *Journal of Experimental Botany* 58, 3017–3024.

Ramiro Arnedo Semillas (2016) Rootstock catalogue. Available at: http://www.ramiroarnedo.com/index.php/en/semillas-profesional-en/290-portainjertos-graft (accessed 29 June 2016).

Reid, R.J., Hayes, J.E., Post, A., Stangoulis, J.C.R. and Graham, R.D. (2004) A critical analysis of the causes of boron toxicity in plants. *Plant, Cell and Environment* 25, 1405–1414.

Riga, P. (2015) Effect of rootstocks on growth, fruit production and quality of tomato plants grown under low temperature and light conditions. *Horticulture, Environment, and Biotechnology* 56, 626–638.

Rijk Zwaan (2016) Rootstock catalogue. Available at: http://www.rijkzwaan.nl/wps/wcm/connect/RZ+NL/Rijk+Zwaan/Products_and_Services/Products/Crops/Rootstock (accessed 29 June 2016).

Rivero, R.M., Ruiz, J.M., Sanchez, E. and Romero, L. (2003a) Does grafting provide tomato plants an advantage against H_2O_2 production under conditions of thermal shock? *Physiologia Plantarum* 117, 44–50. 2003.

Rivero, R.M., Ruiz, J.M. and Romero, L. (2003b) Role of grafting in horticultural plants under stress conditions. *Food, Agriculture and Environment* 1, 70–74.

Rivero, R.M., Ruiz, J.M. and Romero, L. (2005) Iron metabolism in tomato and watermelon plants: influence of grafting. *Journal of Plant Nutrition* 27, 2221–2234.

Rogers, E.D. and Benfey, P.N. (2015) Regulation of plant root system architecture: implications for crop advancement. *Current Opinion in Biotechnology* 32, 93–98.

Romani, M. (2014) Boron toxicity in vegetable crops. PhD thesis, University of Pisa, Pisa, Italy.

Romero, L., Belakbir, A., Ragala, L. and Ruiz, J.M. (1997) Response of plant yield and leaf pigments to saline conditions: effectiveness of different rootstocks in melon plants (*Cucumis melo* L.). *Soil Science and Plant Nutrition* 43, 855–863.

Roosta, H.R. and Karimi, H.R. (2012) Effects of alkali-stress on ungrafted and grafted cucumber plants: using two types of local squash as rootstocks. *Journal of Plant Nutrition* 35, 1843–1852.

Rouphael, Y., Cardarelli, M., Rea, E. and Colla, G. (2008) Grafting of cucumber as a means to minimize copper toxicity. *Environmental and Experimental Botany* 63, 49–58.

Rouphael, Y., Cardarelli, M., Schwarz, D., Franken, P. and Colla, G. (2012) Effects of drought on nutrient uptake and assimilation in vegetable crops. In: Aroca, R. (ed.) *Plant Responses to Drought Stress: From Morphological to Molecular Features*. Springer, Berlin/Heidelberg, pp. 171–195.

Rouphael, Y., Cardarelli, M. and Colla, G. (2015) Role of arbuscular mycorrhizal fungi in alleviating the adverse effects of acidity and aluminium toxicity in zucchini squash. *Scientia Horticulturae* 188, 97–105.

Rouphael, Y., Rea, E., Cardarelli, M., Bitterlich, M., Schwarz, D., Colla, G. (2016) Can adverse effects of acidity and aluminum toxicity be alleviated by appropriate rootstock selection in cucumber? *Frontiers in Plant Science* 7, 1283.

Ruiz, J.M., Belakbir, A., López-Cantarero, I. and Romero, L. (1997) Leaf-macronutrient content and yield in grafted melon plants. A model to evaluate the influence of rootstock genotype. *Scientia Horticulturae* 71, 227–234.

Salehi, R., Kashi, A., Lee, J.M., Babalar, M., Delshad, M., Lee, S.G. and Huh, Y.C. (2010) Leaf gas exchanges and mineral ion composition in xylem sap of Iranian melon affected by rootstocks and training methods. *HortScience* 45, 766–770.

Salinas, J. (2002) Molecular mechanisms of signal transduction in cold acclimation. In: Scheel, D. and Wasternack, C. (eds) *Plant Signal Transduction*. Oxford University Press, London, pp. 116–139.

Sánchez-Rodriguez, E., Romero, L. and Ruiz, J.M. (2013) Role of grafting in resistance to water stress in tomato plants: ammonia production and assimilation. *Journal of Plant Growth Regulation* 32, 831–842.

Sánchez-Rodriguez, E., Leyva, R., Constán-Aguilar, C., Romero, L. and Ruiz, J.M. (2014) How does grafting affect the ionome of cherry tomato plants under water stress? *Soil Science and Soil Nutrition* 2, 145–155.

Sánchez-Rodriguez, E., Romero, L. and Ruiz, J.M. (2016) Accumulation of free polyamines enhances the antioxidant response in fruits of grafted tomato plants under water stress. *Journal of Plant Physiology* 190, 72–78.

Santa-Cruz, M.M., Martinez-Rodriguez, F., Perez-Alfocea, R., Romero-Aranda, R. and Bolarin, M.C. (2002) The rootstock effect on the tomato salinity response depends on the shoot genotype. *Plant Science* 162, 825–831.

Savvas, D., Papastavrou, D., Ntatsi, G., Ropokis, A., Olympios, C., Hartmann, H. and Schwarz, D. (2009) Interactive effects of grafting and manganese supply level on growth, yield and nutrient uptake by tomato. *HortScience* 44, 1978–1982.

Savvas, D., Colla, G., Rouphael, Y. and Schwarz, D. (2010) Amelioration of heavy metal and nutrient stress in fruit vegetables by grafting. *Scientia Horticulturae* 127, 156–161.

Savvas, D., Ntatsi, G., Moiras, N., Tsakalidis, A., Ropokis, A. and Liopa-Tsakalidi, A. (2012) Impact of grafting and rootstock on the responses of cucumber to heavy metal stress. *Acta Horticulturae* 960, 49–56.

Savvas, D., Ntatsi, G. and Barouchas, P. (2013) Impact of grafting and rootstock genotype on cation uptake by cucumber (*Cucumis sativus* L.) exposed to Cd or Ni stress. *Scientia Horticulturae* 149, 86–96.

Schwarz, D., Rouphael, Y., Colla, G. and Venema, J.H. (2010) Grafting as a tool to improve tolerance of vegetables to abiotic stresses: thermal stress, water stress and organic pollutants. *Scientia Horticulturae* 127, 162–171.

Schwarz, D., Öztekin, G.B., Tüzel, Y., Brückner, B. and Krumbein, A. (2013) Rootstocks enhance tomato growth and quality characteristics at low potassium supply. *Scientia Horticulturae* 149, 70–79.

Seguel, A., Cumming, J.R., Klugh-Stewart, K., Cornejo, P. and Borie, F. (2013) The role of arbuscular mycorrhizas in decreasing aluminium phototoxicity in acidic soils: a review. *Mycorrhiza* 23, 167–183.

Seiler, G.J. (1998) Influence of temperature on primary and lateral root growth of sunflower seedlings. *Environmental and Experimental Botany* 40, 135–146.

Semillas Fitó (2016) Seed catalogue. Available at: http://www.semillasfitoworld.com/en/products/vegetable_seeds/rootstock/rootstock_tomato.htm (accessed 29 June 2016).

Semiz, G.D and Suarez, D.L. (2015) Tomato salt tolerance: impact of grafting and water composition on yield and ion relations. *Turkish Journal of Agriculture and Forestry* 39, 876–886.

Shao, H.-B., Chu, L.-Y., Ruan, C.-J., Li, H., Guo, D.-G. and Li, W.-X. (2010) Understanding molecular mechanisms for improving phytoremediation of heavy metal-contaminated soils. *Critical Reviews in Biotechnology* 30, 23–30.

Shibasaki, K., Uemura, M., Tsurumi, S. and Rahman, A. (2009) Auxin response in *Arabidopsis* under cold stress: underlying molecular mechanisms. *Plant Cell* 21, 3823–3838.

Si, Y., Dane, F., Rashotte, A., Kang, K. and Singh, N.K. (2010) Cloning and expression analysis of the *Ccrboh* gene encoding respiratory burst oxidase in *Citrullus colocynthis* and grafting onto *Citrullus lanatus* (watermelon). *Journal of Experimental Botany* 61, 1635–1642.

Silberbush, M. and Barber, S. (1983) Sensitivity of simulated phosphorus uptake to parameters used by a mechanistic-mathematical model. *Plant Soil* 74, 93–100.

Sivaguru, M., Liu, J. and Kochian, L.V. (2013) Targeted expression of SbMATE in the root distal transition zone is responsible for sorghum aluminum resistance. *The Plant Journal* 76, 297–307.

Slovak, R., Ogura, T., Satbhai, S.B., Ristova, D. and Busch, W. (2016) Genetic control of root growth: from genes to networks. *Annals of Botany* 117, 9–24.

Sorgonà, A., Abenavoli, M.R., Gringeri, P.G. and Cacco, G. (2006) A comparison of nitrogen use efficiency definitions in *Citrus* rootstocks. *Scientia Horticulturae* 109, 389–393.

Stazi, S., Cassaniti C., Marabottini R.,Giuffrida F. and Leonardi, C. (2016) Arsenic uptake and partitioning in grafted tomato plants. *Horticulture, Environment, and Biotechnology* 57, 241–247.

Sun, S., Yong-Sheng, T., Dan-Dan, L., Yu, W. and Guo-Ming, X. (2009) Effects of different rootstocks on chilling-tolerance of grafted watermelon seedlings. *Chinese Journal of Ecology* 28, 1561–1566.

Suzuki, N. and Mittler, R. (2006) Reactive oxygen species and temperature stresses: a delicate balance between signaling and destruction. *Physiologia Plantarum* 126, 1, 45–51.

Syngenta (2016) Arnold: Robust rootstock at: http://www3.syngenta.com/country/nl1/nl/Documents/SEFolders/TestimonialRootstockArnold.pdf (accessed 21 November 2016).

Tachibana, S. (1982) Comparison of effects of root temperature on the growth and mineral nutrition of cucumber and figleaf gourd. *Journal of the Japanese Society for Horticultural Science* 51, 299–308.

Takatsuka, H. and Umeda, M. (2014) Hormonal control of cell division and elongation along differentiation trajectories in roots. *Journal of Experimental Botany* 65, 10, 2633–2643.

Takii (2016) Vegetable rootstocks catalogue. Available at: http://www.takii.eu/ws/?tekst=Vegetables (accessed 29 June 2016).

Theocharis, A., Clement, C. and Barka, E.A. (2012) Physiological and molecular changes in plants grown at low temperatures. *Planta* 235, 1091–1105.

Top Seeds Ibérica (2016) Rootstock catalogue. Available at: http://www.topseedsiberica.com/productos/portainjertos (accessed 29 June 2016).

Troeh, F.R. and Thompson, L.M. (2005) *Soils and Soil Fertility*, 6th edn. Blackwell, Ames, Iowa.

Truco, M.J., Randall, L.B., Bloom, A.J. and St Clair, D.A. (2000) Detection of QTLs associated with shoot wilting and root ammonium uptake under chilling temperatures in an interspecific backcross population from *Lycopersicon esculentum* × *L. hirsutum*. *Theoretical and Applied Genetics* 101, 1082–1092.

Tsadilas, C.D. (1997) Soil contamination with boron due to irrigation with treated municipal waste water. In: Bell, R.W. and Berkasem, B. (eds) *Boron in Soil and Plants*. Kluwer Academic Publishers, Dordrecht, The Netherlands, pp. 265–270.

van der Ploeg, A. and Heuvelink, E. (2005) Influence of sub-optimal temperature on tomato growth and yield: a review. *Journal of Horticultural Science and Biotechnology* 80, 652–659.

Venema, J.H., Linger, P., van Heusden, A.W., van Hasselt, P.R. and Brüggemann, W. (2005) The inheritance of chilling tolerance in tomato (*Lycopersicon* spp.). *Plant Biology* 7, 118–130.

Venema, J.H., Dijk, B.E., Bax, J.M., van Hasselt, P.R. and Elzenga, J.T.M. (2008) Grafting tomato (*Solanum lycopersicum*) onto the rootstock of a high-altitude accession of *Solanum habrochaites* improves suboptimal-temperature tolerance. *Environmental and Experimental Botany* 63, 359–367.

Wachsman, G., Sparks, E.E. and Benfey, P.N. (2015) Genes and networks regulating root anatomy and architecture. *New Phytologist* 208, 26–38.

Wahb-Allah, M.A. (2014) Effectiveness of grafting for the improvement of salinity and drought tolerance in tomato (*Solanum lycopersicum*). *Asian Journal of Crop Science* 6, 112–122.

Wahid, A., Gelani, S., Ashraf, M. and Foolad, M. (2007) Heat tolerance in plants: an overview. *Environmental and Experimental Botany* 61, 199–223.

Wang, W., Vinocur, B. and Altman, A. (2003) Plant responses to drought, salinity and extreme temperatures: towards genetic engineering for stress tolerance. *Planta* 218, 1–14.

Wang, S., Yang, R., Cheng, J. and Zhao, J. (2007) Effect of rootstocks on the tolerance to high temperature of eggplants under solar greenhouse during summer season. *Acta Horticulturae* 761, 357–360.

Warschefsky, E.J., Klein, L.L., Frank, M.H., Chitwood, D.H., Londo, J.P., von Wettberg, E.J.B., et al. (2016) Rootstocks: diversity, domestication, and impacts on shoot phenotypes. *Trends in Plant Science* 21, 418–437.

Wei, G.P., Zhu, Y.L., Liu, Z.L., Yang, L.F. and Zhang, G.W. (2007) Growth and ionic distribution of grafted eggplant seedlings with NaCl stress. *Acta Botanica Boreal-Occident Sinica* 27, 1172–1178.

White, P.J. and Brown, P.H. (2010) Plant nutrition for sustainable development and global health. *Annals of Botany* 105, 1073–1080.

Wilkinson, S., Clephan, A.L. and Davies, W.J. (2001) Rapid low temperature-induced stomatal closure occurs in cold-tolerant *Commelina communis* leaves but not in cold-sensitive tobacco leaves, via a mechanism that involves apoplastic calcium but not abscisic acid. *Plant Physiology* 126, 1566–1578.

Wu, X., Yao, X., Chen, J., Zhu, Z., Zhang, H. and Zha, D. (2014) Brassinosteroids protect photosynthesis and antioxidant system of eggplant seedlings from high-temperature stress, *Acta Physiologiae Plantarum* 36, 251–261.

Yamaguchi, H., Fukuoka, H., Arao, T., Ohyama, A., Nunome, T., Miyatake, K. and Negoro, S. (2010) Gene expression analysis in cadmium-stressed roots of a low cadmium accumulating solanaceous plant, *Solanum torvum*. *Journal of Experimental Botany* 61, 423–437.

Yang, M., Zhang, W., Dong, H., Zhang, Y., Lv, K., Wang, D. and Lian, X. (2013) OsNRAMP3 is a vascular bundles-specific manganese transporter that is responsible for manganese distribution in rice. *PLoS ONE* 8, e83990.

Yau, S.K. and Ryan, J. (2008) Boron toxicity tolerance in crops: a viable alternative to soil amelioration. *Crop Science* 48, 854–865.

Yetisir, H. and Uygur, V. (2010) Response of grafted watermelon onto different gourd species to salinity stress. *Journal of Plant Nutrition* 33, 315–327.

Yetisir, H., Caliskan, M.E., Soylu, S. and Sakar, M. (2006) Some physiological and growth responses of watermelon [*Citrullus lanatus* (Thunb.) Matsum. And Nakai] grafted onto *Lagenaria siceraria* to flooding. *Environmental and Experimental Bot*any 58, 1–8.

Zamir, D. and Gadish, I. (1987) Pollen selection for low temperature adaptation in tomato. *Theoretical and Applied Genetics* 74, 545–548.

Zhang, G.W., Liu, Z.L., Zhou, J.G. and Zhu, Y.L. (2008a) Effects of $Ca(NO_3)_2$ stress on oxidative damage, antioxidant enzymes activities and polyamine contents in roots of grafted and non-grafted tomato plants. *Plant Growth Regulation* 56, 7–19.

Zhang, S.P., Gu, X.F. and Wang, Y. (2008b) Effect of low temperature stress on the physiological and biochemical indexes in cucumber seedling grafted on bur cucumber (*Sicyos angulatus* L.). *Acta Horticulturae* 771, 243–247.

Zhang, Z.-K., Li, H., Zhag, Y., Huang, Z.-J., Chen, K. and Liu, S.-Q. (2010) Grafting enhances copper tolerance of cucumber through regulating nutrient uptake and antioxidative system. *Agricultural Sciences in China* 9, 1758–1770.

Zhen, A., Bie, Z., Huang, Y, Liu, Z. and Li, Q. (2010) Effects of scion and rootstock genotypes on the anti-oxidant defense systems of grafted cucumber seedlings under NaCl stress. *Soil Science and Plant Nutrition* 56, 263–271.

Zhou, Y.H., Yu, J.Q., Huang, L.F. and Nogues, S. (2004) The relationship between CO_2 assimilation, photosynthetic electron transport and water–water cycle in chill-exposed cucumber leaves under low light and subsequent recovery. *Plant, Cell and Environment* 27, 1503–1514.

Zhou, Y.H., Huang, L.F., Zhang, Y., Shi, K., Yu, J.Q. and Nogues, S. (2007) Chill-induced decrease in capacity of RuBP carboxylation and associated H_2O_2 accumulation in cucumber leaves are alleviated by grafting onto figleaf gourd. *Annals of Botany* 100, 839–848.

Zhou, Y., Zhou, J., Huang, L., Ding, X., Shi, K. and Yu, J. (2009) Grafting of *Cucumis sativus* onto *Cucurbita ficifolia* leads to improved plant growth, increased light utilization and reduced accumulation of reactive oxygen species in chilled plants. *Journal of Plant Research* 122, 529–540.

Zhu, S.N. and Guo, S.R. (2009) Effects of grafting on K^+, Na^+ contents and distribution of watermelon (*Citrullus vulgaris* Schrad.) seedlings under NaCl stress. *Acta Horticulturae Sinica* 36, 814–820.

Zhu, J., Bie, Z., Huang, Y. and Han, X. (2008) Effect of grafting on the growth and ion concentrations of cucumber seedlings under NaCl stress. *Soil Science and Plant Nutrition* 54, 895–902.

Zhu, J., Zhang, K.-X., Wang, W.-S., Gong, W., Liu, W.-C., Chen, H.-G., Xu, H.-H. and Lu, Y.-T. (2015) Low temperature inhibits root growth by reducing auxin accumulation via *ARR1/12*. *Plant and Cell Physiology* 56, 727–736.

Ziaf, K., Loukehaich, R., Gong, P.J., Liu, H., Han, Q., Wang, T., Li, H. and Ye, Z. (2011) A multiple stress-responsive gene *ERD15* from *Solanum pennellii* confers stress tolerance in tobacco. *Plant and Cell Physiology* 52, 1055–1067.

Zijlstra, S. and den Nijs, A.P.M. (1987) Effects of root systems of tomato genotypes on growth and earliness, studied in grafting experiments at low temperature. *Euphytica* 36, 693–700.

8 Quality of Grafted Vegetables

Cherubino Leonardi,[1*] Marios C. Kyriacou,[2] Carmina Gisbert,[3] Golgen B. Oztekin,[4] Isabel Mourão[5] and Youssef Rouphael[6]

[1]*Università degli Studi di Catania, Catania, Italy;* [2]*Agricultural Research Institute, Nicosia, Cyprus;* [3]*Universitat Politècnica de València, Valencia, Spain;* [4]*Ege University, Bornova-Izmir, Turkey;* [5]*Instituto Politécnico Viana do Castelo, Ponte de Lima, Portugal;* [6]*University of Naples Federico II, Portici, Italy*

8.1 What is Quality?

The quality of fruit and vegetables is commonly defined as 'the set of characteristics of a product able to satisfy the needs of the consumer and to determine its value' (Peri, 2004). Objective qualitative criteria for fresh vegetables are established in Europe and North and Central America. These standards are quality features that dictate the market value of vegetables; they take into consideration external quality attributes, such as size, shape and colour, as well as freshness and the absence of defects and decay (Schnitzler and Gruda, 2002). Critical concentrations of pesticides in plant tissues are not included in the same standards, and important internal quality attributes, such as texture, flavour and health-promoting compounds, are also not considered. In addition, flavour compounds such as soluble carbohydrates, aroma volatiles and organic acids are not taken into account, although these can promote the saleability of the product. Recently, consumer demand for these quality parameters has greatly increased, together with requirements for a higher content of minerals and bioactive compounds (Schnitzler and Gruda, 2002; Gruda, 2005).

A distinction can be made between 'structural and functional quality': the first refers to the objective intrinsic characteristics of a product (e.g. sugar content of the fruit), while the second concerns the perception of these features by the user of the product (e.g. sweet taste of the fruit) (Huyskens-Keil and Schreiner 2003; Gruda, 2005). Indeed, from a product perspective, quality attributes are inherent and can be quantified objectively, whereas from a consumer perspective,

* cherubino.leonardi@unict.it

which defines quality in terms of user satisfaction, quality is a much less tangible and less quantifiable concept (Shewfelt, 1999).

With reference to vegetables produced by grafted crops, quality is a broad and complex issue, involving different target customer groups. The question of quality should therefore be addressed with respect to all factors that contribute to its manifestation. Key intermediaries in the management and standardization of quality are the farm operator, the trader/distributor and the consumer, whose views are sometimes conflicting. However, the consumer remains the economic subject who is more involved in the perception of quality; the fulfilment of consumer demands is therefore the essential aspect in a production system that from market-oriented has become consumer-oriented. For the consumer, in addition to traits common to all intermediaries of the chain (e.g. external appearance, integrity), the concept of quality involves factors resulting from sensory stimuli (e.g. taste, touch) and from expectations (real or virtual) associated with healthiness and the contents of health-promoting substances.

In this chapter, we will review the recent and relevant literature about the effects of grafting on vegetable fruit quality. We will focus on product-oriented quality and objective criteria. The effects of grafting on the main qualitative traits will be presented for the most commonly grafted crops: watermelon (*Citrullus lanatus* (Thunb.) Matsum. & Nakai), melon (*Cucumis melo* L.), cucumber (*Cucumis sativus* L.), tomato (*Solanum lycopersicum* L.), aubergine (*Solanum melongena* L.) and pepper (*Capsicum annuum* L.). Finally, possible mechanisms involved in the effects of grafting on fruit quality will be discussed. The chapter will conclude by identifying the prospects for future research.

8.2 Rootstock Effects on Fruit Quality

8.2.1 Appearance

Fruit size

Fruit size is studied both because it is one of the most frequently respected parameters by consumers and because it is correlated with yield. Provided there is no incompatibility, significant yield increases as a result of grafting have been widely reported for numerous rootstock–scion combinations (Yetisir and Sari, 2003; Yetisir *et al.*, 2003; Huitrón *et al.*, 2007; Cushman and Huan, 2008). Augmented yield reflects enhanced vigour imparted by the rootstock, notwithstanding actual yield variation across seasons and for different climatic and soil conditions (Maynard, 2001; Goreta *et al.*, 2005; Huitrón *et al.*, 2007; Huitrón-Ramírez *et al.*, 2009). More frequently, the effect on yield level is related to the variation in fruit mass, as grafted vegetable plants are characterized by a vigorous root system able to enhance water and nutrient uptake, photosynthesis and consequently productivity (Colla *et al.*, 2006a,b). However, in some cases, increased yield has been attained by an increase in the number of fruits rather than an increase in mean fruit mass (Savvas *et al.*, 2011).

In watermelon, the increase in yield as a result of grafting may be expressed as the increase in fruit mass, which is the main tendency for large-fruited scions, or as the increase in the number of fruits per plant, which is the main trend observed in small-fruited scions (Colla *et al.*, 2006b; Alan *et al.*, 2007; Alexopoulos *et al.*, 2007; Cushman and Huan, 2008; Proietti *et al.*, 2008). In these studies, the reported change in fruit mass as a result of grafting ranged from the absence of effect to an increase ranging from 8 to 55%. A significant decrease in fruit mass usually reflects rootstock–scion incompatibility.

Increased mass of melon fruit resulting in higher yield was reported for *Cucumis melo* var. *inodorus* H. Jacq. 'Incas' grafted on to 'RS841' rootstock. Grafting 'Piel de Sapo' on this same rootstock, Fita *et al.* (2007) reported a similar fruit mass as the control but higher numbers of fruit per plant. An increase in fruit mass of 10% was also reported by Verzera *et al.* (2014) when 'RS841' was used as the rootstock. However, in this study, a similar fruit mass as the control was observed using other tested rootstocks: *Cucurbita* hybrids 'P360' and 'AS10' and the *C. melo* rootstocks 'Sting' and 'Eloisa'. Similarly, Traka-Mavrona *et al.* (2000) found no differences in this parameter with respect to the non-grafted plants for 'Kokkini Banana' and 'Thraki' melon when grafted on to the *Cucurbita* hybrids 'TZ-148' or 'Mamouth'.

In cucumber, Goreta Ban *et al.* (2014) obtained greater fruit mass when plants were grafted on to interspecific hybrid (*Cucurbita maxima* Duch. × *Cucurbita moschata* Duch.) rootstocks 'RS841' and 'Strong Tosa' than when grafted on to courgette (*Cucurbita pepo* L.) rootstocks 'Emphasis' and 'Friend', or in the non-grafted plants. Colla *et al.* (2012, 2013) demonstrated a higher marketable yield in cucumber (*C. sativus*) 'Akito' and 'Ekron' grafted on to three commercial rootstocks, 'PS1313', 'P360' (*C. maxima* × *C. moschata*) and 'Affyne' (*C. sativus*). In the former experiment, the higher marketable yield was due to a larger fruit size, not due to a higher number of fruits per plant.

Fruit mass has likewise been reported to increase significantly by grafting tomato plants compared with the non-grafted plants (Lee, 1994; Passam *et al.*, 2005; Pogonyi *et al.*, 2005; Khah *et al.*, 2006; Theodoropoulou *et al.*, 2007; Tüzel *et al.*, 2009; Turhan *et al.*, 2011; Moncada *et al.*, 2013). For instance, Pogonyi *et al.* (2005) reported that when 'Lemance F_1' was grafted on to 'Beaufort' rootstock, the higher yield was due mainly to higher fruit mass, which is in agreement with the results of Theodoropoulou *et al.* (2007) with 'Star Fighter' grafted on to the same rootstock. However, the effectiveness of grafting on fruit mass or size depends on grafting combinations (Oda *et al.*, 1996; Romano *et al.*, 2000; Leonardi and Giuffrida, 2006; Khah *et al.*, 2006; Schwarz *et al.*, 2013). Accordingly, the rootstocks 'Brigeor' and 'Maxifort' had no effect on the fruit size of cherry tomato 'Classy', whereas increased fruit mass was observed when 'Piccolino' was grafted on to the same rootstocks (Schwarz *et al.*, 2013). Romano *et al.* (2000) found a higher fruit mass when tomato plants ('Rita') were grafted on to rootstocks 'Beaufort', 'Heman', 'Joint', 'P1614' and 'RS1427', whereas other rootstocks, namely 'Energy', 'Firefly', 'Linea9243' and 'Nico', induced a lower fruit mass compared with non-grafted plants.

Similar fruit mass and total number of fruits per plant were observed in aubergine by Gisbert *et al.* (2011b) when 'Black Beauty' was grafted on to *Solanum torvum* Sw., *Solanum incanum* L. × *S. melongena* and *S. melongena* × *Solanum aethiopicum* L. rootstocks. A decrease in yield as a consequence of a lower number of fruits per plant was reported in aubergine 'Mission Bell' F_1 when grafted on to the tomato rootstocks 'PG3', Beaufort' and 'Energy' (Leonardi and Giuffrida, 2006).

For pepper, Gisbert *et al.* (2010) reported a reduction in fruit mass of two commercial cultivars, 'Coyote' and 'Almuden', that were grafted on to the pepper hybrids 'Charlot' and 'Foc'. The reduction in mean fruit mass in 'Coyote'- and 'Almuden'-grafted plants was 11 and 21%, respectively, when compared with fruit from non-grafted plants. Leal-Fernández *et al.* (2013) carried out an experiment on sweet pepper 'Triple Star', either non-grafted or grafted on to the chili pepper 'CM334', and also on to the commercial rootstocks 'Tresor', 'AR96030' 'AR96023', AR96028' and 'AR96029'. The authors concluded that the grafting combination 'Triple Star'–'AR96029' (scion–rootstock) exhibited the highest fruit mean mass in comparison with non-grafted plants. Similarly, two additional studies indicated that grafting can increase fruit size in pepper (Jang *et al.*, 2012; Doñas-Uclés *et al.*, 2014).

Fruit shape

Fruit shape index, defined as the ratio of the two diameters (i.e. equatorial and longitudinal), is a characteristic potentially affected by grafting. However, despite the significant effect of grafting on watermelon fruit mass, its effect on fruit shape is usually non-significant or circumstantial and minimal (Colla *et al.*, 2006b, 2011; Alan *et al.*, 2007; Rouphael *et al.*, 2008; Soteriou and Kyriacou, 2015). This reflects the strong genotypic effect of the scion on this morphological characteristic. In melon, it was also reported that the fruit shape index was unaffected by grafting combinations (Colla *et al.*, 2006a). In contrast, Colla *et al.* (2013) observed that the shape index in grafted cucumber plants ('Ekron'–'Affyne' and 'Ekron'–'P360') was higher by 4% compared with non-grafted plants ('Ekron').

Turhan *et al.* (2011) also reported that the shape index of tomato fruit from non-grafted plants ('Yeni Talya', 'Swanson' and 'Beril') was significantly lower than that of fruits obtained from plants grafted on to 'Beaufort' and 'Arnold' rootstocks. However this result is not applicable to several tomato fruit genotypes and typologies due to the significant genotypic effect on fruit shape. In aubergine, some rootstock–scion combinations affect fruit shape index. For instance, Gisbert *et al.* (2011a) found a greater length in 'Cristal F_1' fruits from plants grafted on to *S. incanum* × *S. melongena* and in plants grafted on to two *S. melongena* rootstocks. However, fruit shape index was only modified in fruits from plants grafted on to *S. incanum* × *S. melongena* and on to *S. incanum*. No significant difference for this parameter was observed when this interspecific hybrid was used as the rootstock for 'Black Beauty' (Gisbert *et al.* 2011b). Grafting aubergine 'Black Bell' on to *S. torvum* also resulted in longer fruits (+4%) than fruit from self-grafted plants (Cassaniti *et al.*, 2011).

Colour

Grafting may affect watermelon pulp colour as much as it affects lycopene content, although limited discrepancy between the two variables has been noted in certain graft combinations (Davis and Perkins-Veazie, 2005). Recent work has demonstrated that ripening events underlying colour development are delayed as a result of grafting watermelon on to interspecific *Cucurbita* hybrids. Red colour

(a^*) peaked later in grafted plants, concomitant with a peak in lycopene content, and was higher than in non-grafted plants (Soteriou *et al.*, 2014).

Several studies have reported colour variations as a consequence of grafting melons on to *Cucurbita* interspecific hybrids. Thus, modified colour (CIELAB values of brightness (L^*) and the redness/yellowness ratio (a^*/b^*) was found in melon fruit ('Cyrano'; *Cucumis melo* var. *cantaloupensis* Naudin) from plants grafted on to 'P360' compared with those of plants grown with their own roots (Colla *et al.*, 2006a). Concerning cucumber, Colla *et al.* (2012) reported that the Hunter colour values, in particular L^* and a^*/b^*ratio, measured on the skin colour of cucumber were higher by 2% and lower by 5%, respectively, in grafted plants compared with non-grafted plants grown under protected cultivation conditions.

According to studies conducted by Oztekin (2009) and Schwarz *et al.* (2013), tomato fruit colour parameters were not affected by the scion–rootstock combinations. However, other researchers indicated that the skin colour of tomato fruits showed variation with grafting. Ulukapi and Onus (2005) reported that the colour parameter L^* did not change between grafted and non-grafted plants, but the a^* colour component was significantly lower in grafted tomato fruits compared with non-grafted plants grown under glasshouse conditions in autumn.

In aubergine ('Birgah') grafted on to *S. torvum*, Moncada *et al.* (2013) found that the fruits were darker and with a less vivid colour with respect to fruit from non-grafted plants. However, in other studies that used *S. torvum* as the rootstock, this effect was not observed (Cassaniti *et al.*, 2011; Gisbert *et al.* 2011b). This difference between results may be due to the difficulty of distinguishing an objective ripening stage in aubergine.

Exocarp characteristics

The effect of grafting on exocarp characteristics has been considered mainly on watermelon fruit, as rind thickness in this produce can be considered a characteristic of commercial interest wherein significant effects in response to grafting have been observed. Thickening of the rind was observed in response to grafting on to *C. maxima* × *C. moschata* and *Lagenaria siceraria* (Molina) Standl. commercial rootstocks (Alexopoulos *et al.*, 2007; Proietti *et al.*, 2008; Kyriacou and Soteriou, 2015), and more so on *C. maxima* and *C. moschata* landraces (Yetisir *et al.*, 2003). However, this effect was not observed with *C. pepo* (squash) rootstock (Davis and Perkins-Veazie, 2005), nor has it been universal with *C. maxima* × *C. moschata* and *L. siceraria* rootstocks (Alan *et al.*, 2007). No difference in scion fruit rind thickness was observed between *C. maxima* × *C. moschata*, *L. siceraria*, *C. moschata*, *Sicyos angulatus* L. and *Citrullus lanatus* var. *citroides* (L.H. Bailey) Mansf. rootstocks (Huitrón *et al.*, 2007). Although grafting indeed presents a general tendency to thicken the rind of watermelon fruit, this effect is usually rather minimal and difficult to assess, as rind thickness along the circumference of the fruit is irregular; moreover, where the effect of grafting was examined in several rootstock–scion combinations, the rootstock effect on rind thickness was minimal compared with the effect of the scion cultivar (Kyriacou and Soteriou, 2015).

For other grafted vegetables, pericarp thickness in pepper was also affected by grafting, where the values recorded in grafted plants 'Palermo/AR40' was significantly higher by 5% than those recorded for non-grafted plants (Doñas-Uclés *et al.*, 2014).

Absence of defects

Defects and abnormalities that impair fruit quality have occasionally been associated with grafting. In watermelon, these most commonly include the formation of yellow bands in the flesh bordering the rind, internal cracking of the flesh known as hollow heart, and discoloured and abnormally compact pith (Ryu *et al.*, 1973; Lee, 1994; Yamasaki *et al.*, 1994; Davis *et al.*, 2008a; Soteriou and Kyriacou, 2015). The association of these defects with grafting has been circumstantial and probably influenced by adverse environmental conditions or rootstock–scion incompatibility. Most reports on grafting-related effects on watermelon quality, particularly in relation to *C. maxima* × *C. moschata* and *L. siceraria* commercial rootstocks, make no mention of apparent defects or abnormalities.

In melon, vitrescence of fruits has been reported as a consequence of grafting (Rouphael *et al.* 2010). In the work of Jang *et al.* (2014), vitrescence was observed in 89% of melon ('Homerunstar'; *inodorus* group) from plants grafted on to 'Shintosa' whereas 50% of fruit from non-grafted plants showed this physiological disorder. Similar or lower percentages of vitrescence were observed in fruits from plants grafted on to experimental *Cucumis* rootstocks.

The incidence of blossom end-rot (BER) with grafting in tomato is related to rootstock genotype and rootstock–scion combination. Krumbein and Schwarz (2013) reported that the rootstocks 'Brigeor' and 'Maxifort' were able to significantly reduce the incidence of BER for 'Classy' at two light levels, but for 'Piccolino' the reduction of BER was observed only at 100% light conditions. In addition, for 'Piccolino' grafted on to 'Maxifort' rootstock at 100% light, the effect on BER was not significant under low potassium (K) supply, despite the antagonistic effect between cations, which could increase the uptake and mobility of calcium (Ca). Takasu *et al.* (1996) claimed that BER increased with grafting and that the effect was greater in hydroponics than in soil culture, which could be due to the faster plant growth in soil-less culture systems. Similarly, Oda *et al.* (1996) suggested that the incidence of BER increased if tomato was grafted on to aubergine rootstock, due to a lower transport of Ca. It can be concluded that the incidence of BER is reduced by grafting when Ca uptake and transport by rootstock is improved, which increases fruit Ca concentration and consequently firmness, and leads to strengthening of cell walls and cellular integrity. However, the alleviation is dependent on the rootstock–scion combination.

8.2.2 Texture

Firmness is one of the typical attributes used to describe fruit texture (*see* Plate 8). Fruit firmness is related to Ca concentration, but more complicated mechanisms, such as water relations, transpiration and wax layers, cell-to-cell adhesion, cell-wall architecture and solubilization, as well as cell-wall protein status, have been shown to contribute to fruit firmness (Huxham *et al.*, 1999; Saladie *et al.*, 2007).

The relative contributions of the various mechanisms combined with grafting are not well understood and merit further investigation.

Grafting can influence firmness in a significant way. Besides the genotypic effect of the scion, pulp firmness also appears to be highly influenced by the type of rootstock used (Yetisir *et al.*, 2003; Cushman and Huan, 2008; Bruton *et al.*, 2009). Exploitation of suitable rootstocks may therefore prove a valuable tool for improving this quality trait in the field. Enhanced pulp firmness may additionally predispose the fruit to a longer postharvest life (Kyriacou and Soteriou, 2015).

Interspecific (*C. maxima* × *C. moschata*) hybrid rootstocks most consistently tend to increase watermelon pulp firmness (Yamasaki *et al.*, 1994; Davis *et al.*, 2008a; Bruton *et al.*, 2009; Huitrón-Ramírez *et al.*, 2009; Soteriou *et al.*, 2014; Soteriou and Kyriacou, 2015). This effect has been shown to prevail even on mini triploid cultivars (e.g. 'Extazy') characterized by an inherently outstanding firmness, in which case it may be considered undesirable (Soteriou and Kyriacou, 2015). Reported increases in the scion's pulp firmness on various commercial *C. maxima* × *C. moschata* rootstocks range from 7 to 36% and from 14 to 55% for diploid and triploid scions, respectively (Huitrón-Ramírez *et al.*, 2009; Soteriou and Kyriacou, 2015). Increases in watermelon pulp firmness have also been reported for *Cucurbita* hybrid rootstocks and for the less common *Cucurbita ficifolia* Bouché, whereas *L. siceraria* rootstocks have rarely been reported to increase firmness, as they usually have no effect on this quality trait (Yetisir *et al.*, 2003; Cushman and Huan, 2008; Bruton *et al.*, 2009). A decrease in firmness as a result of grafting, or rootstock × scion interaction, has been reported mainly in relation to gourd (*L. siceraria*) and pumpkin (*Cucurbita argyrosperma* Huber) rootstocks (Davis and Perkins-Veazie, 2005).

In melon, fruit firmness was modified as a consequence of grafting for some rootstock–scion combinations; for example, firmness increased in 'Cyrano' fruit when grafted on to 'P360' (by 19–32%) (Colla *et al.*, 2006a) and decreased in fruit of the *inodorus* melon 'Homerunstar' grafted on to 'Shintosa' (Jang *et al.* 2014). Similar firmness was observed by Crinó *et al.* (2007) in fruit of 'Incas' non-grafted and grafted on to *Cucurbita* hybrids ('RS841', 'P360', 'ES 99-13' and 'Elsi') or *C. melo* rootstocks ('Belimo', 'Energia', 'Griffin' and 'ES liscio'). Zhao *et al.* (2011) reported that, although grafting led to decreased fruit firmness, 'Strong Tosa' caused less reduction than 'Tetsukabuto' rootstock.

In cucumber, Hwang *et al.* (1992) demonstrated that fruits from plants grafted on to 'Andong' (*S. angulatus*) tended to be firmer than those from cucumbers grafted on to 'Heukjong' (*C. ficifolia*). Morishita (2001) reported that the flesh of 'Kema' or 'Kifujin New Type' cucumbers grafted on to the bloomless rootstock 'Big Ben Kitora' became softer than the flesh of non-grafted cucumbers. The same author also reported that there was no difference in firmness between 'Encore' fruits from 'Shintosa'-grafted plants and those from bloomless rootstock 'Big Ben Kitora'-grafted plants. However, Jang *et al.* (2012) and Colla *et al.* (2013) indicated that fruit firmness in cucumber and pepper remained unchanged in both grafted and non-grafted plants.

Fruit firmness seems not to be influenced by grafting in tomato. Fruits obtained from plants grafted on to 'Heman' and 'Primavera' were not firmer than the fruits from self- or non-grafted tomato 'Big Red', even if fruit Ca content was increased by rootstocks (Khah *et al.*, 2006). Ulukapi and Onus (2005) and Schwarz *et al.* (2013) observed similar results in grafted and non-grafted plants.

Flesh firmness was negatively affected by grafting aubergine ('Tsakoniki') on to *S. torvum* and *Solanum sisymbriifolium* Lam. (Arvanitoyannis *et al.*, 2005). A reduction in firmness (−13%) was also reported in the fruit of aubergine 'Black Bell' grafted on to *S. torvum* (Cassaniti *et al.*, 2011).

8.2.3 Organoleptic compounds and relationship to sensory properties

Sugar content and acidity

Perception of vegetable flavour is a composite of sensory responses in the nose and mouth to aroma and taste (sweetness, acidity, bitterness, pungency or astringency). Acids, sugars, volatiles and other compounds all contribute to flavour, although flavour quality is largely determined by the sugar and acid composition of the fruit. The high contents of sugars and acids are considered parameters of good taste and flavour (Kader, 2008). Flavour and aroma compounds can be affected by grafting vegetables and also by the type of rootstock used (Davis *et al.*, 2008a; Rouphael *et al.*, 2010). Some of these compounds (e.g. sugars, acids, water-soluble pectins, amino acids, pigments) are expressed as total soluble solids (TSS); therefore TSS, expressed as degrees Brix (°Bx), associated with the refractive index of the fruit juice, is commonly considered when analyses of quality are performed. Acidity is evaluated as titratable acid content, or titratable acidity (TA), expressed as the percentage equivalent of the predominant acid (usually citric or malic) in the juice or simply by pH determination.

In the case of watermelon, consumers' occasional impression of reduced sweetness as a result of grafting mostly derives from the negative connotation for sweetness attached to 'pumpkin rootstocks' (Soteriou and Kyriacou, 2015). Reduced TSS have been actually reported for diploid watermelon scions grafted on to cushaw pumpkin (*C. argyrosperma*) and squash (*C. pepo*) rootstocks (Davis and Perkins-Veazie, 2005). The more common types of commercially used rootstocks, *C. maxima* × *C. moschata* and *L. siceraria*, generally do not to seem to affect the TSS of the fruit (Yetisir *et al.*, 2003; Colla *et al.*, 2006b; Davis *et al.*, 2008a; Proietti *et al.*, 2008; Rouphael *et al.*, 2008; Huitrón-Ramírez *et al.*, 2009; Soteriou and Kyriacou, 2015). Exceptionally, depression of the TSS by these rootstocks has been reported, but its magnitude was very limited and did not compromise fruit quality (Alan *et al.*, 2007; Alexopoulos *et al.*, 2007; Cushman and Huan, 2008). Concerning specific sugars, while no significant effect of *C. maxima* × *C. moschata* and *L. siceraria* rootstocks on watermelon fruit glucose, fructose or sucrose content has been identified in some studies (Colla *et al.*, 2006b; Proietti *et al.*, 2008; Soteriou *et al.*, 2014), other studies have reported a limited decrease in total sugars as a result of reduced hexose accumulation during initial fruit development and reduced sucrose accumulation near commercial maturity (Miguel *et al.*, 2004; Liu *et al.*, 2006; Kyriacou and Soteriou, 2015).

Watermelon is a fruit of rather insipid TA, in the range of 0.7–1.2 g l^{-1} malic acid equivalents and pH 5.5–5.8, depending on harvest maturity and shelf life (Kyriacou and Soteriou, 2012; Candir *et al.*, 2013; Soteriou *et al.*, 2014).

Watermelon flavour, however, can be influenced by even limited changes in acidity, resulting from the ripening process, grafting and postharvest storage (Soteriou *et al.*, 2014). The TA of watermelon pulp declines and its pH increases linearly with ripening, up to about 45 days after anthesis, while grafting on *C. maxima* × *C. moschata* increases the TA and reduces the pH of the pulp (Colla *et al.*, 2006b; Proietti *et al.*, 2008; Soteriou *et al.*, 2014). Thus, the sensation of lower sweetness, sometimes perceived in grafted watermelon, may derive in part from elevated acidity. Increased acidity in response to grafting may be an additional indication of retarded maturation in grafted fruit, although, unlike sugar content, acidity declines linearly with ripening and appears steadily higher in grafted watermelon throughout the ripening period (Soteriou *et al.*, 2014).

Similar TSS concentrations have been obtained in melon fruit from grafted and non-grafted melon plants when using different interspecific hybrids of *Cucurbita* (Verzera *et al.*, 2014). However, lower TSS were found in some combinations. For instance, in melon fruits of 'Cyrano' (*C. melo* var. *cantaloupensis*) grafted on to 'P360', a decrease in respect to non-grafted plants was observed (Colla *et al.*, 2006a). Sugar content was greatly diminished (fruits presented 2–3 Bx) when *Cucurbita* spp. were used as the rootstock of 'Earl's Favorite' melon. In contrast, higher TSS content was obtained with respect to non-grafted plants in melon 'Homerunstar' (*C. melo* var. *inodorus*) grafted on to 'Shintosa' (Jang *et al.*, 2014), as well as in melon 'Incas' (*C. melo* var. *inodorus*) grafted on to 'AS10' (Verzera *et al.*, 2014). When *C. melo* rootstocks ('Energia' and 'Sting') were used for grafting the melon 'Incas', similar TSS were obtained (Verzera *et al.*, 2014). However, when melon 'Proteo' was grafted on to *Benincasa hispida* or *Cucumis metuliferus*, lower TSS were measured in fruits from grafted plants compared with those of non-grafted fruits. Recently, Liu *et al.* (2015) reported that melon fruit ('Elizabeth' from plants grafted on to *C. pustulatus* Hook. f. had similar TSS and TA to fruit from self-grafted plants. Both parameters varied for different rootstocks depending on the growing season (Jang *et al.*, 2014). For the same rootstock, 'P360', juice pH and acidity of melons grafted on to this pumpkin interspecific hybrid was reported to be lower than those from non-grafted 'Cyrano' (Colla *et al.*, 2006a), whereas similar TA was observed with 'Incas' by Verzera *et al.* (2014). In the latter study, higher TA was obtained in fruit from plants grafted on to the rootstocks 'Sting' (*C. melo*) and 'AS10' (*Cuburbita* hybrid).

Significant decreases in TSS content and fructose concentration were obtained in cucumber fruits collected from many farmers in cucumber-growing areas using different rootstocks. However, Lee *et al.* (1999) proposed the use of other rootstocks such as 'Andong' (*S. angulatus*) to minimize the reduction in flavour compounds. Huang *et al.* (2009) showed that, under saline conditions (60 mM NaCl), TSS and TA were significantly higher in cucumber fruits of plants grafted on to 'Figleaf Gourd' (*C. ficifolia*) and 'Chaofeng Kangshengwang' (*L. siceraria*), compared with self-grafted ones. Similarly, Colla *et al.* (2012) demonstrated that the TSS in the cucumber fruits of grafted plants were significantly higher (7%) in comparison with those from non-grafted plants, whereas the opposite was observed for TA with a decrease of 5% in fruits harvested from grafted plants.

Results concerning the variation in tomato fruit TSS and TA are contradictory. Some workers have reported that fruit quality characteristics were not affected by grafting or the choice of rootstock (Khah *et al.*, 2006; Savvas *et al.*, 2011). Mourão *et al.* (2013) with 'Valoásis M40' grafted on to rootstock 'Maxifort' reported similar soluble solids content (SSC) and TA in the fruits of two- and three-stemmed grafted plants, in spring season crops grown in the greenhouse. Other authors found that grafting, for some scion–rootstock combinations, can increase or not affect TA and decrease TSS (Turhan *et al.*, 2011; Nicoletto *et al.*, 2013; Schwarz *et al.*, 2013). The lower SSC of grafted fruits is in agreement with other studies (Lee, 1994; Fernandez-Garcia *et al.*, 2004; Pogonyi *et al.*, 2005; Balibrea *et al.*, 2006; Alexopoulos *et al.*, 2007; Qaryouti *et al.*, 2007; Martínez-Ballesta *et al.*, 2010; Turhan *et al.*, 2011; Nicoletto *et al.*, 2013). When tomato plants were grafted on to scarlet aubergine rootstocks, the increased SSC was suggested to be due to the effect of water deficiency in the shoot, which caused lower plant growth and yield and decreased fruit water content (Oda *et al.*, 1996).

Grafting aubergine 'Faselis' on to *S. torvum* reduced the TSS. However, similar TSS were found when 'Pala' was used as the scion. In the first combination, a lower pH in fruit was also obtained. When using the tomato hybrids 'Heman' and 'Primavera' as rootstocks of aubergine ('Rima'), no significant differences in pH, TSS and TA were observed among fruits from grafted and non-grafted or self-grafted plants (Khah, 2011).

In a recent study, López-Marín *et al.* (2013) studied the behaviour (yield and fruit quality) of 'Herminio' plants either non-grafted or grafted on to three commercial pepper rootstocks, 'Atlante', 'Creonte' and 'Terrano', under non-shaded and shaded conditions. There were no significant differences in TSS and TA among the grafting combinations grown in shade conditions, whereas 'Herminio' showed the highest values in non-shaded conditions. Similarly, no differences were observed in TSS, juice pH and electrical conductivity from peppers grafted on to different rootstocks, in particular 'Emphasis', 'S1', 'Strong Tosa', 'Friend', 'Romanesco Zucchini' and 'RS841'. However, the exception was that TA was slightly higher for grafted in comparison with non-grafted plants. Moreover, Colla *et al.* (2008) found no difference in TA and juice pH when two pepper cultivars, 'Edo' and 'Lux', were grafted on to five commercial *Capsicum* rootstocks, 'Snooker', 'Tresor', 'RX360', DRO8801' and '97.9001', cultivated under greenhouse conditions.

Aroma volatiles

Aroma is a complex mixture of a large number of volatile compounds, whose composition is specific to species and often to the variety of fruit (Sanz *et al.*, 1997; Schwab *et al.*, 2008). Although different fruits often share many aromatic characteristics, each fruit has a distinctive aroma that depends on the combination of volatiles, and the concentration and perception threshold of individual volatile compounds (Tucker, 1993). Aroma volatiles contribute to the flavour of both fruits and vegetables (Krumbein *et al.*, 2004; Rouphael *et al.*, 2010).

The most abundant alcohols identified in the aroma profile of mini watermelons include (Z)-3-nonen-1-ol (fresh melon), (Z,Z)-3,6-nonadien-1-ol (pumpkin, cucumber), hexanol (flower, green), nonanol (herbaceous) and (Z)-6-nonen-1-ol (green melon, pumpkin-like) (Kemp, 1975; Yajima *et al.*, 1985; Dima *et al.*, 2014). Among identified aldehydes, the most abundant were (Z)-2-nonenal (honeydew melon, fruity), hexanal (green), (E,Z)-2,6-nonadienal (cucumber, green), nonanal (melon, orange peel), (Z)-6-nonenal (honeydew melon, fruity), 6-methyl-5-hepten-2-one (flower) and (E)-6-nonenal (earthy) (Dima *et al.*, 2014). Grafting midi watermelon cultivars (about 6 kg) on to *C. maxima* × *C. moschata* and *L. siceraria* rootstocks increased fruit content in several aroma volatiles, including (E)-2-nonenal (cucumber, fat) and, most importantly, (Z,Z)-3,6-nonadien-1-ol (pumpkin and cucumber) (Petropoulos *et al.*, 2014). Fredes *et al.* (2016) reported that the volatile profiles of watermelon fruits were influenced by the rootstocks. In fruit harvested at a similar mature stage, those from plants grafted on to *Cucurbita* hybrids showed the highest levels of (Z)-6-nonen-1-ol, which confers pumpkin-like odours. However, fruits harvested from plants grafted on to citron melons had high amounts of the volatile compounds associated with melon- and cucumber-like aromas ((Z)-6-nonenal and (E,Z)-2,6-nonadienal).

Volatiles also play an important role in determining melon fruit flavour. The most important aroma compounds include amino acid-derived compounds, lipid-derived compounds, phenolic derivatives, and mono- and sesquiterpenes (Schwab *et al.*, 2008). A decrease (20–95% less than the control) in the content of the key aroma esters ethyl 2-methylbutanoate and ethyl butanoate in most of the grafted samples when 'Polifemo', 'AS10', 'RS841', 'P360' and 'Energia' were used as rootstocks for the melon 'Proteo' was reported by Verzera *et al.* (2014). A decrease in octyl acetate, isobuthyl acetate and butyl acetate was reported by Yarsi *et al.* (2012) in melon ('Galia C8') grafted on to 'RS841'; the content of 2-methyl butyl acetate was similar for both non-grafted and grafted melon fruits and that of ethyl hexanoate higher in grafted plants. This last compound was positively correlated with fruity and sweet taste.

In a recent study, Guler *et al.* (2013) found substantial differences in the volatile compounds in the peel and flesh of cucumber 'Cengelköy' grafted on to bottle gourd (*L. siceraria*) rootstocks. Grafting caused an increase in the alcohol content and a decrease in the aldehyde content of the peel and flesh of the cucumber. The percentage compositions of aldehydes and alcohols identified in the flesh of cucumbers harvested from plants grafted on to '33-41' were similar to those of non-grafted plants. The authors concluded that the former rootstock could be considered a potential and promising candidate with regard to the major volatile organic compounds identified in cucumber.

Around 400 volatiles have been identified in fresh tomato (Petro-Turza, 1986). However, only odour-active volatiles are able to contribute to the flavour, such as (Z)-3-hexenal, hexanal, 1-octen-3-one, methional, 1-penten-3-one and 3-methylbutanal (Buttery *et al.*, 1989; Krumbein and Auerswald, 1998; Tandon *et al.*, 2000; Mayer *et al.*, 2002). Krumbein and Schwarz (2013) showed for cherry tomato 'Piccolino' and 'Classy' grafted on to rootstocks 'Brigeor' and 'Maxifort' that, in general, grafting increased the concentration of the three aroma volatiles methyl salicylate, guaiacol and eugenol, with oily, sweet and spicy odours, respectively, but the concentration of the three aroma volatiles with almondy odour (benzaldehyde), violet-like odour (β-ionone) and tomato-like flavour (geranylacetone) were decreased by grafting.

Hence, further research is required to test whether and how changes in flavour compounds in tomato fruits of plants grown under different growing conditions, for different rootstock–scion combinations, can influence consumer acceptance.

8.2.4 Health-promoting substances

Carotenoids

Although grafting in general tends to augment lycopene content in watermelon fruit, controversy among reports is perhaps indicative of widespread rootstock–scion interactions. An increase in lycopene content by 10–40% was obtained on *C. maxima* × *C. moschata* rootstocks (Perkins-Veazie *et al.*, 2007; Proietti *et al.*, 2008; Soteriou *et al.*, 2014; Kyriacou and Soteriou, 2015). An increase in the range of 11–27% was reported on selected *L. siceraria* genotypes (Candir *et al.*, 2013), and also for triploid watermelon grafted on to *C. argyrosperma* (cushaw pumpkin) and *C. pepo* L. (Davis and Perkins-Veazie, 2005). An absence of any significant effect was reported in some studies (Bruton *et al.*, 2009; Soteriou and Kyriacou, 2015), but a decrease was reported for particular *L. siceraria* and *C. argyrosperma* rootstock–scion combinations (Candir *et al.*, 2013). More recent work has demonstrated that watermelon lycopene content may be affected foremost by ripening and less so by grafting (Soteriou *et al.*, 2014). The peak in lycopene content, as well as in the chroma of the pulp, appears earlier in non-grafted than in grafted fruit as a result of delayed ripening behaviour in the latter. Differences in harvest maturity may in fact confound the effect of grafting, thus explaining the apparent absence of effect or even a decrease in response to grafting. In addition, lycopene synthesis events are carried over to the postharvest period where they appear to be temperature controlled and linked to changes in pulp colour (Perkins-Veazie and Collins, 2006).

The carotenoid profile of melon ‘Proteo’ (*Cucumis melo* var. *reticulatus*) was also improved when grafted on to pumpkin hybrid rootstocks (Condurso *et al.*, 2012). Recently, Verzera *et al.* (2014) and Zhou *et al.* (2014) reported that the use of pumpkin hybrid rootstocks resulted in an improvement of the qualitative and quantitative carotenoid profile. An increase in lutein (5.5–13.7 mg kg^{-1} fresh mass) was produced in fruit samples from plants grafted on to ‘Elsi’, ‘P360’, ‘RS841’ and ‘AS10’, and about eightfold more β-carotene than the control and about 56% more α-carotene than the control where noted in fruits grafted on to ‘P360’ (Verzera *et al.*, 2014). In this work, fruit from the combination ‘Liyu’–‘Nanzhen No. 1’ contain fourfold more β-carotene than fruit from non-grafted plants (Zhou *et al.*, 2014). However, not all the combinations reported in the above studies had similar consequences. Thus, when Verzera *et al.* (2014) used ‘Energy’ and ‘Sting’ as rootstocks, a decrease in the amount of β-carotene (about 55% less than the control) was obtained.

Carotenoid (lycopene and β-carotene) content of tomato fruit can be influenced by grafting, but the rootstock–scion combination seems to play an important role. For example, six out of seven rootstocks investigated, including ‘Maxifort’ and ‘Beaufort’, decreased the lycopene concentration of the tomato ‘Jeremy’ grown over 3 years, while the aubergine rootstock ‘Madonna’ increased the lycopene concentration (Mišković *et al.*, 2008). According to many authors, lycopene concentration in tomato fruits also decreases with grafting (Mišković *et al.*, 2008; Helyes *et al.*, 2009). Mohammed *et al.* (2009) reported the same results with tomato ‘Cecilia’ grafted on to ‘Beaufort’ and ‘Heman’, as did Krumbein and Schwarz (2013) with ‘Classy’ grafted on to ‘Brigeor’. Nevertheless, other authors reported that grafting had no effect on carotenoids (Neocleous, 2010; Turhan *et al.*, 2011). Khah *et al.* (2006) found no differences in the lycopene content between non-grafted tomato 'Big Red' and one grafted on to 'Heman' and 'Primavera' rootstocks under open-field and greenhouse conditions, as did Vinkovic-Vrcek *et al.* (2011) with the 'Tamaris' grafted on to 'Heman', 'Efiato' and 'Maxifort'.

Chávez-Mendoza *et al.* (2013) evaluated the influence of the scion and the rootstock on pepper lycopene and β-carotene. The two bell pepper cultivars 'Fascinato' and 'Jeanette' were grafted on to the rootstock 'Terrano'. Grafting increased β-carotene but not lycopene content, and also enhanced antioxidant activity.

Ascorbic acid

Grafting watermelon on to *C. maxima* × *C. moschata* rootstock has been shown to increase total ascorbate content by 7.3%, mainly due to an increase in the oxidized dehydroascorbate form (Proietti *et al.*, 2008). However, grafting on *L. siceraria* did not change total ascorbate content, while grafting on *C. maxima* resulted in a decrease (Qin *et al.*, 2014). Based on available reports, the effect of grafting on ascorbic acid (vitamin C) content of watermelon fruit seems marginal; moreover, the actual content is highly influenced by culture conditions and ripening stage, although the effect of the latter appears rather erratic (Leskovar *et al.*, 2004; Tlili *et al.*, 2011).

In cucumber, Huang *et al.* (2009) showed that vitamin C content was positively affected by grafting, as an enhancement in ascorbic acid content of the fruits was obtained by grafting on to 'Figleaf Gourd' and 'Chaofeng Kangshengwang' under both saline and non-saline nutrient solution conditions.

Tomato fruit contains significant amounts of ascorbic acid, and studies showed that fruit content was strongly reduced by grafting (Fernandez-Garcia *et al.*, 2004; Arvanitoyannis *et al.*, 2005; di Gioia *et al.*, 2010; Vinkovic-Vrcek *et al.*, 2011). For example, Qaryouti *et al.* (2007) found that fruit vitamin C content was reduced in soil cultivation when tomato plants were grafted on to 'Heman' and 'Spirit' rootstocks. Moreover, the ascorbic acid content can change according to the scion–rootstock combination, as shown by Turhan *et al.* (2011) with tomato plants grafted on to 'Beaufort' rootstock, which exhibited higher ascorbic acid content compared with the same plants grafted on to 'Arnold' rootstock.

Studying bell pepper 'Fascinato' and 'Jeanette' grafted on to the rootstock 'Terrano', Chávez-Mendoza *et al.* (2013) observed an increase in vitamin C content. However, the effect of grafting on pepper vitamin C content was not confirmed by López-Marín *et al.* (2013), who observed that results differed in relation to the rootstock adopted and the light availability at crop level. No differences in vitamin C content were reported by Gisbert *et al.* (2010) when two commercial hybrid peppers were grafted on to two developed rootstocks. As expected, differences in vitamin C content were observed between the scions in agreement with the high genotypic dependence of this trait reported in pepper.

Phenolic compounds

The more abundant flavonoids in tomato fruits are the hydroxycinnamic acids and their derivatives (Gómez-Romero *et al.*, 2010), as well as naringenin, chalcone and rutin (quercetin-3-*O*-rutinoside) (Slimestad *et al.*, 2008; Sánchez-Rodríguez *et al.*, 2012), which are natural antioxidants. Steward *et al.* (2000) indicated that the choice of cultivar was a major factor contributing to the total content of phenolics in tomatoes when grown under similar environmental conditions, but the flavonoids content of tomato also varies with both abiotic and agronomic factors (Tomas-Barberan and Espin, 2001). Sánchez-Rodríguez *et al.* (2012), combining grafting and water stress with a reciprocal cherry tomato grafting between drought-tolerant 'Zarina' and drought-sensitive 'Josefina' found that 'Zarina' as rootstock under water-stress conditions showed the highest value in total flavonoids, hydroxycinnamic acids and rutin. However, Vinkovic-Vrcek *et al.* (2011) reported that grafting significantly reduced the total phenolics content of tomato 'Tamaris' grafted on to cvs. 'Heman', 'Efiato' and 'Maxifort', and no significant differences were found among the different rootstocks. Nicoletto *et al.* (2013) found a higher phenolic acid content for the tomato 'Profitto' grafted on to 'Beaufort' rootstocks compared with non-grafted plants, but this was not observed with the rootstock 'Big Power'.

Rootstocks have little or no effect on the fruit phenolic content of aubergines (Gisbert *et al.*, 2011b; Moncada *et al.*, 2013; Marsic *et al.*, 2014). Gisbert *et al.* (2011b) obtained a higher phenolic content only in fruits of aubergine 'Cristal F_1' grafted on to *Solanum macrocarpon* L. rootstock, but no modification in phenolics content was observed in fruit of the rest of the tested combinations, which included intraspecific and interspecific hybrids. The work of Moncada *et al.* (2013) confirmed that rootstocks have no effect on phenolics and indicated the main influence of the scion genotype. However, Marsic *et al.* (2014) obtained inconsistent results for the same rootstock–scion combinations (aubergine cultivars grafted on to the tomato rootstock 'Maxifort') in two consecutive trials, concluding that the browning potential of fruit pulp was highly dependent on variety and partly on grafting combination. Differences observed by Marsic *et al.* (2014) in some combinations and trials, and those observed by Gisbert *et al.* (2011b) in plants grafted on to *S. macrocarpon*, may be related to responses to incompatibility.

Concerning pepper, no significant effects of grafting were observed on fruit phenolic content (Chávez-Mendoza *et al.*, 2013; López-Marín *et al.*, 2013).

Minerals

Grafting watermelon on *C. pepo* rootstocks has been shown to improve nutrient uptake capacity of the scion (Pulgar *et al.*, 2000). Grafting mini watermelon scion on to *C. maxima* × *C. moschata* rootstock was shown to affect fruit mineral composition, mainly by increasing K and magnesium (Mg) concentrations in the pulp (Yamasaki *et al.*, 1994; Rouphael *et al.*, 2008). Under increased NaCl salinity conditions, grafting on both *C. maxima* × *C. moschata* and *L. siceraria* rootstocks maintained lower sodium (Na) concentrations in the fruit than in the non-grafted control (Colla *et al.*, 2006b).

The mineral contents of melon fruit were influenced by rootstock and varied with the growing season (Jang *et al.*, 2014). These authors observed that the phosphorus (P) and Mg contents of fruit grafted on to 'Shintosa' were lower than those grafted on to *C. melo* rootstocks or non-grafted melons. In fruit of

'Khatooni', nitrogen (N), P and K were higher in fruit from plants grafted on to 'Ace' or 'Shintoza' (Salehi *et al.*, 2014).

Khah *et al.* (2006) investigated the mineral content of tomato fruits from 'Big Red' grafted on to hybrid rootstocks 'Heman' and 'Primavera'. The fruit contents of copper (Cu), manganese (Mn) and iron (Fe) were not significantly different between grafted, non-grafted and self-grafted plants, either in the greenhouse or in the open-field, but the Ca content of the fruits on plants grafted on to 'Heman' were higher in the greenhouse. The absorption of Ca could be strongly associated with the higher rate of absorption of water and minerals from the soil by roots of the rootstock 'Heman'. Similar results were observed by Tsouvaltzis *et al.* (2004) with tomato 'Sacos' grafted on to 'Primavera' rootstock. As with the incidence of BER, the Ca fruit content appeared to be dependent on the rootstock genotype and rootstock–scion combination.

In aubergine, different amounts of N, P and Ca are necessary to obtain the same quantity of commercial products (Leonardi and Giuffrida, 2006). In plants of 'Black Bell' grafted on to *S. torvum*, Cassaniti *et al.* (2011) reported an increase in the uptake of N, P, K and Ca with respect to the control. These differences were reflected in the fruit mineral contents. However, in the study by Gisbert *et al.* (2011b) where mineral content was analysed in fruit ('Black Beauty') from non-grafted, self-grafted and plants grafted on to different interspecific hybrids or *S. torvum*, differences among treatments were not found.

With regard to pepper, Sánchez-Torres *et al.* (2016) showed the importance of choosing the best rootstock–scion combinations from a nutritional point of view. The sweet peppers 'Coyote' and 'Almuden' were grafted on to 'Charlot' and 'Foc' rootstocks, and higher levels of P and K were found in the fruit of 'Coyote' grafted on to 'Charlot' (17.2% higher P than for the control and the highest fruit K concentration: 226.70 mg per 100 g fresh mass).

8.2.5 Contaminants

Root systems have different mechanisms for transporting nutrients and metals from the soil to the shoots. Therefore, differences among rootstocks with respect to metal and contaminant translocation can be expected. Watermelon grafted on to *C. maxima* × *C. moschata* rootstock demonstrated increased uptake of soilborne organochlorine pesticides and increased fruit content of these contaminants, although their actual levels remained 6–12 times below acceptable levels (Isleyen *et al.*, 2012).

Edelstein *et al.* (2006) determined the effects of plant grafting on heavy-metal and trace mineral concentrations in melon fruits ('Arava') grafted on to 'TZ-148' rootstocks and non-grafted plants, and found lower concentrations of B, zinc (Zn), strontium (Sr), Mn, Cu, titanium (Ti), chromium (Cr), nickel (Ni) and cadmium (Cd) in the grafted plants.

Drins have been classified as a group of persistent organic pollutants because of their high toxicity, high bioaccumulation and persistency in the environment. Otani and Seike (2007) investigated the influence of three *Cucurbita* spp. rootstocks ('Yuyuikki-black', 'Shintosa-1gou' and 'Hikaripower-gold') on dieldrin concentration in grafted cucumber fruits grown in contaminated soils. Dieldrin concentration in cucumber fruits grafted on to 'Yuyuikki-black' (*Cucurbita mixta* Pang.) decreased by 50–70% compared with those grafted on to 'Shintosa-1gou' (*C. maxima* × *C. moschata*), and by 30–50% compared with those grafted on 'Hikaripower-gold' (*C. moschata*). Selection of low-uptake rootstocks would be an effective tool for reducing the drins concentration in edible parts of fruit vegetables under adverse chemical soil conditions.

Tomato is characterized by rather low rates of heavy-metal translocation to the fruit (Angelova *et al.*, 2009). However, the impact of grafting on the uptake or translocation of heavy metals in tomato has not been investigated adequately.

Grafting on to *S. torvum* reduced aubergine fruit Cd concentration by 63–75% in Cd-polluted soil and unpolluted soil compared with grafting on to *S. melongena* and *Solanum integrifolium* Lam. (Arao *et al.*, 2008). These authors conclude that grafting on to *S. torvum* is a practical method for reducing the Cd concentrations of aubergine fruit, which is a problem in Japan. Grafting in aubergines may improve the fruit quality by reducing the amount of oxalic acid in the fruit. In aubergine grafted on to *Datura inoxia* P. Mill., scopolamine and atropine accumulated in the fruit at levels high enough to cause poisoning (Calvo-Asensio *et al.*, 2014).

8.3 Effects of Grafting on Ripening and Postharvest Behaviour

The effect of grafting on fruit quality is also mediated by variation in ripening behaviour and postharvest behaviour. Recent work has demonstrated that grafting watermelon on *C. maxima* × *C. moschata* rootstock delays ripening events associated with physicochemical changes in fruit composition. Important quality parameters, such as the concentration of TSS and non-structural carbohydrates, were not affected by grafting but almost entirely by harvest maturity (Soteriou *et al.*, 2014). Lycopene content and colour intensity, in contrast, were increased but also delayed in development as a result of grafting, and proved highly ripening dependent (Fig. 8.1); characteristics such as pulp firmness and TA appeared to be increased by grafting throughout ripening (Soteriou *et al.*, 2014). Overall fruit quality of grafted watermelon, as opposed to fruit from non-grafted plants, may therefore benefit from delayed harvest. Rootstock-mediated shifting of flower initiation has been implicated in grafting effects on fruit maturation (Satoh, 1996; Sakata *et al.*, 2007). However, while grafting on widely used *C. maxima* × *C. moschata* rootstocks may confer a limited tendency for delayed anthesis, tagging flowers at anthesis and monitoring fruit age in days past anthesis has demonstrated that potential delay in maturation relates predominantly to postanthesis ripening events (Soteriou *et al.*, 2014; Soteriou and Kyriacou, 2015; Kyriacou and Soteriou, 2015). Retarded ripening derives mainly from overburdened source–sink relationships as a result of increased crop load on grafted plants. The effect of grafting on more or less vigorous rootstocks could also be hypothesized for tomato; in this case, some of the effect (e.g. on TSS or carotenoids) should be confirmed.

Grafting performed on *C. maxima* × *C. moschata* rootstock improved watermelon fruit postharvest performance mainly by sustaining higher pulp lycopene levels, more intense chroma and higher pulp firmness (Kyriacou and Soteriou,

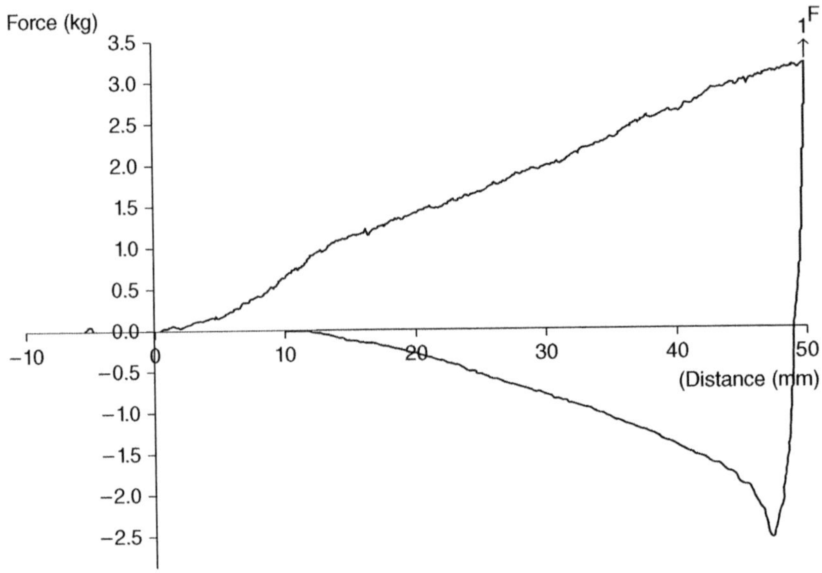

Fig. 8.1. Watermelon pulp firmness in relation to depth of penetration. (Courtesy of M.C. Kyriacou, Agricultural Research Institute, Nicosia, Cyprus.)

2015). Lycopene content peaked 7 days postharvest at 25°C and was further increased by grafting on *C. maxima* × *C. moschata* rootstocks. However, postharvest decline in watermelon sweetness, resulting mainly from a decrease in the levels of reducing sugars, may occasionally offset these effects, especially in seasons of rootstock-supported prolific crop and harvest performed before full commercial maturity.

Zhao *et al.* (2011) studied grafting effects on postharvest quality of melon and found that grafting led to decreased fruit firmness of 'Athena' melon grafted on to 'Strong Tosa' and 'Tetsukabuto'. They further demonstrated that grafting effects on postharvest ripening and quality of 'Athena' muskmelon can vary markedly with the rootstocks used. Grafting with 'Strong Tosa' achieved the same fruit shelf life as non-grafted 'Athena', whereas self-grafting with 'Athena' and use of 'Tetsukabuto' rootstock resulted in a decline of shelf life by 6 days for the early harvest and 3 days for the late harvest under 1-methylcyclopropene treatment. In solanaceous fruits, postharvest quality is also affected by grafting. Fruits of aubergine 'Tsakoniki' grafted on to *S. torvum* and *S. sisymbriifolium* rootstocks and stored under modified-atmosphere packaging showed a reduction in vitamin C content and firmness (Arvanitoyannis *et al.*, 2005).

8.4 Biophysiological Processes Affecting Fruit Quality

Only a few studies dealing with the effect of grafting on product quality have hypothesized or clearly demonstrated a cause–effect relationship between biophysiological processes and the variation of a specific qualitative trait.

However, in this chapter we are not considering the effect that may derive from a more or less evident degree of resistance that a specific rootstock may give to a plant. In this case, the effect on quality may derive from different photosynthesis rates of grafted plants resistant to specific soilborne diseases (Jang *et al.*, 2014). We are not even considering all the aspects related to scion/rootstock compatibility, which, for instance, may affect the development of vascular bundles at the graft union and in turn could result in water deficiency in the shoot and fruit (Oda *et al.*, 1996). Furthermore, we have to consider that the effect on product quality in some cases may actually derive from limitations in the methodological approach adopted. Disparity with respect to the TSS between fruits of grafted and non-grafted scions may actually relate to sampling size and method. In crops bearing limited number of fruits per plant, such as watermelon, recurrent harvests of experimental plots may drastically alter the photosynthetic source–sink balance and confound actual effects on the TSS.

Grafting may affect the TSS indirectly through its effect on scion vigour, timing of flowering, fruit load, yield and, ultimately, fruit maturation, as TSS are highly dependent on fruit maturity at harvest (Yamasaki *et al.*, 1994; Rouphael *et al.*, 2010; Soteriou and Kyriacou, 2015). In this respect, grafting may be considered a high-input production method, with a prevalent tendency at least in watermelon for increasing fruit number on small-fruited scions and fruit mass on large-fruited ones, hence overburdening source–sink relationships, and potentially suppressing the TSS (Davis *et al.*, 2008b; Soteriou and Kyriacou, 2015). The effect of crop load on mineral concentrations of grafted watermelon/bottle gourd and watermelon/squash xylem exudates was non-significant at a young fruit stage and became apparent only on watermelon/squash plants as the fruit matured, mainly as increased P concentrations in plants with increased crop load (Yamasaki *et al.*, 1994). While particular rootstock–scion combinations may adversely impact fruit sweetness, combinations resulting in TSS lower than 10%, proposed as a minimum for consumer acceptability (Kader, 1999; Maynard *et al.*, 2002), are uncommon.

Quadratic regression analysis of total sugar content with fruit age has revealed that an apparent reduction in total sugar content of watermelon fruit in response to grafting derives largely from retarded maturation; grafting delays the peak in total sugars to over 40 days after fruit setting as opposed to just under 35 days in non-grafted fruit (Fig. 8.2; Soteriou *et al.*, 2014). Notwithstanding the effects of cultural practice and prevailing climatic conditions on flowering and fruit setting, a sugar peak delay in response to grafting is impacted foremost by rootstock vigour; hence, scion genotypes of low sugar content should preferably be matched to less vigorous rootstocks.

In tomato, the lower SSC of fruits harvested from grafted plants could be due to an affected sucrose-metabolizing enzyme or an increase in fruit water content, or the vigorous rootstocks may act as an additional sink for assimilates and thus reduce assimilate flow to the fruits (Balibrea *et al.*, 2006; Xu *et al.*, 2006; Martínez-Ballesta *et al.*, 2010).

Another mechanism that could be involved is related to the capability of a vigorous root system to enhance the uptake of nutrients such as K (Ruiz and Romero, 1999; Leonardi and Giuffrida, 2006; Albacete *et al.*, 2009). K is positively related

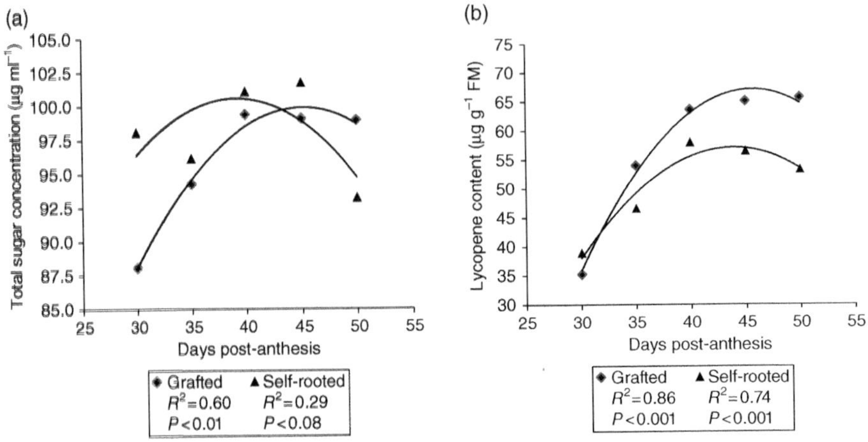

Fig. 8.2. Regression of total sugar concentration (a) and lycopene content (b) with maturity for watermelon fruit harvested at 30, 35, 40, 45 and 50 days postanthesis from plants grown self-rooted (non-grafted) or grafted on to *C. maxima* × *C. moschata* 'TZ-148' rootstock. Data points represent means of four replicates consisting of six sample fruits each. FM, fresh mass. (Adapted from Soteriou *et al.*, 2014.)

to the acid concentration in tomato fruits, and plays a role in the maintenance of the electroneutrality of acids in the fruit. However, K transport depends not only on the rootstock but also on other growing conditions, such as the current K concentration in the root zone and climatic factors (Albacete *et al.*, 2009). Interestingly, differences in K concentration were found not to be significant between fruits of self-grafted and grafted plants, but differences in the fruit TA were significant. Therefore, the reasons for the increase in TA in fruits of grafted plants demand further investigation (Leonardi and Giuffrida, 2006).

As mentioned above for tomato, one of the consequences of grafting for some scion–rootstock combinations is the variation of carotenoid content, which in turn affect the carotenoid-derived volatiles such as geranylacetone, β-cyclocitral and β-ionone, which are responsible for tomato flavour, such as the violet-like odour (β-ionone) and the tomato-like flavour (geranylacetone) (Krumbein and Schwarz, 2013).

In tomato, the lower ascorbic acid content could be explained by the higher plant/shoot biomass in grafted plants compared with non-grafted ones or by the fact that grafted plants were initially subjected to stress following the grafting operation. Ascorbic acid is known to control cell differentiation (Arrigoni, 1994) and to promote callus division and growth (Tabata *et al.*, 2001). The decrease in content of total vitamin C in the fruits of grafted plants could therefore be a result of the redistribution or accumulation of vitamin C in other parts of grafted plants (Wadano *et al.*, 1999).

Hormonal signals in the rootstock–scion interaction may modify many of the steps in cell-wall metabolism, such as monomer distribution, side-chain modifications, de-esterification and polymer solubilization of cell-wall and middle-lamella structures (Brummell and Harpster, 2001; Aloni *et al.*, 2010).

8.5 Conclusions

Vegetable quality depends on the interaction of genetic, climatic, cultural and postharvest factors. Finding the best combinations of these factors to maximize vegetable quality will be a challenge in the future. Among the different factors, the use of specific rootstocks for plant grafting plays an important role. Selection of appropriate rootstock for a certain scion is therefore crucial for optimal yield but also for fruit quality. Therefore, besides the possibility of better management of soilborne diseases and the improvement of abiotic stress tolerance, grafting can be also an important tool for the improvement of fruit quality.

However, in some cases, grafting may influence negatively product quality; therefore, suitable cultural practices have to be considered to minimize the negative effects, which may arise from the adoption of certain rootstocks expressing tolerance to specific stresses or determining a higher plant vigour.

The current scientific literature abounds with about 140 papers dealing with the effect of grafting on product quality. Most research concerns visual quality and compounds responsible for taste (i.e. sugars, Brix), whereas only a few studies have investigated parameters related to health or flavour-related compounds (Table 8.1).

Table 8.1. Number of published papers specifically related to various grafted vegetables and their different fruit quality characteristics.

Characteristics	Water melon	Melon	Cucumber	Tomato	Aubergine	Pepper
Unit mass	6	10	9	13	6	9
Shape index	4	1	3	1	3	5
Defects incidence	5			5		2
Colour	4	4	2	3	3	
Exocarp characteristics	8					1
Firmness	10	2	5	3	2	1
Total soluble solids	15	7	5	17	1	3
Titratable acidity	5	2	5	11		2
Juice pH	4		3	3	2	1
Total sugars	6		1	7		
Glucose	6					
Fructose	6					
Sucrose	6					
Aroma volatiles	1	2	1	4		
Vitamins	2		1	7		2
Polyphenols				3	2	2
Carotenoids	10	2		8		1
Mineral content	4	1	3	4	1	
Antioxidant activity						2
Alkaloids					1	
Heavy metals				1	2	
Organic pollutants			1			

The total number of articles used for preparing the table were: 34 for watermelon, 23 for melon, 10 for cucumber, 43 for tomato, 20 for aubergine and 9 for pepper.

A limited number of studies have demonstrated or hypothesized the rootstock characteristics or the biophysiological mechanisms involved in mediating the effects of rootstocks on product quality. Many of the biochemical and molecular mechanisms remain relatively unknown. A better comprehension of the cause–effect relationships between these mechanisms and product quality could be the basis of specific breeding programmes.

The scarce information therefore does not allow us to draw a clear picture concerning the effects of grafting on fruit quality. This is particularly true if we consider that the effects can be different and sometimes conflicting (Table 8.2); this may come from the masking effect of specific growing conditions (e.g. salinity, rhizosphere volume), as well as from the scion–rootstock interactions.

Considering the above, our understanding of the effect of grafting on product quality is far from complete. However, it is clear that the scion/rootstock combination represents one key aspect that should be considered not only in facing specific climatic and agronomic challenges but also in relation to the effects on product quality. Further research is needed to elucidate the effect of grafting on some specific parameters that are assuming increasing importance (e.g. aroma

Table 8.2. Effect of grafting on different quality characteristics of various annual fruits.

Quality characteristics	Water melon	Melon	Cucumber	Tomato	Aubergine	Pepper
Unit mass	+=	+−=	+	+−=	+=	+−
Shape index	=	=	+	−=	+=	+
Defect incidence	−=			−+		
Colour	+=	+−=	+	=−	+=	
Exocarp characteristics	+=					+
Firmness	+=−	+−	+−	=	−=	=
Total soluble solids	−=	+−=		+−=	−=	=
Titratable acidity	−=		+−	−=		+=
Juice pH	−=	+−=	+−	=	−=	=
Total sugars	−=		=	+−		
Glucose	−=		=			
Fructose	−=					
Sucrose	−=					
Aroma volatiles	+	+−=		+−		
Vitamins	+=−		+	−		+
Polyphenols				+−	+−=	=
Carotenoids	+=−	+−=		−+=		+=
Citrulline	+					
Mineral content	+	−		+−	=	+=
Antioxidant activity						+
Alkaloids					+	
Heavy metals			+=	−	−	
Organic pollutants						

+, Increased; −, decreased; =, no significant effects. More than one symbol is used if different results were observed in different studies; the first symbol indicates the general trend.

volatiles, amino acid content). Furthermore, attention should also be focused on new species that are going to be grafted in the future (e.g. artichoke). An integrated approach based on the cooperation between quality specialists, agronomists, physiologists and breeders is therefore needed for a fruitful selection of suitable rootstocks and the production of high-quality fruits from grafted vegetable crops.

References

Alan, O., Ozdemir, N. and Gunen, Y. (2007) Effect of grafting on watermelon plant growth, yield and quality. *Journal of Agronomy* 6, 362–365.

Albacete, A., Martínez-Andújar, C., Ghanem, M.E., Acosta, M., Sánchez-Bravo, J., Asins, M.J., Cuartero, J., Lutts, S., Dodd, I.C. and Pérez-Alfocea, F. (2009) Rootstock-mediated changes in xylem ionic and hormonal status are correlated with delayed leaf senescence, and increased leaf area and crop productivity in salinized tomato. *Plant, Cell and Environment* 32, 928–938.

Alexopoulos, A.A., Kondylis, A. and Passam, H.C. (2007) Fruit yield and quality of watermelon in relation to grafting. *Journal of Food Agriculture and Environment* 5, 178–179.

Aloni, B., Cohen, R., Karni, L., Aktas, H. and Edelstein, M. (2010) Hormonal signaling in rootstock–scion interactions. *Scientia Horticulturae* 127, 119–126.

Angelova, V.R., Babrikov, T.D. and Ivanov, K.I. (2009) Bioaccumulation and distribution of lead, zinc, and cadmium in crops of Solanaceae family. *Communications in Soil Science and Plant Analysis* 40, 2248–2263.

Arao, T., Takeda, H. and Nishihara, E. (2008) Reduction of cadmium translocation from roots to shoots in eggplant (*Solanum melongena*) by grafting onto *Solanum torvum* rootstock. *Soil Science and Plant Nutrition* 54, 555–559.

Arrigoni, O. (1994) Ascorbate system in plant development. *Journal of Bioenergetics and Biomembranes* 26, 407–419.

Arvanitoyannis, I.S., Khah, E.M., Christakou, E.C. and Bletsos, F.A. (2005) Effect of grafting and modified atmosphere packaging on eggplant quality parameters during storage. *International Journal of Food Science and Technology* 40, 311–322.

Balibrea, M.E., Martinez-Andujar, C., Cuartero, J., Bolarín, M.C. and Perez-Alfocea, F. (2006) The high fruit soluble sugar content in wild *Lycopersicon* species and their hybrids with cultivars depends on sucrose import during ripening rather than on sucrose metabolism. *Functional Plant Biology* 33, 279–288.

Brummell, D.A. and Harpster, M.H. (2001) Cell wall metabolism in fruit softening and quality and its manipulation in transgenic plants. *Plant Molecular Biology* 47, 311–340.

Bruton, B.D., Fish, W.W., Roberts, W. and Popham, T.W. (2009) The influence of rootstock selection on fruit quality attributes of watermelon. *Open Food Science Journal* 3, 15–34.

Buttery, R.G., Teranishi, R., Flath, R.A. and Ling, L.C. (1989) Fresh tomato aroma volatiles – composition and sensory studies. In: Teranishi, R., Buttery, R.G. and Shahidi, F. (eds) *Flavour Chemistry: Trends and Developments*. ACS Symposium Series No. 388. American Chemical Society, Washington, DC, pp. 213–222.

Calvo-Asensio, I., Prohens, J. and Gisbert, C. (2014) Vigor for in vitro culture traits in *S. melongena* × *S. aethiopicum* hybrids with potential as rootstocks for eggplant. *Scientific World Journal* 2014, 702071.

Candir, E., Yetişir, H., Karaca, F. and Üstün, D. (2013) Phytochemical characteristics of grafted watermelon on different bottle gourds (*Lagenaria siceraria*) collected from the Mediterranean region of Turkey. *Turkish Journal of Agriculture and Forestry* 37, 443–456.

Cassaniti, C., Giuffrida, F., Scuderi, D. and Leonardi, C. (2011) Effect of rootstock and nutrient solution concentration on eggplant grown in a soilless system. *Journal of Food, Agriculture and Environment* 9, 252–256.

Chávez-Mendoza, C., Sánchez, E., Carvajal-Millán, E., Munõz-Márquez, E. and Guevara-Aguillar, A. (2013) Characterization of the nutraceutical quality and antioxidant activity in bell pepper in response to grafting. *Molecules* 18, 15689–15703.

Colla, G., Rouphael, Y., Cardarelli, M., Massa, D., Salerno, A. and Rea, E. (2006a) Yield, fruit quality and mineral composition of grafted melon plants grown under saline conditions. *Journal of Horticultural Science and Biotechnology* 81, 146–152.

Colla, G., Rouphael, Y., Cardarelli, M. and Rea, E. (2006b) Effect of salinity on yield, fruit quality, leaf gas exchange, and mineral composition of grafted watermelon plants. *HortScience* 41, 622–627.

Colla, G., Rouphael, Y., Cardarelli, M., Temperini, O., Rea, E., Salerno, A. and Pierandrei, F. (2008) Influence of grafting on yield and fruit quality of pepper (*Capsicum annuum* L.) grown under greenhouse conditions. *Acta Horticulturae* 782, 359–363.

Colla, G., Rouphael, Y., Mirabelli, C. and Cardarelli, M. (2011) Nitrogen-use efficiency traits of mini-watermelon in response to grafting and nitrogen-fertilization doses. *Journal of Plant Nutrition and Soil Science* 174, 933–941.

Colla, G., Rouphael, Y., Rea, E. and Cardarelli, M. (2012) Grafting cucumber plants enhance tolerance to sodium chloride and sulfate salinization. *Scientia Horticulturae* 135, 177–185.

Colla, G., Rouphael, Y., Jawad, R., Kumar, P., Rea, E. and Cardarelli, M. (2013) The effectiveness of grafting to improve NaCl and $CaCl_2$ tolerance in cucumber. *Scientia Horticulturae* 164, 380–391.

Condurso, C., Verzera, A., Dima, G., Tripodi, G., Crinò, P., Paratore, A. and Romano, D. (2012) Effects of different rootstocks on aroma volatile compounds and carotenoid content of melon fruits. *Scientia Horticulturae* 148, 9–16.

Crinò, P., Lo Bianco, C., Rouphael, Y., Colla, G., Saccardo, F. and Paratore, A. (2007) Evaluation of rootstock resistance to Fusarium wilt and gummy stem blight and effect on yield and quality of a grafted 'Inodorus' melon. *HortScience* 42, 521–525.

Cushman, K.E. and Huan, J. (2008) Performance of four triploid watermelon cultivars grafted onto five rootstock genotypes: yield and fruit quality under commercial growing conditions. *Acta Horticulturae* 782, 335–337.

Davis, A.R. and Perkins-Veazie, P. (2005) Rootstock effects on plant vigor and watermelon fruit quality. *Cucurbit Genetics Cooperative Reports* 28–29, 39–42.

Davis, A.R., Perkins-Veazie, P., Hassell, R., Levi, A., King, S.R. and Zhang, X. (2008a) Grafting effects on vegetable quality. *HortScience* 43, 1670–1672.

Davis, A.R., Webber, C.L., Perkins-Veazie, P., Ruso, V., López Galarza, S. and Sakata, Y. (2008b) A review of production systems on watermelon quality. In: Pitrat, M. (ed.) *Proceedings of the IXth EUCARPIA Meeting on Genetics and Breeding of Cucurbitaceae*, 21–24 May. INRA, Avignon, France, pp. 515–520.

di Gioia, F., Serio, F., Buttaro, D., Ayala, O. and Santamaria, P. (2010) Vegetative growth, yield, and fruit quality of 'Cuore di Bue', an heirloom tomato, as influenced by rootstock. *Journal of Horticultural Science and Biotechnology* 85, 477–482.

Dima, G., Tripodi, G., Condurso, C. and Verzera, A. (2014) Volatile constituents of mini-watermelon fruits. *Journal of Essential Oil Research* 26, 323–327.

Doñas-Uclés, F., Jiménez-Luna, M.M., Gongora-Corral, J.A., Pérez-Madrid, D., Verde-Fernandez, D. and Camacho-Ferre, F. (2014) Influence of three rootstocks on yield and commercial quality of 'Italian sweet' pepper. *Ciência e Agrotecnologia* 38, 538–545.

Edelstein, M., Ben-Hur, M. and Plaut, Z. (2006) Response of grafted and non-grafted melon plants to effluent and boron under greenhouse conditions. *Acta Horticulturae 782:*

IV International Symposium on Seed, Transplant and Stand Establishment of Horticultural Crops; Translating Seed and Seedling, pp. 325–328.

Fernandez-Garcia, N., Martinez, V. and Carvajal, M. (2004) Effect of salinity on growth, mineral composition, and water relations of grafted tomato plants. *Journal of Soil Science and Plant Nutrition* 167, 616–622.

Fita, A., Pico, B., Roig, C. and Nuez, F. (2007) Performance of *Cucumis melo* ssp. *agrestis* as a rootstock for melon. *Journal of Horticultural Science and Biotechnology* 82, 184.

Fredes, A., Roselló, S., Beltrán, J., Cebolla-Cornejo, J., Pérez-de-Castro, A., Gisbert, C. and Picó, M.B. (2016) Quality of watermelons grafted onto citron melon rootstock. *Journal of the Science of Food and Agriculture* 10.1002/jsfa.7915 [Epub ahead of print].

Gisbert, C., Sánchez-Torres, P., Dolores Raigón, M. and Nuez, F. (2010) *Phytophtora capsici* resistance evaluation in pepper hybrids: agronomic performance and fruit quality of pepper grafted plants. *Journal of Food Agriculture and Environment* 8, 116–121.

Gisbert, C., Prohens, J. and Nuez, F. (2011a) Performance of eggplant grafted onto cultivated and wild materials and hybrids of eggplant and tomato. *International Journal of Plant Production* 5, 367–380.

Gisbert, C., Prohens, J., Raigón, M.D., Stommel and J., Nuez, F. (2011b) Eggplant relatives as sources of variation for developing new rootstocks: effects of grafting on yield and fruit apparent quality and composition. *Scientia Horticulturae* 128, 14–22.

Gómez-Romero, M., Segura-Carretero, A. and Fernández-Gutiérrez, A. (2010) Metabolite profiling and quantification of phenolic compounds in methanol extracts of tomato fruit. *Phytochemistry* 71, 1848–1864.

Goreta, S., Perica, S., Dumicic, G., Bucan, L. and Zanic, K. (2005) Growth and yield of watermelon on polyethylene mulch with different spacings and nitrogen rates. *HortScience* 40, 366–369.

Goreta Ban, S., Zanic, K., Dumiciv, G., Raspudic, E., Vuletin Selak, G. and Ban, D. (2014) Growth and yield of grafted cucumbers in soil infested with root-knot nematodes. *Chilean Journal of Agricultural Research* 74, 29–34.

Gruda, N. (2005) Impact of environmental factors on product quality of greenhouse vegetables for fresh consumption. *Critical Reviews in Plant Sciences* 24, 227–247.

Guler, Z., Karaca, F. and Yetisir, H. (2013) Volatile compounds in the peel and flesh of cucumber (*Cucumis sativus* L.) grafted onto bottle gourd (*Lagenaria siceraria*) rootstocks. *Journal of Horticultural Science and Biotechnology* 88, 123–128.

Helyes, L., Lugasi, A., Pogonyi, A. and Pek, Z. (2009) Effect of variety and grafting on lycopene content of tomato (*Lycopersicon lycopersicum* L. Karsten) fruit. *Acta Alimentaria* 38, 27–34.

Huang, Y., Tang, R., Cao, Q. and Bie, Z. (2009) Improving the fruit yield and quality of cucumber by grafting onto the salt tolerant rootstock under NaCl stress. *Scientia Horticulturae* 122, 26–31.

Huitrón, M.V., Diaz, M., Dianez, F. and Camacho, F. (2007) The effect of various rootstocks on triploid watermelon yield and qualify. *Journal of Food, Agriculture and Environment* 5, 344–348.

Huitrón-Ramírez, M.V., Ricárdez-Salinas, M. and Camacho-Ferre, F. (2009) Influence of grafted watermelon plant density on yield and quality in soil infested with melon necrotic spot virus. *HortScience* 44, 1838–1841.

Huxham, M.I., Jarvis, M.C., Shakespeare, L., Dover, C.D., Johnson, D., Knox, J.P. and Seymour, G.B. (1999) Electron-energy-loss spectroscopic imaging of calcium and nitrogen in the cell walls of apple fruits. *Planta* 208, 483–443.

Huyskens-Keil, S. and Schreiner, M. (2003) Quality of fruits and vegetables. *Journal of Applied Botany and Food Quality* 77, 147–151.

Hwang, J.M., Lee, W.H. and Hwang, G.S. (1992) Yield performance test of cucumber (*Cucumis sativus* L.) grafted with the rootstocks, *Sicyos angulatus* L. grown in the farm plastic house. *Korean Journal of Horticultural Science and Technology* 10, 34–35.

Isleyen, M., Sevim, P. and White, J.C. (2012) Accumulation of weathered *p,p′*-DDTs in grafted watermelon. *Journal of Agricultural and Food Chemistry* 60, 1113–1121.

Jang, Y., Cho, M., Um, Y., Ko, K. and Chun, C. (2012) Effect of grafting on growth and incidence of *Phytophtora* blight and bacteria wilt pepper (*Capsicum annuum* L.). *Horticulture, Environment, and Biotechnology* 53, 9–19.

Jang, Y., Huh, Y.C., Dong-Kum, P., Mun, B., Lee, S. and Um, Y. (2014) Greenhouse evaluation of melon rootstock resistance to Monosporascus root rot and vine decline as well as of yield and fruit quality in grafted 'Inodorus' melons. *Korean Society for Horticultural Science* 32, 614–622.

Kader, A.A. (1999) Fruit maturity, ripening and quality relationships. *Acta Horticulturae* 485, 203–208.

Kader, A.A. (2008) Flavor quality of fruits and vegetables. *Journal of the Science of Food and Agriculture* 88, 1863–1868.

Kemp, T.R. (1975) Identification of some volatile compounds from *Citrullus vulgaris*. *Phytochemistry* 14, 2637–2638.

Khah, E.M. (2011) Effect of grafting on growth, performance and yield of aubergine (*Solanum melongena* L.) in greenhouse and open-field. *International Journal of Plant Production* 5, 359–366.

Khah, E.M., Kakava, E., Mavromatis, A., Chachalis, D. and Goulas, C. (2006) Effect of grafting on growth and yield of tomato (*Lycopersicum esculentum* Mill) in greenhouse and open-field. *Journal of Applied Horticulture* 8, 3–7.

Krumbein, A. and Auerswald, H. (1998) Characterization of aroma volatiles in tomatoes by sensory analysis. *Nahrung Food* 42, 395–399.

Krumbein, A. and Schwarz, D. (2013) Grafting: a possibility to enhance health-promoting and flavour compounds in tomato fruits of shaded plants? *Scientia Horticulturae* 149, 97–107.

Krumbein, A., Peters, P. and Brückner, B. (2004) Flavour compounds and a quantitative descriptive analysis of tomatoes (*Lycopersicon esculentum* Mill.) of different cultivars in short-term storage. *Postharvest Biology and Technology* 32, 15–28.

Kyriacou, M.C. and Soteriou, G.A. (2012) Compositional, optical and physical quality change postharvest in grafted watermelon [*Citrullus lanatus* (Thunb) Matsum & Nakai] cultivars. *Acta Horticulturae* 934, 985–991.

Kyriacou, M.C. and Soteriou, G.A. (2015) Quality and postharvest behavior of watermelon fruit in response to grafting on inter-specific cucurbit rootstocks. *Journal of Food Quality* 38, 21–29.

Leal-Fernández, C., Godoy-Hernández, H., Núñez-Colín, C.A., Anaya-López, J.A., Villalobos-Reyes, S. and Castellanos, J.Z. (2013) Morphological response and fruit yield of sweet pepper (*Capsicum annuum* L.) grafted onto different commercial rootstocks. *Biological Agriculture and Horticulture* 29, 1–11.

Lee, J.M. (1994) Cultivation of grafted vegetables, I. Current status, grafting methods, and benefits. *HortScience* 29, 235–239.

Lee, J.M., Bang, H.J. and Ham, H.S. (1999) Quality of cucumber fruit as affected by rootstock. *Acta Horticulturae* 483, 117–123.

Leonardi, C. and Giuffrida, F. (2006) Variation of plant growth and macronutrient uptake in grafted tomatoes and eggplants on three different rootstocks. *European Journal of Horticultural Science* 71, 97–101.

Leskovar, D.I., Bang, H., Crosby, K.M., Maness, N., Franco, J.A. and Perkins-Veazie, P. (2004) Lycopene, carbohydrates, ascorbic acid and yield components of diploid and triploid watermelon cultivars are affected by deficit irrigation. *Journal of Horticultural Science and Biotechnology* 79, 75–81.

Liu, H.Y., Zhu, Z.J., Diao, M. and Guo, Z.P. (2006) Characteristic of the sugar metabolism in leaves and fruits of grafted watermelon during fruit development. *Plant Physiology Communications* 42, 835–840.

Liu, B., Ren, J., Zhang, Y., An, J., Chen, M., Chen, H., Xu, Ch. and Ren, H. (2015) A new grafted rootstock against root-knot nematode for cucumber, melon, and watermelon. *Journal of Agriculture and Food Chemistry* 62, 10504–10514.

López-Marín, J., González, A., Pérez-Alfocea, F., Egea-Gilabert, C. and Fernandez, J.A. (2013) Grafting is an efficient alternative to shading screens to alleviate thermal stress in greenhouse-grown sweet pepper. *Scientia Horticulturae* 149, 39–46.

Marsic, N.K., Mikulik-Petkovsek, M. and Stampar, F. (2014) Grafting influences phenolic profile and carpometric traits of fruits of greenhouse-grown eggplant (*Solanum melongena* L.). *Journal of Agricultural and Food Chemistry* 62, 10504–10514.

Martínez-Ballesta, M.C., Alcaraz-Lopez, C., Muries, B., Mota-Cadenas, C. and Carvajal, M. (2010) Physiological aspects of rootstock scion interactions. *Scientia Horticulturae* 127, 112–118.

Mayer, F., Takeoka, G., Butery, R., Nam, Y., Naim, M., Bezman, Y. and Rabinowich, H. (2002) Aroma of fresh field tomatoes. In: Cadwallader, K.R. and Weenen, H. (eds) *Freshness and Shelf Life of Foods*. ACS Symposium Series No. 836. American Chemical Society, Washington, DC, pp. 144–161.

Maynard, N.D. (2001) *Watermelons Characteristics, Production, and Marketing*. American Society of Horticultural Science Press, Alexandria, Virginia.

Maynard, D.N., Dunlap, A.M. and Sidoti, B.J. (2002) Sweetness in diploid and triploid watermelon fruit. *Cucurbit Genetics Cooperative Report* 25, 32–35.

Miguel, A., Maroto, J.V., San Bautista, A., Baixauli, C., Cebolla, V., Pascual, B., López-Galarza, S. and Guardiola, J.L. (2004) The grafting of triploid watermelon is an advantageous alternative to oil fumigation. *Scientia Horticulturae* 103, 9–17.

Mišković, A., Ilin, Z. and Marković, V. (2008) Effect of different rootstock type on quality and yield of tomato fruits. *Acta Horticulturae* 807, 619–622.

Mohammed, S.T.M., Humidan, M., Boras, M. and Abdalla, O.A. (2009) Effect of grafting tomato on different rootstocks on growth and productivity under glasshouse conditions. *Asian Journal of Agricultural Research* 3, 47–54.

Moncada, A., Miceli, A., Vetrano, F., Mineo, V., Planeta, D. and D'Anna, F. (2013) Effect of grafting on yield and quality of eggplant (*Solanum melongena* L.). *Scientia Horticulturae* 149, 108–114.

Morishita, M. (2001) Origin and characteristics of the local cucumber variety 'Kema' in the Osaka district. *Bulletin of Osaka Prefectural Agricultural and Forestry Research Center* 37, 27–34 (in Japanese with English abstract).

Mourão, I., Brito, L.M. and Coutinho, J. (2013) Evaluation of organic fertilization with composts in the production of grafted tomatoes with different pruning systems. In: Vallez, G., Cambier, P., Bacheley, H., Cheviron, N., Formisano, S., Lepeuple, A.S., Revallier, A. and Houot, S. *Proceedings of the 15th RAMIRAN International Conference of the FAO ESCORENA*, 3–5 June, p. S4.11 (poster).

Neocleous, D. (2010) Yield, nutrients and antioxidants of tomato in response to grafting and substrate. *International Journal of Vegetable Science* 16, 212–221.

Nicoletto, C., Tosini, F. and Sambo, P. (2013) Effect of grafting and ripening conditions on some qualitative traits of 'Cuore di bue' tomato fruits. *Journal of the Science of Food and Agriculture* 93, 1397–1403.

Oda, M., Nagata, M., Tsuji, K. and Sasaki, H.S.O. (1996) Effects of scarlet eggplant rootstock on growth, yield, and sugar content of grafted tomato fruits. *Journal of the Japanese Society for Horticultural Science* 65, 531–536.

Otani, T. and Seike, N. (2007) Rootstock control of fruit dieldrin concentration in grafted cucumber (*Cucumis sativus*). *Journal of Pest Science* 32, 235–242.

Oztekin, G.B. (2009) [Response of tomato rootstocks to salinity stress]. PhD thesis, Ege University, Graduate School of Natural and Applied Science, Bornova-İzmir, Turkey.

Passam, H.C., Stylianoy, M. and Kotsiras, A. (2005) Performance of eggplant grafted on tomato and eggplant rootstocks. *European Journal of Horticultural Science* 70, 130–134.

Peri, C. (2004) *Qualità Nelle Aziende e Nelle Filiere Agroalimentari*. Hoepli, Milano, Italy.

Perkins-Veazie, P. and Collins, J.K. (2006) Carotenoid changes of intact watermelons after storage. *Journal of Agriculture and Food Chemistry* 54, 5868–5874.

Perkins-Veazie, P., Zhang, X., Lu, G. and Huan, J. (2007) Grafting increases lycopene in seedless watermelon. *HortScience* 42, 959.

Petro-Turza, M. (1986) Flavor of tomato and tomato products. *Food Reviews International* 2, 309–351.

Petropoulos, S.A., Olympios, C., Ropokis, A., Vlachou, G., Ntatsi, G., Paraskevopoulos, A. and Passam, H.C. (2014) Fruit volatiles, quality, and yield of watermelon as affected by grafting. *Journal of Agricultural Science and Technology* 16, 873–885.

Pogonyi, A., Pek, Z., Helyes, L. and Lugasi, A. (2005) Effect of grafting on the tomato's yield, quality and main fruit components in spring forcing. *Acta Alimentaria* 34, 453–462.

Proietti, S., Rouphael, Y., Colla, G., Cardarelli, M., De Agazio, M., Zacchini, M., Rea, E., Moscaterllo, S. and Battistelli, A. (2008) Fruit quality of mini-watermelon as affected by grafting and irrigation regimes. *Journal of the Science of Food and Agriculture* 88, 1107–1114.

Pulgar, G., Villora, G., Moreno, D.A., and Romero, L. (2000) Improving the mineral nutrition in grafted watermelon plants: nitrogen metabolism. *Biologia Plantarum* 43, 607–609.

Qaryouti, M.M., Qawasmi, W., Hamdan, H. and Edwan, M. (2007) Tomato fruit yield and quality as affected by grafting and growing system. *Acta Horticulturae* 741, 199–206.

Qin, Y., Yang, C., Xia, J., He, J., Ma, X., Yang, C., Zheng, Y., Lin, X., He, Z., Huang and Z. Yan, Z. (2014) Effects of dual/threefold rootstock grafting on the plant growth, yield and quality of watermelon. *Notulae Botanicae Horti Agrobotanici Cluj-Napoca* 42, 495–500.

Romano, D., Paratore, A. and Vindigni, G. (2000) Caratteristiche dei frutti di pomodoro in rapporto a diversi portinnesti. In: *Proceedings of the Workshop on 'Applicazione di Tecnologie Innovative per il Miglioramento dell'Orticoltura Meridionale'*. Consiglio Nazionale delle Ricerche, Rome, pp. 113–114.

Rouphael, Y., Cardarelli, M. and Colla, G. (2008) Yield, mineral composition, water relations, and water use efficiency of grafted mini-watermelon plants under deficit irrigation. *HortScience* 43, 730–736.

Rouphael, Y., Schwarz, D., Krumbein, A. and Colla, G. (2010) Impact of grafting on product quality of fruit vegetables. *Scientia Horticulturae* 127, 172–179.

Ruiz, J.M. and Romero, L. (1999) Nitrogen efficiency and metabolism in grafted melon plants. *Scientia Horticulturae* 81, 113–123.

Ryu, J.S., Choi, K.S. and Lee, S.S. (1973) Effect of grafting stocks on growth, quality, and yields of watermelon. *Journal of the Korean Society for Horticultural Science* 13, 45–49.

Sakata, Y., Ohara, T. and Sugiyama, M. (2007) The history and present state of grafting of cucurbitaceous vegetables in Japan. *Acta Horticulturae* 731, 159–170.

Saladie, M., Matas, A.J., Isaacson, T., Jenks, M.A., Goodwin, S.M., Niklas, K.J., Ren, X.L., Labavitch, J.M., Shackel, K.A., Fernie, A.R., Lytovchenko, A., O'Neill, M.A., Watkins, C.B. and Rose, J.K.C. (2007) A reevaluation of the key factors that influence tomato fruit softening and integrity. *Plant Physiology* 144, 1012–1028.

Salehi, R., Kashi, A., Lee, J. and Javanpour, R. (2014) Mineral concentration, sugar content and yield of Iranian 'Khatooni' melon affected by grafting, pruning and thinning. *Journal of Plant Nutrition* 37, 1255–1268.

Sánchez-Rodríguez, E., Ruiz, J.M., Ferreres, F. and Moreno, D.A. (2012) Phenolic profiles of cherry tomatoes as influenced by hydric stress and rootstock technique. *Food Chemistry* 134, 775–782.

Sánchez-Torres, .P, Raigón, M.D., Gammoudi, N. and Gisbert, C. (2016) Effects of grafting combinations on the nutritional composition of pepper fruit. *Fruits* 71, 249–256.

Sanz, C., Olias, J.M. and Perez, A.G. (1997) Aroma biochemistry of fruits and vegetables. In: *Phytochemistry of Fruit and Vegetables*. Oxford University Press, New York, pp. 125–155.

Satoh, S. (1996) Inhibition of flowering of cucumber grafted on rooted squash rootstocks. *Physiologia Plantarum* 97, 440–444.

Savvas, D., Savva, A., Ntatsi, G., Ropokis, A., Karapanos, I., Krumbein, A. and Olympios, A. (2011) Effects of three commercial rootstocks on mineral nutrition, fruit yield, and quality of salinized tomato. *Journal of Soil Science and Plant Nutrition* 174, 154–162.

Schnitzler, W.H. and Gruda, N. (2002) Hydroponics and product quality. In: Savvas, D. and Passam, H.C. (eds) *Hydroponic Production of Vegetables and Ornamentals*. Embryo Publications, Athens, pp. 373–411.

Schwab, W., Davidovich-Rikanati, R. and Lewinsohn, E. (2008) Biosynthesis of plant-derived flavor compounds. *Plant Journal* 54, 712–732.

Schwarz, D., Oztekin, G.B., Tüzel, Y., Brückner, B. and Krumbein, A. (2013) Rootstocks can enhance tomato growth and quality characteristics at low potassium supply. *Scientia Horticulturae* 149, 70–79.

Shewfelt, R. (1999) What is quality? *Postharvest Biology and Technology* 15, 197–200.

Slimestad, R., Fossen, T. and Verhuel, M.J. (2008) The flavonoids of tomatoes. *Journal of Agricultural and Food Chemistry* 56, 2436–2441.

Soteriou, G.A. and Kyriacou, M.C. (2015) Rootstock mediated effects on watermelon field performance and fruit quality characteristics. *International Journal of Vegetable Science* 21, 344–362.

Soteriou, G.A., Kyriacou, M.C., Siomos, A.S. and Gerasopoulos, D. (2014) Evolution of watermelon fruit physicochemical and phytochemical composition during ripening as affected by grafting. *Food Chemistry* 165, 282–289.

Steward, A.J., Bozonnet, S., Mullen, W., Jenkins, G.I., Lean, M.E.J. and Crozier, A. (2000) Occurrence of flavonols in tomatoes and tomato-based products. *Journal of Agricultural and Food Chemistry* 48, 2663–2669.

Tabata, K., Oba, K., Suzuki, K. and Esaka, M. (2001) Generation and properties of ascorbic acid-deficient transgenic tobacco cells expressing antisense RNA for L-galactono-1,4-lactone dehydrogenase. *Plant Journal* 27, 139–148.

Takasu, E., Wu, D.L. and Imai, H. (1996) Effects of grafting on blossom end rot incidence in summer tomato production. *TVIS Newsletter* 1, 22–23.

Tandon, K.S., Baldwin, E.A. and Shewfelt, R.L. (2000) Aroma perception of induviduel volatile compounds in fresh tomatoes (*Lycopersicon esculentum* Mill.) as affected by the medium of evaluation. *Postharvest Biology and Technology* 20, 261–268.

Theodoropoulou, A., Giotis, C., Hunt, J., Gilroy, J., Toufexi, E., Liopa-Tsakalidis, A., Markellou, A., Lueck, L., Seal, C. and Leifert, C. (2007) Effect of variety choice and use of resistant rootstock on crop yield and quality parameters of tomato plants grown in organic, low input and conventional production systems/growth media. In: Niggli, U., Leifert, C., Alföldi, T., Lück, L. and Willer, H. (eds) *Proceedings of the International Congress of the European Integrated Project Quality Low Input Food (QLIF): Improving Sustainability in Organic and Low Input Food Production Systems*, 20–23 March. University of Hohenheim, Stuttgart, Germany, pp. 177–180.

Tlili, I., Hdider, C., Lenucci, M.S., Ilahy, R., Jebari, H. and Dalessandro, G. (2011) Bioactive compounds and antioxidant activities during fruit ripening of watermelon cultivars. *Journal of Food Composition and Analysis* 24, 923–928.

Tomas-Barberan, F.A. and Espin, J.C. (2001) Phenolic compounds and related enzymes as determinants of quality in fruits and vegetables. *Journal of the Science Food and Agriculture* 81, 853–876.

Traka-Mavrona, E., Koutsika-Sotiriou, M. and Pritsa, T. (2000) Response of squash (*Cucurbita* spp.) as rootstock for melon (*Cucumis melo* L.). *Scientia Horticulturae* 83, 353–362.

Tsouvaltzis, P.I., Siomos, A.S. and Dogras, K.C. (2004) The effect of the two tomatoes grafting on the performance, earliness and fruit quality. In: *Proceedings of the 21st Pan-Hellenic Congress of the Greek Society for Horticultural Science*, 8–10 October 2003, Vol. 11. Ioannina, Greece, pp. 51–55.

Tucker, G.A. (1993) Introduction. In: Seymour, G.B., Taylo, R.J.E. and Tucker, G.A. (eds) *Biochemistry of Fruit Ripening*. Chapman & Hall, London, pp. 1–51.

Turhan, A., Ozmen, N., Serbeci, M.S. and Seniz, V. (2011) Effects of grafting on different rootstocks on tomato fruit yield and quality. *Horticultural Science* 38, 142–149.

Tüzel, Y., Duyar, H., Oztekin, G.B. and Gül, A. (2009) [Effects of tomato rootstocks on plant growth, temperature sum requirements, yield and quality in different planting dates]. *Journal of Ege University Faculty of Agriculture* 46, 79–92.

Ulukapi, K. and Onus, A.N. (2005) Effect of grafted tomato (cv. 191 F1) on yield and fruit quality. In: *GAP IVth Agricultural Congress*, Sanliurfa, Turkey, pp. 1314–1317.

Verzera, A., Dima, G., Tripodi, G., Condurso, C., Crinò, P. and Romano, D. (2014) Aroma and sensory quality of honeydew melon fruits (*Cucumis melo* L. subsp. *melo* var. inodorus H. Jacq.) in relation to different rootstocks. *Scientia Horticulturae* 169, 118–124.

Vinkovic-Vrcek, I., Samobor, V., Bojic, M., Medic-Saric, M., Vukobratovic, M., Erhatic, R., Horvat, D. and Matotan, Z. (2011) The effect of grafting on the antioxidant properties of tomato (*Solanum lycopersicum* L.). *Spanish Journal of Agricultural Research* 9, 844–851.

Wadano, A., Azeta, M., Itotani, S., Kanda, A., Iwaki, T., Taira, T., Fujii, Y., Nishiura, Y., Murase, H. and Honami, N. (1999) Change of ascorbic acid level after grafting of tomato seedlings. *Zeitschrift für Naturforschung* 54, 830–833.

Xu, C.Q., Li, T.L. and Qui, H.Y. (2006) Effects of grafting on development, carbohydrate content, and sucrose metabolizing enzymes activities of muskmelon fruit. *Acta Horticulturae* 33, 773–778.

Yajima, I., Sakakibara, H., Ide, J., Yanai, T. and Hayashi, K. (1985) Volatile flavor components of watermelon (*Citrullus vulgaris*). *Agricultural and Biological Chemistry* 49, 3145–3150.

Yamasaki, A., Yamashita, M. and Furuya, S. (1994) Mineral concentrations and cytokinin activity in the xylem exudate of grafted watermelons as affected by rootstocks and crop load. *Journal of the Japanese Society for Horticultural Science* 62, 817–826.

Yarsi, G., Sari, N., Sari, N., Solmaz, I. and Aras, V. (2012) Determination of aroma compounds of grafted and ungrafted Galia C8 melon cultivar in greenhouse growing. In: Sari, N., Solmaz, I. and Aras, V. (eds) *Cucurbitaceae 2012. Proceedings of the Xth EUCARPIA Meeting on Genetics and Breeding of Cucurbitaceae*, Antalya, Turkey, 15–18 October. Ziraat Fakultesi, University of Cukurova, Turkey, pp. 415–420.

Yetisir, H. and Sari, N. (2003) Effect of different rootstock on plant growth, yield and quality of watermelon. *Australian Journal of Experimental Agriculture* 43, 1269–1274.

Yetisir, H., Sari, N. and Yücel, S. (2003) Rootstock resistance to fusarium wilt and effect on watermelon fruit yield and quality. *Phytoparasitica* 31, 163–169.

Zhao, X., Guo, Y., Huber, D.J. and Lee, J. (2011) Grafting effects on postharvest ripening and quality of 1-methylcyclopropene-treated muskmelon fruit. *Scientia Horticulturae* 130, 581–587.

Zhou, X., Wu, Y., Chen, S., Chen, Y. and Zhang, W. (2014) Using *Cucurbita* rootstock to reduce Fusarium wilt incidence and increase fruit yield and carotenoid content in oriental melons. *Horticultural Science* 49, 1365–1369.

9 Practical Applications and Speciality Crops

Amnon Koren,[1*] Eyal Klein,[1] J. Anja Dieleman,[2] Jan Janse,[2] Youssef Rouphael,[3] Giuseppe Colla[4] and Isabel Mourão[5]

[1]*Hishtil Nurseries, Moshav Nehalim, Israel;* [2]*Wageningen UR Greenhouse Horticulture, Wageningen, The Netherlands;* [3]*University of Naples Federico II, Portici, Italy;* [4]*University of Tuscia, Viterbo, Italy;* [5]*Instituto Politécnico Viana do Castelo, Ponte de Lima, Portugal*

9.1 Establishment of Grafted Transplants under Mediterranean Climate Conditions

Plantlet establishment is the part of horticultural sciences that focuses on the parameters and conditions enabling maximal throughput from propagation material, optimal yield from the crop and best horticultural performance under commercial growing conditions.

The major factors affecting plantlet establishment are seed health and vigour, plantlet quality and health, and the environmental conditions under which the transplantation is performed, namely abiotic and biotic stresses. Implementation of the best conditions at the transplanting stage and adjustment of agrotechniques enable the genetic potential of the seed to be realized (Grassbaugh and Bennett, 1998).

The Mediterranean climate is characterized by a long, hot summer with little precipitation, followed by a relatively short winter that is either dry with little rain or rich in precipitation from short rainstorms (Bolle, 2003; Alpert *et al.*, 2006). During most of the year, the level of radiation is high, which makes the summer very hot and confers moderate temperatures during most of the winter (Pardossi *et al.*, 2004). In the Mediterranean environment, the traditional planting dates start at the end of the warm summer season due to the unfavourable temperatures inside the vegetable growing greenhouses during the summer (up to 45–48°C), which are usually equipped with only rudimentary ventilation systems and no

* amkor52@gmail.com

active cooling. The planting dates vary according to region and crop. In Italy and France, transplanting takes place mainly in winter time (November–March) for Cucurbitaceae (i.e. melon and watermelon), although in France there is a small area of melon transplanting in the summer. The main tomato transplanting in Sicily occurs in both the summer (July–August) and winter (December–January); in the centre and north of Italy, transplanting occurs from January until March/April. In France, tomato transplanting starts in December, mainly in rockwool soil-less substrate, and goes on until March/April in both soil-less substrate and in soil. In Israel, current planting dates have been achieved by extending the growing season, usually starting in the warm summer season (July/August), characterized by extremely hot air and soil temperatures (Arava valley and Jordan valley). As cropping progresses, the climatic conditions become moderate to cold with the transition to winter. The transition from winter to spring is rapid, with many warm days and high levels of radiation ranging between 15 and 30 MJ m^{-2} day^{-1} from March to June (Castilla and Baeza, 2013; Ityel *et al.*, 2014). In many Mediterranean habitats, there is a trend towards growing under insect-proof nets or in greenhouses to avoid aphids, other insects and virus vectors, such as *Bemisia tabaci* (Castilla, 2002). The general trend towards reduced pesticide application is promoting the use of physical barriers, such as dense screens, which contribute to higher temperatures and humidity in the greenhouse.

This section describes the factors involved in grafted transplant establishment in Mediterranean habitats, as is practised on a large scale in Israel, Turkey, Spain, Italy, France and Greece and on a smaller scale in many other countries in the region.

9.1.1 Factors affecting the establishment of grafted plants

Seeds

The seed is one of the most important components in the production of grafting. Unlike regular seedling production, which usually includes a single seed per seedling, grafted transplant production requires two seeds for each plantlet – one for the scion and the other for the rootstock. This does not necessarily increase demand for seeds because grafted compared with non-grafted crops are grown at lower planting densities, often with two main stems per rootstock (*see* Chapter 3, this volume). However, the grafting procedure requires vigorous rootstock seeds, which should include well defined structural features and high uniformity (Mavi *et al.*, 2006; Kubota and McClure, 2008). For example, for grafting cucurbitaceous plants using a diagonal cut, which is also known as one-cotyledon splice grafting (*see* Chapter 1, this volume), vigorous seedlings are needed with good cotyledon quality. In this method, the apical meristem and one cotyledon of the rootstock are removed by an approximate 65° cut; the scion is cut above the cotyledons at the same angle and then attached to the rootstock at the remaining cotyledon (Hassell *et al.*, 2008). Less vigorous seeds or cotyledon disorders will not suit this process. It should be noted that the transition to automatic grafting machinery requires uniformity and consistency of high-quality plant materials – both scion and rootstock seedlings (Chiu *et al.*, 2010). Seeds with high and uniform germination performance are a key factor for high throughputs (Kubota and McClure, 2008).

In recent years, seeds have been the main vector for several epidemic outbreaks in crops (de León *et al.*, 2011; Dutta *et al.*, 2014). Bacterial pathogens such as *Clavibacter michiganensis* and *Acidovorax citrulli* in vegetables, tobamoviruses such as cucumber green mottle mosaic virus (CGMMV), tobacco mosaic virus (TMV), tomato mottle virus and pepper mild mottle virus (PMMV) and viroids are easily contaminated via plant and soil by mechanical means, in addition to seedborne 'vertical' transmission. Their spread within the nursery and by the nursery product to the growers can be proliferated by the activities associated with grafting, such as sorting, cutting of the rootstock and the scion, and attaching them together. Seed health issues are intensified in grafting, and the required standard is higher than in normal production of non-grafted seedlings.

There is a vital need to improve standard tests for the detection of seedborne diseases (Shang *et al.*, 2011; Lecoq and Desbiez, 2012). For example, during 2013, a severe epidemic of CGMMV occurred in many of the watermelon plots in Israel. This was found to be because the seed inspections, which had been performed in the best seed-producing laboratories around the world, had used a kit that was not suitable for identifying the specific strain of the virus. The test results for virus infection in the seeds were negative, whereas in fact the seeds were infected. This situation led to a large-scale epidemic in the fields and greenhouses. Using serological detection of CGMMV in the inner parts of infected seeds may be more accurate for infected batch identification (Reingold *et al.*, 2015).

Development of more effective testing methods that are well suited to larger samples of seeds is essential to avoid epidemics such as that caused by CGMMV. This is particularly true in the case of tomatoes, which are highly sensitive to *C. michiganensis* (Mathis *et al.*, 2015). These bacteria are transmitted by the seeds both internally and on the seedcoat. Even extremely low rates of 25 colony-forming units in one seed in 10,000 might cause disease outbreaks in nurseries, especially in grafted transplants. The existing methods to identify the presence of bacteria in tomato seed lots, although sensitive (OEPP/EPPO, 2013), are still insufficient. This is because the representative sample size needed for a seed batch requires tens of thousands of seeds, placing economic difficulties in its implementation. Therefore, *C. michiganensis* epidemics are relatively common in many Mediterranean countries, mainly in grafted tomato production plots; detection in symptomless plantlets has not yet been fully validated.

Plantlets

Prior to grafting, seedlings should have the correct proportions among their components: rootstock size (i.e. height and diameter) and scion size (Hassell *et al.*, 2008). Seedlings that are too elongated are weak, exposed to burns during summer (heat, radiation) and might display further negative effects during cropping. However, components that are too short are difficult to graft correctly and their general development might be too slow during the winter.

The seedling should be healthy and free of diseases. To ensure disease-free seedlings, especially from viruses, which are transferred mainly by aphids and other insect vectors, a 50-mesh screen (50 openings per square inch) is essential for greenhouses in the Mediterranean region (Harel *et al.*, 2013). The hot season, when the insects are most active, creates a major challenge, as the heat load in the greenhouse is maximal, and professional skills are required to enable proper development of the grafted transplants.

Although good cotyledon quality is important for Cucurbitaceae, French growers absolutely require that plants of melon, grafted on to squash or melon, are provided without the cotyledon of the rootstock in order to avoid problems with *Oidium* spp. infection (powdery mildew). In Israel, in recent years, due to the threat of tobamoviruses transmitted by mechanical means, handling and direct touching of the plantlet after one-cotyledon grafting is minimal and therefore the cotyledon is not removed. Powdery mildew is instead treated by chemical and cultural methods.

The structure of the root system is important for plantlet establishment (*see* Plate 19). It is affected by nursery management as well as field practices such as irrigation methods (Sánchez-Rodríguez *et al.*, 2014). Root air pruning is common in the plug transplant production (Cantliffe, 2008). Root tips that are exposed to dry air by holes at the bottom of the trays become dried. This generates hormonal changes in the root system, and increases splitting and the production of lateral roots instead of taproots (Leskovar and Stoffella, 1995). Twisting roots adversely affect plantlet establishment, and enhance root vulnerability to soilborne pathogens such as *Macrophomina phaseolina*, as well as other pathogens. The nursery should avoid conditions that enhance root curling, especially in the Cucurbitaceae. Root architecture can also be adjusted by using pressed peat plugs or Ellepots (plugs made from degradable paper cylinders, filled with growth medium; http://ellepot.dk/ellepotsystem.html, accessed 24 November 2016), as an alternative method of air pruning. Pressed peat is used for watermelon in Italy and for melon in France; currently, more than 60% of these young plants are produced in pressed peat plugs. In Israel, the Ellepot system is more common for grafted Cucurbitaceae. The combination of a healthy root system and irrigation methods that are adjusted to the field conditions contributes to deep root systems and normal plant development.

The collapse of grafted transplants in the field shortly after planting may be related to nursery practices. Grafted watermelons may collapse shortly after transplanting if the scion and rootstock are not compatible by genotype, age, or stalk diameter. Low tolerance to environmental conditions and biotic stress may result from improper healing and/or hardening off process. Grafted plants may be infected by soilborne pathogens even though a resistant rootstock is being used because of the development of susceptible roots from the scion. For example, internal rooting in cucumber happens when the scion roots into an internal hole of the rootstock stem (*see* Plate 20). Usually, this problem appears in cucurbit plant species grafted by using the hole insertion method (Hassell *et al.*, 2008), but also when the one-cotyledon grafting method is used. This may lead to infection with soilborne pathogens that the rootstock is apparently resistant to, such as *Fusarium oxysporum* f. sp. *radices-cucumerinum*, which causes crown and root rot in cucumber and melon (Pavlou *et al.*, 2002). Similarly, a bypass may occur when the scion roots directly in the growth medium after transplanting. This phenomenon, which is quite common in tomato and other Solanaceae but less so in the Cucurbitaceae, occurs when the rootstock is too short for grafting or when the grafted position is too low on the rootstock.

Logistics and transportation

Under the Mediterranean climate, transporting plantlets from the nursery to the field is a sensitive practice, and problems during planning, transport or at some other station along the way to the field might negatively affect plantlet establishment (Kubota and Kroggel, 2006). Insufficient irrigation before packing, transportation over long distances under non-cooled conditions or overheating immediately prior to transplanting can all impede plantlet establishment, and exposure to high temperatures, even for a few hours, may be deleterious to the plantlets. Vegetables, mainly cucumbers, accumulate ethylene under heat stress, which enhances plant ageing and inhibits establishment. Transplanting plantlets with low turgor or at high temperatures has similar negative effects. Long exposure of plants to low temperature may also reduce turgor, inhibit water absorption and decrease establishment success (Justus and Kubota, 2010). Because of the need to protect the plants from pests during transportation, different coverings are used, and these tend to intensify temperature build-up and heat stress. Shortened delivery times, maintaining a suitable temperature before planting and minimizing waiting times all facilitate the development of plantlets in the field.

Plot selection

The high management expense of greenhouses, net-houses and tunnels, and a scarcity of additional land suitable for agricultural use, have resulted in intensive use of available lands and continuous cropping on the same plots. The main limiting factor of this practice is the build-up of phytotoxic chemical residues and of soilborne pathogens, mainly fungi and nematodes but also viruses. The phase-out of methyl bromide for soil disinfection has led to applications of different soil fumigants and other pesticides and herbicides, some of which cause further residual build-up, requiring an awareness of the possibility that any residue might have a detrimental effect and verification of acceptable chemical levels before planting (Colla *et al.*, 2012).

Plot preparation

An area designated for planting grafted transplants should be totally cleared of weeds and previous crop residues (Katan, 2000). However, due to high costs, time investment and labour shortages, there has been a trend in recent years towards taking short cuts in plot preparation. Some farmers pulverize the previous crop's residues and spread them above the plots or incorporate them into the ground. Shallow and partial cultivation and no-till farming have also become popular agrotechnology practices. Although there are some benefits to these methods (Triplett and Dick, 2008), planting plantlets into residues of previous crops may inhibit plantlet establishment, mainly when the soil contains raw residues such as fibrous root systems, leaves and fruit, and may also increase the potential for hazards, such as pathogens that have remained in the previous crop's residues, as well as chemicals and salinity. For short periods, no-till cultivation may be used if sanitation, residue removal and washing are applied before planting. In the long term, any cultivation should be adjusted to soil type and season. In soils low in organic matter, compost amendments can improve soil structure (Mekki *et al.*, 2014). In heavy soils, high beds are recommended to improve drainage.

In warm areas and seasons, soil solarization is recommended wherever possible to control soilborne pests (Gamliel and Katan, 2012). Soil solarization consists of covering moist soil with a layer of transparent plastic and exposure to sunlight for a few weeks in the summer. Solarization can be used alone, or in combination with other chemical or biological agents as an integrated pest management programme in high-value horticultural crops grown in greenhouses and open fields. Successful solarization requires effective soil preparation and the use of suitable tarping and plastics technology (van Bruggen *et al.*, 2016). Cold and moderate seasons limit soil heating and reduce heating deeper in the soil. Other limiting factors are the long period (30–60 days) required for an effective control process, although this can be shorter if combined with chemical soil disinfection, and the need for plastics disposal. Interactions with grafted transplants should be considered (Fallik *et al.*, 2016): suitable plot preparation, consideration of the history of soilborne diseases, organic inputs and fertilizers, and reduction of soil salinity will support the grafted plant in its first stages in the field and enable successful cropping.

There exist some improved methods for coping with soilborne pathogens (Jarvis, 1989). For example, application of an intermediate medium before transplanting has been found to be effective against tobamovirus infection in infested soils (Antignus *et al.*, 2005). This medium (clean soil, compost or inert medium) can be applied in pits or in strips along the plot and the plantlets are transplanted into it, thereby preventing infestation of the freshly injured roots by viruses. The new roots that break through the intermediate medium without transplanting injuries will be less sensitive to viral infection. This technique was successfully applied for grafted melons with a sensitive rootstock, which were transplanted in CGMMV-infested soil, and for grafted peppers that were sensitive to PMMV and were planted in infested soil (Antignus, 2012; Reingold *et al.*, 2016).

The increased threat of soilborne pathogens, especially fungi and nematodes, in recent few years and depletion of permitted chemical soil disinfectants have contributed to increasing use of soil-less growth media such as rockwool and coco peat slabs. In Italy, increasing numbers of hectares of soil-less tomato are cultivated in coco peat slabs every year, and this system is also increasing in France, replacing soil cultivation. Soil-less growth on rockwool is also increasing but less rapidly. Soil-less substrates are usually highly conducive to soilborne diseases (Diara *et al.*, 2012), mainly when used the first time; therefore, plantlets grafted on to resistant rootstocks are recommended. An additional advantage of the grafted transplants is enhanced vigour and improved establishment in soil-less growing systems (Patakioutas *et al.*, 2015).

Planting

The grafted transplant should be transplanted carefully and with specific attention. As most the growth media include hydrophobic substrates, it should be confirmed that the plantlet has been well irrigated before transplanting. Otherwise, absorption of water from the soil will be less effective. It is recommended that the rootstock be squeezed gently before transplanting. If it discharges some water, then the irrigation is regarded as sufficient. If the rootstock is only damp or even dry, flooding the bottom of the trays (while keeping the foliage dry) in a cool water tank for a few minutes will wet the rootstock before transplanting. Deep planting and burial of the grafted point in wet soil creates a direct connection between the scion and the soil and encourages bypass of the rootstock by direct rooting. The size of the receiving hole for transplanting should be appropriate to the rootstock size to avoid air spaces, which lead to partial development of the new root system (*see* Plate 21). However, intensive compaction around the rootstock, twisted plantlet roots or insufficient wetting during transplanting causes significant delay in plantlet establishment. A certain distance between the ground and grafting point should be maintained, and covering the grafting point with soil, plastic mulch or other moisture-retaining materials is to be avoided. The grafted transplants should not be placed on the ground before planting, to avoid infection with pathogens such as tobamoviruses (Antignus *et al.*, 2005). These viruses can be present in plant residues, soil and organic particles, and any direct contact between them and the plantlet can lead to later infestation of the crop. Therefore, planting should be done directly from the tray into the transplanting holes, followed by gentle covering and pressing around the rootstock to avoid large air spaces.

9.1.2 Abiotic stress

Immediately after transplanting, the plantlet can be exposed to abiotic stresses, such as temperature (hot or cold ambient or soil temperature, high radiation), water availability (drought or salt) or flooding (lack of oxygen in the root zone) (Schwarz *et al.*, 2010). As a result, plantlet sensitivity to biotic stresses can also increase (Atkinson and Urwin, 2012). For example, exposure to hot or cold temperatures causes damage to the root membrane and leakage (Mahajan and Tuteja, 2005), thereby increasing plantlet sensitivity to pathogens such as *Pythium* spp. (Pivonia *et al.*, 2012).

Exposure to stresses at the establishment stage can affect the structure of the plant later on. For example, high radiation, either direct or reflected from the ground, increases plantlet height in a short time as an avoidance response to excess radiation (Takaichi *et al.*, 2000).

Heat and radiation

Heat stress is the most common abiotic stress during summer or autumn plantings in the Mediterranean region (Harel *et al.*, 2013). It is enhanced by high radiation and long days. Heat stress can be followed by osmotic stress due to water shortages and turgor loss. Plastic mulching of the soil and poor structural ventilation make the problem worse (*see* Plate 22). The mulching film colour also has a significant effect on stress potential (Lamont, 1995): black or transparent coverings absorb the radiation, increase soil and ambient temperatures around the plantlet, and may suppress new root development or even burn the plant where it is in direct contact with the covering (*see* Plate 23). Drip-irrigation pipes deployed under the mulch also absorb and emit heat and may harm the plantlets at points of direct contact.

Use of small planting holes in the mulch or covering of these holes with soil after planting can cause heat build-up in the rootstock, as soil cooling is limited under the mulch (*see* Plate 24). Heat is usually lost through evaporation, but if the soil is mulched, and the holes in the mulch are too narrow or covered, the high temperatures around the roots can become detrimental. If the ground is covered, large holes of more than 10 cm in diameter should be made to allow the soil around the plantlet to cool off by evaporation. Shading can also have a similar effect, but its combination with evaporation is most effective. Understanding the different stress factors in the different cropping sites and seasons will help in developing further techniques to reduce stress, such as additional plastic film mulching to avoid low temperatures, or application of non-woven fabric sheets or shading nets to reduce high radiation.

The grafted transplant is sensitive to high ambient and soil temperatures (Garibaldi and Minuto, 2003; Cohen *et al.*, 2007). As the grafted transplant is undergoing graft healing during the days before planting, the water-transport system between the scion and the rootstock is still incomplete and hence highly sensitive. For example, certain botanical varieties of *Cucumis melo* (melon), grafted on to *Cucurbita* (pumpkin) rootstocks, are sensitive to high soil temperature, which may cause physiological incompatibility at an early stage (Aloni *et al.*, 2008) or a late stage, during fruit ripening (Soteriou *et al.*, 2016). This phenomenon might be reflected by poor root development, early flowering and inhibited vegetative growth. Reduction of soil temperature by adequate mulch or shading successfully prevents the negative effects of heat on these grafted transplants. However, shading should be used properly and for a limited time during plantlet establishment. A constant and continuous reduction in radiation may decrease plant functioning, fitness, vigour and root-system development. Hence, any action taken after transplanting should be adjusted according to accumulated experience throughout the growing season and according to the developmental stage of the plant.

Cold temperature

Low temperature causes a delay in the development and growth of the plantlets (Korkmaz and Dufault, 2001). Temperatures that are too low will delay the appearance of the first inflorescence. In general, vegetable transplants are very sensitive to low temperatures (Jouyban *et al.*, 2013), and this is particularly noticeable in grafted cucumber and grafted melon (Ahn *et al.*, 1999). During the winter, it is recommended that the plantlets be protected using a plastic cover or non-woven fabric sheets, even if the plantlets are planted indoors. In temperate regions in the Mediterranean area, these non-woven fabric sheets are used mainly during the cold season as row covers. They have different purposes, such as to conserve warmth, stimulate germination and early growth, protect plants from frost injury and improve the quality of the crops.

The date of planting is critical in terms of plantlet development. In Israel, planting in the middle of winter – late November to mid-December – inhibits plantlet establishment and slows down plantlet development. Before or after mid-winter, moderate stress is expected, and plantlets will become established and develop faster than those planted in the middle of winter. Hence, one means of avoiding plantlet growth retardation is transplanting before or after the lowest winter temperatures.

Low oxygen

Poorly drained soil or suboptimal irrigation-management strategies can cause a temporary excess of water; this situation may result in low-oxygen stress in the roots, followed by toxicity and death (Patel *et al.*, 2014). In sensitive crops such as cucumber, growth and development are retarded, but toxicity will only become apparent in the shoots at a later stage (van Dongen and Licausi, 2014). Therefore, it is important to identify the various abiotic stress factors and avoid them in order to prevent consequent damage by biotic stresses.

Salinization

Salinization is often associated with irrigated areas that are characterized by low rainfall and high evapotranspiration (Postiglione, 2003; Rengasamy, 2006). Excessive fertilization and irrigation with water containing high levels of salt dramatically aggravate the problem (Balliu *et al.*, 2012). Although grafted plants are relatively tolerant to high salinity (Edelstein *et al.*, 2005), it can influence normal functioning. High salinity levels, as well as high concentrations of specific ions in the irrigation water, reduce the yield potential in melon plants (Edelstein *et al.*, 2005); however, grafting on to a suitable rootstock can improve fruit quality under stress conditions (Colla *et al.*, 2006). Water shortages and high management costs have led many farmers to stop soil washing by sprinklers to remove salts until after tillage. The alternative of washing through conventional drip-irrigation systems with dripper intervals of 20–50 cm is not sufficient, as the salts accumulate at the margins of the wetting area and in the upper layer, instead of being flushed into the deeper soil layers. Soil tillage increases salt accumulation and fertilizer residues in the root zone from deeper layers. Ineffective or partial implementation of washing can result in high salt concentrations in the plantlet root zone, reduced water availability and increased ion toxicity, mainly from sodium and chloride ions (Shalhevet, 2004). It is therefore recommended to wash the soil with sprinklers, but if drip irrigation is used, it should include short dripper spacing or double the rate of dripping into a bed. Transplanting should only be performed in the washed zone, and planting at the bed margins where the salts concentrate to their highest level should be avoided.

The combination of salinization and high temperature has a weakening effect on grafted transplants such as cucumber or melon, which are particularly sensitive crops (Al-Sadi *et al.*, 2010); this combination makes them highly vulnerable to pathogens, mainly *Pythium* spp. If transplanting is carried out indoors and under hot climatic conditions, it is recommended that the ground at the bottom of the rootstock is well packed while leaving its upper part uncovered, that is, the hypocotyl and the adjacent roots at the crown (roots at the top of the plug) should left exposed. This will help avoid salt damage to the lower part of the stem, which is the most salt-sensitive part of the grafted transplant. This method should be performed carefully and only in the hot season, and is not be intended for application during the winter.

9.1.3 Biotic stress

Weeds and self-seeding

Self-seeded plants or residues from the prior crop, as well as weeds, can become hosts for pests such as mites, thrips, mealybugs and whiteflies (Jarvis, 1989; Wisler and Norris, 2005). In some cases, they could be a source of acquired virus, or other inoculum of plant pathogens. Sanitation and removal of previous crop residues are effective for reducing these hazards (Ucko and Dayan, 1986; Katan, 2004; Louws *et al.*, 2010). Chopping and rototilling of crop residues into the plot may allow pathogenic microorganisms to survive, with consequent risk for the next crop.

Viruses

The major viral risk factors in intensive crops include viruses from the family *Tobamovirus* such as CGMMV, TMV and ToMV (Loebenstein and Lecoq, 2012). In recent years, these seedborne viruses, which are spread by crop residues and agricultural equipment, have been found to significantly infest and survive in agricultural land. As already mentioned, an effective method for preventing root infection by viruses is application of a physical buffer before transplanting. Other groups of viruses, transmitted by vectors such as whiteflies or aphids, can damage plantlets after planting (Tomlinsone, 1987). Physical barriers, mainly nets or screens (minimum 50-mesh) prevent pest invasion or enable insect confusion, based on wavelength-dependent insect vision. Such barriers reduce the proliferation and direct damage from these potential viral vectors (Antignus and Ben-Yakir, 2004). UV-absorbing films that significantly reduce transmission of UV-A and UV-B natural solar radiation can interfere with the vision of pests. The exclusion of UV from natural sunlight hampers flight initiation, dispersal, host finding and establishment, hence affecting the population build-up (Gulidov and Poehling, 2013). For example, UV-absorbing plastic roofs showed a pronounced deterrent effect for movement of thrips towards greenhouses, and the UV-absorbing net effectively reduced the number of thrips crossing the net barrier into the greenhouse (Nguyen *et al.*, 2009).

Bacteria

Exclusion of pathogenic inoculum sources before transplanting is essential to decrease bacterial infestation after transplanting (van der Wolf and de Boer, 2015). For example, one of the major sources of *Clavibacter* survival in tomatoes is fruit residues from the previous crop. Keeping the plot clean of residues effectively decreases inoculum density and possible sources for epidemics. Guttation should be considered as an additional source for the secondary spread of *C. michiganensis* (Sharabani *et al.*, 2013). Avoiding touching tomato plants bearing guttation droplets significantly decreased the occurrence of severe bacterial canker epidemics (Shtienberg *et al.*, 2015). Some pathogenic bacteria are encouraged by abiotic stress conditions such as excess water or extreme temperatures. Avoiding these conditions shortly after transplanting decreased the presence of *Erwinia* and other bacteria with the potential to cause diseases (Albajes, 1999; Gnanamanickam, 2007).

Fungal diseases

The primary purpose of grafting vegetables worldwide has been to provide resistance to soilborne diseases (Louws *et al.*, 2010), and its use dramatically increased after the phase-out of the soil disinfectant methyl bromide (King *et al.*, 2008). However, under the hot Mediterranean climate conditions, crop sensitivity to certain pathogens may increase. Breakdown of rootstock resistance (Abdul-Baki and Haroon, 1996), or favoured conditions for disease expression during plantlet establishment, will negatively affect plant health. For example, oomycetes such as *Pythium* spp. are considered to have a broad host range. *Pythium aphanidermatum* favours the very warm conditions (Ben-Yephet and Nelson, 1999) that are unfavourable for proper plantlet establishment, making the latter more sensitive to *Pythium* infection. Grafting cucumber on to different *Cucurbita* rootstocks indeed improved its tolerance to *P. aphanidermatum* infection, but infected plantlet survival was reduced when the temperature and humidity were higher (Al-Mawaali *et al.*, 2012). Insufficient soilborne disease management may result in suppression of growth after transplanting.

Macrophomina phaseolina, which causes vine decline and yield losses in muskmelon, and *Monosporascus cannonballus*, which causes sudden wilt of muskmelon, are both enhanced by hot environmental conditions (Cohen *et al.*, 2000). Verticillium wilt of watermelon was successfully managed by grafting, although disease incidence, fruit yield and quality were affected by soil inoculum density and environmental conditions (Wimer *et al.*, 2015). Soil and water salinity may increase transplant sensitivity to soilborne pathogens (Al-Sadi *et al.*, 2010). Irrigation with saline water significantly increased disease severity in tomato transplants inoculated with *F. oxysporum* f. sp. *radicis-lycopersici*, and mineral fertilization further increased it (Triky-Dotan *et al.*, 2005). In certain cases, grafting improved tomato salinity tolerance (Martorana *et al.*, 2007; di Gioia *et al.*, 2013), but possible interactions with soilborne diseases should be considered.

9.2 Recommendations for the Use of Grafted Plants in Greenhouses: the Case of The Netherlands

Annual fruit crops such as tomato, pepper, cucumber and aubergine are commonly grown in soil-less culture in The Netherlands. Almost 100% of the year-round cultivation of tomato and aubergine is grafted, whereas rootstocks are hardly used in the cultivation of pepper and cucumber. In the latter crops, grafting is economically not feasible due to the lack of production increase. In pepper and cucumber, rootstocks are only used in organic cultivation in soil, in order to prevent problems with soilborne diseases.

The main objectives of using grafted transplants are: (i) to increase production (Rouphael *et al.*, 2010; Flores *et al.*, 2010); (ii) to improve the physiology of the plants, making them more robust to tolerate biotic and abiotic environmental stresses (Lee *et al.*, 2010); and (iii) to impart resistance against a number of soilborne diseases and nematodes (Louws *et al.*, 2010). In order to realize these advantages, the grafted plantlets from the nurseries have to meet the highest quality standards and the cultivation practice has to be adapted to the demands of grafted plants. In this section, a number of recommendations on the cultivation of grafted vegetable plants in high-tech greenhouse systems are presented.

9.2.1 The grafting process

Before grafting takes place, the grower has to decide on the genotypes of the rootstock and scion that will best fit their cultivation strategy. This decision is usually taken based on discussions with a consultant and a seed company, who will have an array of rootstock genotypes available ranging from high to moderate vigour and with resistances to viruses, fungi and nematodes.

For commercial cultivation of grafted plants in large-scale greenhouses, the grafting process usually takes place at specialized nurseries. For a successful graft union formation, the cambium of the rootstock and scion must be well aligned and in close contact. The scion and rootstock plants must therefore have similar stem diameters at the moment of grafting. To realize this, tomato rootstock seeds are sown approximately 1 week earlier than the scion cultivar. Compared with the cultivation of non-grafted plants, seeds of the scion have to be sown approximately 5 days earlier to obtain plants of the desired size on the date of planting, as grafting results in a delay in plant growth and development (Peet and Welles, 2005). Currently, the most commonly used grafting technique is splice grafting in which the rootstock and the scion are cut at matching 45° angles and fixed together with a silicone grafting clip (*see* Chapter 1, this volume) (Lee *et al.*, 2010).

After grafting and a healing schedule in which the vascular connections are established under high humidity conditions at the nursery, the grafted plants are transferred to commercial greenhouses. For successful cultivation in which high production levels and high product quality are realized, the quality of the grafted plantlet is a key factor. High-quality plantlets should be uniform in size and traits, of proper size or height and with a well developed root system and an aboveground plant appearance as agreed with the grower (in terms of thickness of the stem, plant height, developmental stage and leaf area). The grafted plantlets should be completely free of bacterial and viral diseases, although this may not always be easily recognized at the time of transplanting (Lee *et al.*, 2010).

9.2.2 Cultivation system of grafted plants

One of the issues in the use of grafted plants is their price. Grafted plants are 50–100% more expensive than non-grafted plants, which is caused by the fact that two seeds are required per plantlet (one for the rootstock and the other for the scion) and that there are additional labour costs in the production of grafted plants (Morra, 2004), a problem that could be mitigated by the development of automated grafting robots (Kurata, 1994). To reduce the expenses for the grafted plant material per m^2 for the grower, often two stems per plant are kept. Multiple strategies can be used to obtain two stems. The first is to allow the main stem to grow normally, and let a side shoot below the first or second truss become the second stem. Depending on the plant density in the greenhouse, additional shoots may be allowed on all plants at the same time, or in a number of steps, to realize the final stem density. Another strategy to obtain two stems per plant is to decapitate the plant and allow two buds to grow and form the two stems per plant. This decapitation can be done above the cotyledons (*see* Plate 25) so that two equal-sized shoots will be produced from the buds at the base of the cotyledons. However, at the moment of decapitation, the plants are relatively small with little capacity for light interception. Therefore, this is a strategy that is applied mainly in countries with higher light intensities at the moment of planting (e.g. France, Italy), or at sites where artificial light is used. The most commonly used method in countries with lower light intensities in winter, such as The Netherlands and Belgium, is to decapitate above the second leaf (*see* Plate 26) so that the plants have a larger leaf area for light interception after decapitation than when decapitated above the cotyledons. In this case, the buds belonging to the first and second leaf will grow out to form the shoots. Both decapitation methods will set back the plant growth, delaying the shoot appearance by 14 days (decapitation above the cotyledons) to 1 week (decapitation above the second leaf).

9.2.3 Start of cultivation

In general, the start of cultivation of grafted plants requires more skills of the grower than non-grafted plants. Cultivation starts with plants that tend to be more vegetative (larger leaves), with one or more leaves below the first truss, than in non-grafted plants (Peet and Welles, 2005), and with two stems that differ in vigour. The high vigour of the plants demands a high rate of cultivation, which implies that they should be cultivated at a higher average 24 h temperatures. Other measures that can be taken are a higher electrical conductivity level or less irrigation than non-grafted plants to stimulate flowering and fruit set. In general, the early fruit production of grafted plants is equal to or lower than that of the non-grafted plants (Wittemans, 2012). To prevent too great a difference in early production, the growth pattern of grafted plants should be controlled accurately at the beginning of the cultivation.

The rootstocks that are currently used are highly vigorous, and so would produce much leaf mass at the expense of fruit production if their growth pattern was not adjusted by the grower. The grower therefore has to take 'generative' actions, that is, measures that will stimulate the assimilate distribution to the fruits rather than to the leaves. The measures that can be taken are as follows:

- Maintain two stems per plant. This generates a larger aboveground 'sink' compared with the root volume, which suppresses the vegetative growth of the plant.
- Maintain a large difference between the temperature during the day and the temperature at the beginning of the night (Zhang *et al.*, 2010). This is common practice in tomato cultivation, although there is limited evidence that this stimulates assimilate partitioning to the fruits.
- Prevent the plants from using a large part of their assimilates for root growth. This means that the root volume should be restricted during the first (2) weeks of cultivation when the plantlet is not placed on the rock wool slab and is only

allowed to form the root system in a rock wool cube of 10 × 10 cm. In the first weeks after planting on the rockwool slab, restricted root growth and water transport to the leaves is realized by increasing the electrical conductivity of the nutrient solution, and by not watering the plants during the day so that the amount of available water will be less then.

- Leaf removal: during cultivation, young leaves are removed from the top of the shoot to reduce the partitioning of assimilates to the leaves in favour of partitioning of assimilates to the fruits.

9.2.4 Later phases in cultivation

One of the objectives of grafting vegetables is to obtain a yield increase. In general, in crops that are planted in December, early production levels of non-grafted plants are equal to or higher than those of grafted plants (Wittemans, 2012). The advantages of grafting are seen only in the second part of the cropping cycle (after June). Especially in periods when the climate is unfavourable, grafted plants have proven to be able to deal with these conditions better, due to better uptake of water and nutrients (Tachibana, 1982; Ahn *et al.*, 1999). This is the case both in the winter months when light conditions are unfavourable, and in summer when greenhouse temperatures are high. Grafted plants also have the ability to recover better after infection with diseases like pepino mosaic virus. In summer, grafted plants were able to maintain fruit set and fruit quality, whereas non-grafted plants had problems with these processes (Kell and Jaksch, 1998). The production increase is due to the higher mean fruit mass, as well as the rate of truss initiation, which is higher from June onwards for grafted plants compared with non-grafted plants (Wittemans, 2012).

The consequence is that a grafted crop can be cultivated for a longer period than a non-grafted crop (Lee, 1994), which is convenient in western European cultivation where the cropping cycle of a tomato or aubergine crop is 11 months. The advantages of rootstocks are even more pronounced in high-tech greenhouses that have high light transmission, CO_2 supply and light assimilation.

9.3 Role of Grafting in Speciality Crops

9.3.1 Globe artichoke

Cynara cardunculus L., a Mediterranean perennial species within the *Asteraceae* (Compositaeae) family, includes the two cultivated taxa globe artichoke (*C. cardunculus* var. *scolymus* L.) and cardoon (*C. cardunculus* var. *altilis* DC), along with their ancestor the wild cardoon, also called artichoke thistle (*C. cardunculus* var. *sylvestris* (Lam.) Fiori) (Rouphael *et al.*, 2012; Colla *et al.*, 2013). Currently, most commercial grafting is practised in annual fruit crops belonging to the Solanaceae, such as aubergine, pepper and tomato, and the Cucurbitaceae, such as cucumber, melon and watermelon. However, in recent years, grafting has been adopted in other vegetables, in particular artichoke (Ciccarese *et al.*, 2012; Temperini *et al.*,

2013; Trinchera *et al.*, 2013). Reports mentioning Verticillium wilt tolerance in artichoke being affected by grafting began in 2012. Verticillium wilt, caused by the soilborne fungus *Verticillium dahliae*, is spreading in artichoke-growing areas worldwide and is becoming a threat to artichoke production (Cirulli *et al.*, 2010). Ciccarese *et al.* (2012) tested a collection of wild and cultivated cardoons under controlled conditions for resistance to *V. dahliae* by artificial inoculation with a conidial suspension of the fungus. Of 44 populations obtained by open pollination in the field, nine showed a high level of resistance to Verticillium wilt without any symptoms. The authors concluded that the use of artichoke plants grafted on to rootstocks resistant to Verticillium wilt can be an efficient agronomical tool, mainly in areas where the soil is significantly infested (Ciccarese *et al.*, 2012). Similarly, Temperini *et al.* (2013) evaluated the crop performance of seed-propagated artichoke selection 'T3' either non-grafted or grafted on to the cultivated cardoon 'Bianco gigante inerme a foglia intera' in soil infested with *Verticillium* spp. over 3 years. Selection 'T3' was used due to the susceptibility to Verticillium wilt. The total yield was on average higher by 44–53% over the 3 years in grafted compared with non-grafted artichoke plants (Temperini *et al.*, 2013). The increase in total yield in grafted plants was attributed to an increase in both head mean mass and numbers. Moreover, the Verticillium wilt incidence in grafted plants was significantly lower in comparison with non-grafted plants (10 versus 43%). The higher yield and yield components of grafted artichoke plants in infested soils were also recorded in non-infested soil conditions when the seed-propagated artichoke hybrid 'Concerto' was grafted on to the cultivated cardoon variety 'Belgio' (Temperini *et al.*, 2013). In the same work, the authors evaluated the compatibility of the two seed artichoke cultivars 'Romolo' and 'Istar' grafted on to the cultivated cardoon 'Bianco avorio' and wild cardoon, and also identified the most accurate grafting method (splice versus cleft grafting) (*see* Plate 27). Temperini *et al.* (2013) demonstrated that the higher grafting survival rate (82–92%) was recorded when the two cultivars were self-grafted, cross-grafted and grafted on to 'Bianco avorio', whereas a lower affinity rate (39–49%) was found when the wild cardoon was used as rootstock. The grafting technique affected the survival rate, with higher values observed with the splice grafting (78%) compared with cleft grafting (73%) method, indicating that splice grafting is the most suitable method for globe artichoke.

Rootstock–scion compatibility is a key factor in grafted vegetables depending on anatomical, physiological and genetic variables (Edelstein *et al.*, 2004). The mechanism of graft compatibility/incompatibility in artichoke was studied by Trinchera *et al.* (2013), who reported that the degree of affinity between different artichoke scion and cardoon rootstocks was associated with the healing time of the two bionts. Scanning electron microscopy images of longitudinal sections of graft junctions, just 3 days after grafting, revealed the appearance of many interconnection structures between the two grafting components, followed by a vascular rearrangement and callus development during graft union formation. The 'Romolo'–'Sardo' affinity has been associated with the higher capacity of wild cardoon 'Sardo' to produce pectic and carbohydrate materials at the graft interface (Trinchera *et al.*, 2013). The authors also demonstrated that the duration of the early-stage grafting process could be affected not only by the scion–rootstock compatibility, but also by the different growing seasons, being favoured by lower temperatures and a reduced light/dark photoperiod.

9.3.2 Green bean

Grafting of green beans (*Phaseolus vulgaris* L.) has been used in Portugal to prevent biotic and abiotic stresses since 2010 and is currently an important strategy for the production of this crop in greenhouses, in both hydroponic and soil-based production systems. Keen interest in green bean grafting has also arisen because of its environmental sustainability and ease of management, which make it ideal for organic production.

Historically, protected crops of green beans grown in soil have been cultivated under intensive conditions, frequently twice a year, with high application rates of mineral synthetic fertilizers and pesticides and without crop rotation. These practices led to a significant increase in soil salinity and to a higher incidence of soilborne diseases such as those caused by *Fusarium* spp. and by root-knot nematodes (*Meloidogyne* spp.), which have impaired the production of this crop in the main vegetable production areas in Portugal (Rodrigues, 2010). Therefore, grafting became a promising technique to improve nutrient uptake, increase crop tolerance to salinity and suppress soilborne diseases through natural resistance/tolerance of some rootstocks of *Phaseolus coccineus* L. and *P. vulgaris* species.

P. coccineus was cultivated for several thousand years in Mesoamerica and was introduced to Europe in the 17th century (Hernández-Bermejo and León, 1992). Its centre of origin is the Mexican highlands where it grows wild and sometimes at the edge of cultivated bean plots (Delgado-Salinas, 1988). This species has a high degree of allogamy, shown, for example, by different seed colours produced from a single-coloured seed sown (Giurcă, 2009). The wild forms show great phenotypic variation and are currently under active speciation, although crosses between wild and domesticated forms have changed the speciation patterns. However, this species still shows great potential for breeding, and it is recognized that rustic forms of *P. coccineus* are resistant to some viruses (e.g. bean golden mosaic virus), bacteria (e.g. *Pseudomonas syringae* Van Mall; *Xanthomonas phaseoli*) and fungi (e.g. *Colletotrichum lindemuthianum*) (CATIE, 2014). For example, Giurcă (2009) showed that *P. vulgaris* plants had a lower vigour and were more susceptible to attack by *Xanthomonas campestris*, compared with *P. coccineus* species, which showed high vigour and greater resistance to attack by this pathogen.

P. coccineus has a well developed root system with a tuberized taproot, which is rich in starch, and numerous large and fleshy secondary roots (Delgado-Salinas, 1988; Giurcă, 2009; Labuda, 2010; CATIE, 2014), while *P. vulgaris* has a fusiform, weakly developed root system (Giurcă, 2009). *P. coccineus* plants have smaller numbers of stem ramifications but greater growth vigour compared with *P. vulgaris* plants (Giurcă, 2009). There are only a few distinguishable cultivars of *P. coccineus* for the determinate or the indeterminate plant growth habit types (Hernández-Bermejo and León, 1992). In Portugal, the indeterminate type is cultivated traditionally in the north of the country and is called *feijão de 7 anos* or

'7-year bean', due to the vegetative buds in the root crown of this species, allowing regrowth for several consecutive years.

The traits of *P. coccineus* and its botanical proximity to the common bean *P. vulgaris* match the green bean rootstock requirements for both plant breeding programmes (Gurusamy *et al.*, 2010) and crop production. Trials developed in Portugal with the *P. coccineus* cultivars 'Aintree' (P1) and 'White Emergo' (P2) (TozerSeeds) and the '7-year bean' landrace 'Ponte de Lima' (P3) as rootstocks (*see* Plate 28), were conducted in two different locations in north-west Portugal to evaluate the effect of root/shoot genotype combinations on yield and pod quality of two commercial cultivars, 'Oriente' (Vreeken's Zaden) and the standard Portuguese traditional 'Vagem rajada' (Anseme) (I. Mourão, M.L. Moura, L.M. Brito, J. Coutinho and S.R. Costa, 2016, unpublished results). The splice grafted plants could withstand two sets of stems along two training supports but the self- and non-grafted plants only one. In order to maintain the same stem density, self- and non-grafted plants were planted in pairs. Planting densities were equivalent to 3.3 and 2.1 stems m^{-2}, respectively, for the first and second experimental sites.

At the first site, the experiment was conducted with low input of soil mineral nutrients and the crop showed symptoms of vascular wilt caused by the soilborne fungus *F. oxysporum* f. sp. *phaseoli*. At this site, the highest yield was obtained with 'Oriente' shoot grafted on to '7-year bean' root (P3), followed by the scion–rootstock combinations 'Oriente'–P2 and 'Vagem rajada'–P2 and –P3. A similar lower yield was found in the graft combinations of both cultivars grafted on to rootstock P1, in the self-grafted and in the non-grafted plants of both green bean cultivars. At the second site, with absence of disease symptoms, increased mineral nutrient availability and lower crop density, a comparable yield was obtained from plants of both cultivars in self-grafted and non-grafted plants and where they were grafted on to P1 and P3 rootstocks, which was higher than the yield from plants of both cultivars grafted on to rootstock P2. Symptoms of root-knot nematodes were not observed in either site, which agreed with the non-detection of this disease through nematological soil analysis done at planting.

Grafting green beans showed no advantages at the second experimental site, but in the first site, with low nutrient input, the root/shoot genotype affected crop growth and development. 'Oriente' grafted on to rootstock of the '7-year bean' landrace (P3) appeared to be an appropriate strategy for increasing crop tolerance to vascular wilt disease (*F. oxysporum* f. sp. *phaseoli*) and to allow greater absorption of nutrients in the soil, probably due to the better developed root system, compared with the *P. vulgaris* cultivars (Giurcă, 2009). The rootstock 'Aintree' (P1) may have been more sensitive to this soilborne pathogen. Different root/shoot genotype responses to *F. oxysporum* f. sp. *phaseoli* were also reported for the root-rot disease caused by *Fusarium solani* f. sp. *phaseoli*, whose severity in *P. vulgaris* crops has been shown to increase with the environmental factors that stressed the plants (Cichy *et al.*, 2007). Two quantitative trait loci associated with *F. solani* f. sp. *phaseoli* resistance have been identified in the *P. vulgaris* genome (Schneider *et al.*, 2001; Hagerty *et al.*, 2015) and accounted for 9 and 22% of the total genetic variation in the studies described by Hagerty *et al.* (2015). Although these findings improved bean resistance selection and breeding towards a more efficient marker-assisted selection, different mechanisms of resistance were apparently at work under different environmental conditions. Cichy *et al.* (2007), working with different *P. vulgaris* graft combinations, reported that root genotype controlled the expression of root-rot incidence in the absence of soil compaction, but with the addition of a compacted soil layer, the interaction of the root/shoot genotype dictated *F. solani* f. sp. *phaseoli* severity. In addition, quantitative trait loci for root vigour were correlated with *F. solani* f. sp. *phaseoli* resistance, indicating that selection for vigorous root systems can be especially valuable in uncompacted soils. Therefore, crop management and plant breeding efforts that effectively improve root growth can play a critical role in mitigating this disease.

A pot experiment in controlled conditions was conducted to evaluate the resistance/tolerance to *Meloidogyne javanica* of five cultivars of *P. vulgaris* and four cultivars *P. coccineus* (I. Mourão, M.L. Moura, L.M. Brito, J. Coutinho and S.R. Costa, 2016, unpublished results). Plants were inoculated with 5000 eggs and second-stage juveniles of *M. javanica*, with non-inoculated plants serving as a negative control and susceptible tomato 'Tiny Tim' being used as a positive control. Sixty days after inoculation, roots were observed to determine the number of galls and egg masses. None of the tested cultivars was completely resistant to the nematode, but a potential for resistance was detected in two *P. vulgaris* cultivars out of the nine cultivars tested. These two cultivars showed levels of nematode-induced galls and egg masses comparable to cultivars classified as resistant. Soils with nematode problems could benefit from these non-grafted cultivars if they were used as rootstocks because resistance factors to the root-knot nematodes *Meloidogyne incognita* were either localized within roots or not translocated basipetally through the stem graft union in studies conducted by Mullin *et al.* (1991). These authors, using graft combinations between resistant and susceptible *P. vulgaris* cultivars, reported that a resistant rootstock resulted in a resistant response to the root-knot nematodes, and that those combinations in which the rootstock was susceptible had a susceptible response, regardless of the scion component.

Green bean pod quality can also be influenced by the rootstock genotype. In the above-mentioned field experiments in Portugal (I. Mourão, M.L. Moura, L.M. Brito, J. Coutinho and S.R. Costa, 2016, unpublished results), grafting with 'White Emergo' (P2) and the '7-year bean' landrace (P3) rootstocks resulted in an increased phosphorus content of the pods for both cultivars ('Oriente' and 'Vagem rajada') at both experimental sites, as well as increased magnesium content at the first site and calcium content at the second site, compared with self-grafted and non-grafted plants. These higher nutrient contents of the pods indicated increased nutrient uptake and better utilization of these nutrients by grafted plants with P2 and P3 rootstocks. In addition to these nutritional quality attributes, the green pods of both cultivars grafted on to rootstock P3 at the first site showed a 15% increase in mean length per pod compared with self-grafted and non-grafted plants.

Recently, green bean crop production has begun to benefit from grafting, and we believe that improved rootstocks will be able to overcome the main production constraints for both conventional and organic protected crops. Green bean grafting appears to be an appropriate strategy to increase crop tolerance to important soilborne diseases and to increase nutrient uptake, which will allow crop yield increases and reduced use of synthetic chemical solutions. Selection and breeding efforts, together with a better understanding of scion–rootstock combinations and crop management effects on plant growth and development, need further study.

9.4 Conclusions and Future Perspectives on Vegetable Grafting

Vegetable grafting has the potential to improve resistance to biotic and abiotic stresses, as well as to increase yield biomass and quality. As a result of its benefits and value, demand for high-quality grafted transplants by vegetable growers and interest by specialist propagators are expected to increase in the years to come. However, care must be taken to ensure sanitation, Good Manufacturing Practice and appropriate conditions during transport and transplanting to ensure proper establishment of the grafted transplant. Investing effort in sufficient plot preparation before transplanting and elimination of stress factors after transplanting will improve plantlet establishment and enable expression of the proper benefits of grafting. As grafted plants have different behaviours and requirements from non-grafted ones, special cultivation practices (plant density, length of the growing cycle, irrigation and fertilization) have to be adapted to the demands of grafted plants. New frontiers in vegetable grafting have been reached in recent years as specialty crops such as globe artichoke and green bean have been successfully grafted on to specific rootstocks with very promising results. Scientists, extension specialists, propagators and vegetable growers need to work together in the future to integrate this environmentally friendly technology as a key factor for sustainable horticultural production.

Acknowledgements

The authors of each section are: A. Koren and E. Klein (section 9.1); J.A. Dieleman and J. Janse (section 9.2); and Y. Rouphael, G. Colla and I. Mourão (section 9.3).

References

Abdul-Baki, A.A. and Haroon, S.A. (1996) Temperature effects on resistance to *Meloidogyne* spp. in excised tomato roots. *HortScience* 31, 147–149.

Ahn, S.J., Im, Y.J., Chung, G.C., Cho, B.H. and Suh, S.R. (1999) Physiological responses of grafted-cucumber leaves and rootstock roots affected by low root temperature. *Scientia Horticulturae* 81, 397–408.

Albajes, R. (ed.) (1999) *Integrated Pest and Disease Management in Greenhouse Crops*. Springer Science & Business Media, New York.

Al-Mawaali, Q.S., Al-Sadi, A.M., Khan, A.J., Al-Hasani, H.D. and Deadman, M.L. (2012) Response of cucurbit rootstocks to *Pythium aphanidermatum*. *Crop Protection* 42, 64–68.

Aloni, B., Karni, L., Deventurero, G., Levin, Z., Cohen, R., Katzir, N., Lotan-Pompan, M., Edelstein, M., Aktas, H., Turhan, E., Joel, D.M., Horev, C. and Kapulnik, Y. (2008) Possible mechanisms for graft incompatibility between melon scions and pumpkin rootstocks. *Acta Horticulturae* 782, 313–324.

Alpert, P., Baldi, M., Ilani, R., Krichak, S., Price, C., Rodó, X., Saaroni, H., Ziv, B., Kishcha, P., Barkan, J., Mariotti, A. and Xoplaki, E. (2006) Relations between climate variability in the Mediterranean region and the tropics: ENSO, South Asian and African monsoons, hurricanes and Saharan dust. In: Lionello, P., Malanotte-Rizzoli, P. and Boscolo, R. (eds) In: *Mediterranean Climate Variability. Developments in Earth and Environmental* Sciences, Vol. 4. Elsevier, Amsterdam, The Netherlands, pp. 149–177.

Al-Sadi, A.M., Al-Masoudi, R.S., Al-Habsi, N., Al-Said, F.A., Al-Rawahy, S.A., Ahmed, M. and Deadman, M.L. (2010) Effect of salinity on pythium damping-off of cucumber and on the tolerance of *Pythium aphanidermatum*. *Plant Pathology* 59, 112–120.

Antignus, Y. (2012) Control methods of virus diseases in the Mediterranean basin. *Advances in Virus Research* 84, 533–553.

Antignus, Y. and Ben-Yakir, D. (2004) Ultraviolet-absorbing barriers, an efficient integrated pest management tool to protect greenhouses from insects and virus diseases. In: Horowitz, A.R. and Ishaaya, I. (eds) *Insect Pest Management*. Springer, Berlin/Heidelberg, pp. 319–335.

Antignus, Y., Lachman, O., Pearlsman, M. and Koren, A. (2005) Containment of cucumber fruit mottle mosaic virus (CFMMV) infection through roots by planting into a virus-free intermediating medium. *Phytoparasitica* 33, 85–87.

Atkinson, N.J. and Urwin, P.E. (2012) The interaction of plant biotic and abiotic stresses: from genes to the field. *Journal of Experimental Botany* 63, 695–709.

Balliu, A., Sallaku, G. and Babaj, I. (2012) The effect of watermelon (*Citrullus aedulis*) grafting methods on seedling's growth rate and plant stand establishment rate under saline conditions. *Acta Horticulturae* 927, 393–398.

Ben-Yephet, Y. and Nelson, E.B. (1999) Differential suppression of damping-off caused by *Pythium aphanidermatum*, *P. irregulare*, and *P. myriotylum* in composts at different temperatures. *Plant Disease* 83, 356–360.

Bolle, H.J. (ed.) (2003) *Mediterranean Climate: Variability and Trends*. Springer Science & Business Media, New York.

Cantliffe, D.J. (2008) Plug technology: horticultural revolution. *Chronica Horticulturae* 4, 7–10.

Castilla, N. (2002) Current situation and future prospects of protected crops in the Mediterranean region. *Acta Horticulturae* 582, 135–147.

Castilla, N. and Baeza, E. (2013) Greenhouse site selection. In: *Good Agricultural Practices for Greenhouse Vegetable Crops*. FAO Plant Production and Protection Paper 217. UN Food and Agriculture Organization, Rome, pp. 21–34.

CATIE (2014) *Phaseolus coccineus*. Banco de Germoplasma de semillas ortodoxas del Centro Agronómico Tropical de Investigación y Enseñanza (CATIE), Costa Rica. Available at: http://bancodegermoplasma.catie.ac.cr/phaseolus_coccineus.php (accessed 6 July 2015).

Chiu, Y.C., Chen, S. and Chang, Y.C. (2010) Development of a circular grafting robotic system for watermelon seedlings. *Applied Engineering in Agriculture* 26, 1077–1084.

Ciccarese, F., Crinò, P., Raccuia, S.A., Temperini, A. (2012) Use of resistant cardoons as rootstocks for the control of verticillium wilt in globe artichoke. *Acta Horticulturae* 942, 201–206.

Cichy, K.A., Snapp, S.S. and Kirk, W.W. (2007) Fusarium root rot incidence and root system architecture in grafted common bean lines. *Plant Soil* 300, 233–244.

Cirulli, M., Bubici, G., Amenduni, M., Armengol, J., Berbegal, M., Jiménez-Gasco, M.M. and Jiménez-Díaz, R.M. (2010) Verticillium wilt: a threat to artichoke production. *Plant Disease Feature Article*, 94, 1176–1187.

Cohen, R., Pivonia, S., Burger, Y., Edelstein, M., Gamliel, A. and Katan, J. (2000) Towards integrated management of *Monosporascus* wilt of melons in Israel. *Plant Disease* 84, 496–505.

Cohen, R., Burger, Y., Horev, C., Koren, A. and Edelstein, M. (2007) Introducing grafted cucurbits to modern agriculture: the Israeli experience. *Plant Disease* 91, 916–923.

Colla, G., Roupahel, Y., Cardarelli, M. and Rea, E. (2006) Effect of salinity on yield, fruit quality, leaf gas exchange, and mineral composition of grafted watermelon plants. *HortScience* 41, 622–627.

Colla, P., Gilardi, G. and Gullino, M.L. (2012) A review and critical analysis of the European situation of soilborne disease management in the vegetable sector. *Phytoparasitica* 40, 515–523.

Colla, G., Rouphael, Y., Cardarelli, M., Svecova, E., Rea, E. and Lucini, L. (2013) Effects of saline stress on mineral composition, phenolics acids and flavonoids in leaves of artichoke and cardoon genotypes grown in floating system. *Journal of the Science of Food and Agriculture* 93, 1119–1127.

de León, L., Siverio, F., López, M.M. and Rodríguez, A. (2011) *Clavibacter michiganensis* subsp. *michiganensis*, a seedborne tomato pathogen: healthy seeds are still the goal. *Plant Disease* 95, 1328–1338.

Delgado-Salinas, A. (1988) Variation, taxonomy, domestication, and germplasm potentialities in *Phaseolus coccineus*. In: Gepts, P. (ed.) *Genetic Resources of Phaseolus Beans*. Kluwer Academic Publishers, Dordrecht, The Netherlands, pp. 441–463.

Diara, C., Incrocci, L., Pardossi, A. and Minuto, A. (2012) Reusing greenhouse growing media. *Acta Horticulturae* 927, 793–800.

di Gioia, F., Signore, A., Serio, F. and Santamaria, P. (2013) Grafting improves tomato salinity tolerance through sodium partitioning within the shoot. *HortScience* 48, 855–862.

Dutta, B., Gitaitis, R., Smith, S., Langston, D. Jr (2014) Interactions of seedborne bacterial pathogens with host and non-host plants in relation to seed infestation and seedling transmission. *PLoS ONE* 9, e99215.

Edelstein, M., Burger, Y., Horev, C., Porat, A., Meir, A. and Cohen, R. (2004) Assessing the effect of genetic and anatomic variation of *Cucurbita* rootstocks on vigour, survival and yield of grafted melons. *Journal of Horticultural Science and Biotechnology* 79, 370–374.

Edelstein, M., Ben-Hur, M., Cohen, R., Burger, Y. and Ravina, I. (2005) Boron and salinity effects on grafted and non-grafted melon plants. *Plant and Soil* 269, 273–284.

Fallik, E., Alkalai-Tuvia, S., Chalupowicz, D., Zutahy, Y., Zaaroor, M., Beniches, M. and Gamliel, A. (2016) Effects of rootstock and soil disinfection on quality of grafted watermelon fruit (*Citrullus lanatus* L.): a two-year study. *Israel Journal of Plant Sciences* 63, 38–44.

Flores, F.B., Sanchez-Bel, P., Estan, M.T., Martinez-Rodriguez, M.M., Moyano, E., Morales, B., Campos, J.F., Garcia-Abellán, J.O., Egea, M.I., Fernández-Garcia, N., Romojaro, F. and Bolarín, M.C. (2010) The effectiveness of grafting to improve tomato fruit quality. *Scientia Horticulturae* 125, 211–217.

Gamliel, A. and Katan, J. (eds) (2012) *Soil Solarization: Theory and Practice*. American Phytopathological Society, St Paul, Minnesota.

Garibaldi, A. and Minuto, A. (2003) The application of grafting against soilborne pests and diseases of tomato in Italy: present situation and perspectives. In: *Proceedings of International Congress 'Greenhouse tomato – Integrated Crop Production – Organic Production*, Avignon, France, pp. 55–59.

Giurcă, D.M. (2009) Morphological and phenological differences between the two species of the *Phaseolus* genus (*Phaseolus vulgaris* and *Phaseolus coccineus*). *Cercetări Agronomice în Moldova* 42, 39–45.

Gnanamanickam, S.S. (ed.) (2007) *Plant-associated Bacteria.* Springer Science & Business Media, New York.

Grassbaugh, E.M. and Bennett, M.A. (1998) Factors affecting vegetable stand establishment. *Scientia Agricola* 55, 116–120.

Gulidov, S. and Poehling, H.M. (2013) Control of aphids and whiteflies on Brussels sprouts by means of UV-absorbing plastic films. *Journal of Plant Diseases and Protection* 120, 122–130.

Gurusamy, V., Bett, K.E. and Vandenberg, A. (2010) Grafting as a tool in common bean breeding. *Canadian Journal of Plant Science* 90, 299–304.

Hagerty, C.H., Cuesta-Marcos, A., Cregan, P.B., Song, Q., McClean, P., Noffsinger, S. and Myers, J.R. (2015) Mapping *Fusarium solani* and *Aphanomyces euteiches*. Root rot resistance and root architecture quantitative trait loci in common bean. *Crop Science* 55, 1969–1977.

Harel, D., Fadida, H., Gantz, S., Shilo, K. and Yasuor, H. (2013) Evaluation of low pressure fogging system for improving crop yield of tomato (*Lycopersicon esculentum* Mill.): grown under heat stress conditions. *Agronomy* 3, 497–507.

Hassell, R.L., Memmott, F. and Liere, D.G. (2008) Grafting methods for watermelon production. *HortScience* 43, 1677–1679.

Hernández-Bermejo, J.E. and León, J. (1992) Cultivos marginados, otra perspectiva de 1492. *Colleción FAO – Producción y Protección Vegetal* 26, 46–49.

Ityel, E., Avraham, L., Alon, H. and Offenbach, R. (2014) Predicting pepper fruit yield based on temperature and solar radiation. *Journal of Agricultural Science and Technology* 4, 360–367.

Jarvis, W.R. (1989) Managing diseases in greenhouse crops. *Plant Disease* 73, 190–194.

Jouyban, Z., Hasanzade, R. and Sharafi, S. (2013) Chilling stress in plants. *International Journal of Agriculture and Crop Sciences* 5, 2961–2968.

Justus, I. and Kubota, C. (2010) Effects of low temperature storage on growth and transplant quality of non-grafted and grafted cantaloupe-type muskmelon seedlings. *Scientia Horticulturae* 125, 47–54.

Katan, J. (2000) Physical and cultural methods for the management of soil-borne pathogens. *Crop Protection* 19, 725–731.

Katan, J. (2004) Role of cultural practices for the management of soilborne pathogens in intensive horticultural systems. *Acta Horticulturae* 635, 11–18.

Kell, K. and Jaksch, T. (1998) Veredlungsunterlagen bei Tomaten im Vergleich. *Gemüse* 12, 700–704.

King, S.R., Davis, A.R., Liu, W. and Levi, A. (2008) Grafting for disease resistance. *HortScience* 43, 1673–1676.

Korkmaz, A. and Dufault, R.J. (2001) Developmental consequences of cold temperature stress at transplanting on seedling and field growth and yield. I. Watermelon. *Journal of the American Society for Horticultural Science* 126, 404–409.

Kubota, C. and Kroggel, M. (2006) Air temperature and illumination during transportation affect quality of mature tomato seedlings. *HortScience* 41, 1640–1644.

Kubota, C. and McClure, M.A. (2008) Vegetable grafting: history, use, and current technology status in North America. *HortScience* 43, 1664–1669.

Kurata, K. (1994) Cultivation of grafted vegetables II. Development of grafting robots in Japan. *HortScience* 29, 240–244.

Labuda, H. (2010) Runner bean (*Phaseolus coccineus* L.) – biology and use. *Acta Scientiarum Polonorum – Hortorum Cultus* 9, 117–132.

Lamont, W.J. Jr (1995) Plastics: modifying the microclimate for the production of vegetable crops. *HortTechnology* 15, 477–481.

Lee, J.M. (1994) Cultivation of grafted vegetables I. Current status, grafting methods, and benefits. *HortScience* 29, 235–239.

Lee, J.M., Kubota, C., Tsao, S.J., Bie, Z., Hoyos Echevarria, P., Morra, L. and Oda, M. (2010) Current status of vegetable grafting: diffusing, grafting techniques, automation. *Scientia Horticulturae* 127, 93–105.

Lecoq, H. and Desbiez, C. (2012) Virus of cucurbit crops in the Mediterranean region: an ever-changing picture. In: Loebenstein, G. and Lecoq, H. (eds) *Viruses and Virus Diseases of the Vegetables in the Mediterranean Basin*. Academic Press, Norwich, UK, pp. 67–126.

Leskovar, D.I. and Stoffella, P.J. (1995) Vegetable seedling root systems: morphology, development, and importance. *HortScience* 30, 1153–1159.

Loebenstein, G. and Lecoq, H. (eds) (2012) *Viruses and Virus Diseases of the Vegetables in the Mediterranean Basin.* Academic Press, San Diego.

Louws, F.J., Rivard, C.A. and Kubota, C. (2010) Grafting fruiting vegetables to manage soilborne pathogens, foliar pathogens, arthropods and weeds. *Scientia Horticulturae* 127, 127–146.

Mahajan, S. and Tuteja, N. (2005) Cold, salinity and drought stresses: an overview. *Archives of Biochemistry and Biophysics* 444, 139–158.

Martorana, M., Giuffrida, F., Leonardi, C., and Kaya, S. (2007) Influence of rootstock on tomato response to salinity. *Acta Horticulturae* 747, 555–561.

Mathis, R., Fricot, C., Larenaudie, M., Quillévéré, A., Rolland, M., Grimault, V., Olivier, V., Dousset, C., Gentit, P., Germain, R. and Baldwin, T. (2015) *Clavibacter michiganensis* subsp. *michiganensis*: optimization of detection in seed and effect of seed treatment on efficiency of detection methods. *Acta Horticulturae* 1069, 113–118.

Mavi, K., Ermis, S. and Demir, I. (2006) The effect of tomato rootstock seeds in relation to seedling growth. *Asian Journal of Plant Sciences* 5, 940–947.

Mekki, A., Mdhaffar, M. and Sayadi, S. (2014) Advance in Mediterranean soil properties following compost amendment. *International Journal of Agricultural Policy and Research* 2, 373–379.

Morra, L. (2004) Grafting in vegetable crops. In: *International Workshop on the Production in the Greenhouse After the Era of the methyl bromide*. Comiso, Italy, pp. 147–154.

Mullin, B.A., Abawi, G.S., Pastor-Corrales, M.A. and Kornegay, J.L. (1991) Contribution of root and shoot tissues of *Phaseolus vulgaris* to *Meloidogyne incognita* resistance. *Hortscience* 26, 1503–1504.

Nguyen, H.N., Borgemeister, C., Max, J. and Poehling, H.-M. (2009) Manipulation of ultraviolet light affects immigration behavior of *Ceratothripoides claratris* (Thysanoptera: Thripidae). *Journal of Economic Entomology* 102, 1559–1566.

Pardossi, A. Tognoni, F. and Incrocci, L. (2004) Mediterranean greenhouse technology. *Chronica Horticulturae* 44, 28–34.

Patakioutas, G., Dimou, D., Kostoula, O., Yfanti, P., Paraskevopoulos, A., Ntatsi, G. and Savvas, D. (2015) Inoculation of tomato roots with beneficial micro-organisms as a means to control *Fusarium oxysporum* f. sp. *lycopersici* and improve nutrient uptake and yield. *Acta Horticulturae* 1107, 141–148.

Patel, P.K., Singh, A.K., Tripathi, N., Yadav, D. and Hemantaranjan, A. (2014) Flooding: abiotic constraint limiting vegetable productivity. *Advances in Plants and Agriculture Research* 1, 00016.

Pavlou, G.C., Vakalounakis, D.J. and Ligoxigakis, E.K. (2002) Control of root and stem rot of cucumber, caused by *Fusarium oxysporum* f. sp. *radicis-cucumerinum*, by grafting onto resistant rootstocks. *Plant Disease* 86, 379–382.

Peet, M.M. and Welles, G. (2005) Greenhouse tomato production. In: Heuvelink, E. (ed.) *Tomatoes. Crop Production Science in Horticulture*, Vol. 13. CABI Publishing, Wallingford, UK, pp. 257–304.

Pivonia, S., de Cock, A.W.A.M., Levita, R., Etiel, E. and Cohen, R. (2012) Low temperatures enhance winter wilt of pepper plants caused by *Pythium* sp. *Phytoparasitica* 40, 525–531.

OEPP/EPPO (2013) PM 7/42 (2) *Clavibacter michiganensis* subsp. *michiganensis*. *OEPP/EPPO Bulletin* 43, 46–67.

Postiglione, L. (2003) Soil salinization in the Mediterranean. In: Geeson, N.A., Brandt, C.J. and Thorne, J.B. (eds) *Mediterranean Desertification: a Mosaic of Processes and Responses*. Wiley, Chichester, UK, pp. 163–173.

Reingold, V., Lachman, O., Blaosov, E. and Dombrovsky, A. (2015) Seed disinfection treatments do not sufficiently eliminate the infectivity of *Cucumber green mottle mosaic virus* (CGMMV) on cucurbit seeds. *Plant Pathology* 64, 245–255.

Reingold, V., Lachman, O., Belausov, E., Koren, A., Mor, N. and Dombrovsky, A. (2016) Epidemiological study of *Cucumber green mottle mosaic virus* in greenhouses enables reduction of disease damage in cucurbit production. *Annals of Applied Biology* 168, 29–40.

Rengasamy, P. (2006) World salinization with emphasis on Australia. *Journal of Experimental Botany* 57, 1017–1023.

Rodrigues, C. (2010) Plantas enxertadas conquistam portugueses. *Frutas Legumes e Flores* 110, 40–41.

Rouphael, Y., Schwarz, D., Krumbein, A. and Colla, G. (2010) Impact of grafting on product quality of fruit vegetables. *Scientia Horticulturae* 127, 172–179.

Rouphael, Y., Cardarelli, M., Lucini, L., Rea, E. and Colla, G. (2012) Nutrient solution concentration affects growth, mineral composition, phenolic acids and flavonoids in leaves of artichoke and cardoon. *HortScience* 47, 1424–1429.

Sánchez-Rodríguez, E., Leyva, R., Constán-Aguilar, C., Romero, L. and Ruiz, J.M. (2014) How does grafting affect the ionome of cherry tomato plants under water stress? *Soil Science and Plant Nutrition* 60, 145–155.

Schneider, K.A., Grafton, K.F. and Kelly, J.D. (2001) QTL analysis of resistance to Fusarium root rot in bean. *Crop Science* 41, 535–542.

Schwarz, D., Rouphael, Y., Colla, G. and Venema, J.H. (2010) Grafting as a tool to improve tolerance of vegetables to abiotic stresses: thermal stress, water stress and organic pollutants. *Scientia Horticulturae* 127, 162–171.

Shalhevet, J. (2004) Using water of marginal quality for crop production: major issues. *Agricultural Water Management* 25, 233–269.

Shang, H., Xie, Y., Zhou, X., Qian, Y. and Wu, J. (2011) Monoclonal antibody-based serological methods for detection of Cucumber green mottle mosaic virus. *Virology Journal* 8, 228.

Sharabani, G., Manulis-Sasson, S., Borenstein, M., Shulhani, R., Lofthouse, M., Chalupowicz, L. and Shtienberg, D. (2013) The significance of guttation in the secondary spread of *Clavibacter michiganensis* subsp. *michiganensis* in tomato greenhouses. *Plant Pathology* 62, 578–586.

Shtienberg, D., Sharabani, G., Bornstein, M., Shulhani, R., Frenkel, O., Manulis-Sasson, S., Lofthouse, M. and Sofer, M. (2015) Management of bacterial canker of tomatoes: an attainable goal. *Acta Horticulturae* 1069, 151–158.

Soteriou, G.A., Papayiannis, L.C. and Kyriacou, M.C. (2016) Indexing melon physiological decline to fruit quality and vine morphometric parameters. *Scientia Horticulturae* 203, 207–215.

Tachibana, S. (1982) Comparison of effects of root temperature on the growth and mineral nutrition of cucumber cultivars and figleaf gourd. *Journal of the Japanese Society for Horticultural Science* 51, 299–308.

Takaichi, M., Shimaji, H. and Higashide, T. (2000) Effect of red/far-red photon flux ratio of solar radiation on growth of fruit vegetable seedlings. *Acta Horticulturae* 514, 147–156.

Temperini, O., Calabrese, N., Temperini, A., Rouphael, Y., Tesi, R., Lenzi, A., Carito, A. and Colla, G. (2013) Grafting artichoke onto cardoon rootstocks: graft compatibility, yield and Verticillium wilt incidence. *Scientia Horticulturae* 149, 22–27.

Tomlinsone, J.A. (1987) Epidemiology and control of virus diseases of vegetables. *Annals of Applied Biology* 110, 661–681.

Triky-Dotan, S., Yermiyahu, U., Katan, J. and Gamliel, A. (2005) Development of crown and root rot disease of tomato under irrigation with saline water. *Phytopathology* 95, 1438–1444.

Trinchera, A., Pandozy, G., Rinaldi, S., Crinò, P., Temperini, O. and Rea, E. (2013) Graft union formation in artichoke grafting onto wild and cultivated cardoon: an anatomical study. *Journal of Plant Physiology* 170, 1569–1578.

Triplett, G.B. and Dick, W.A. (2008) No-tillage crop production: a revolution in agriculture! *Agronomy Journal* 100, 153–165.

Ucko, O. and Dayan, R. (1986) Prevention of virus diseases in vegetables by sanitation in the Arava Valley. *Phytoparasitica* 14, 159 (abstract).
van Bruggen, A.H.C., Gamliel, A. and Finck, M.R. (2016) Plant disease management in organic farming systems. *Pest Management Science* 72, 30–44.
van Dongen, J.T. and Licausi, F. (eds) (2014) *Low-oxygen Stress in Plants: Oxygen Sensing and Adaptive Responses to Hypoxia.* Plant Cell Monographs 21. Springer, Wien, Austria.
van der Wolf, J. and de Boer, S.H. (2015) Phytopathogenic bacteria. In: Lugtenberg, B. (ed.) *Principles of Plant–Microbe Interactions: Microbes for Sustainable Agriculture*. Springer International Publishing, Cham, Switzerland, pp. 65–77.
Wimer, J., Inglis, D. and Miles, C. (2015) Evaluating grafted watermelon for Verticillium wilt severity, yield, and fruit quality in Washington State. *HortScience* 50, 1625–1630.
Wisler, G.C. and Norris, R.F. (2005) Interactions between weeds and cultivated plants as related to management of plant pathogens. *Weed Science* 53, 914–917.
Wittemans, L. (2012) Kunnen ongeënte planten even rendabel zijn als geënte planten? *ProeftuinNieuws* 17, 38–39.
Zhang, L., Hao, X., Li, Y. and Jiang, G. (2010) Response of greenhouse tomato to varied low pre-night temperatures at the same daily integrated temperature. *HortScience* 45, 1654–1661.

Index

Note: Page numbers in **bold** type refer to **figures**
Page numbers in *italic* type refer to *tables*

Printed and bound by CPI Group (UK) Ltd, Croydon, CR0 4YY

15/09/2026

14962383-0001